CO-ABR-621

Catalysis and Chemical Processes

Catalysis and Chemical Processes

Editors

R. PEARCE
B.Sc., D.Phil.
Senior Research Chemist
I.C.I. Petrochemicals and Plastics Division

W. R. PATTERSON
B.Sc., Ph.D.
Senior Research Chemist
I.C.I. Corporate Laboratory

A HALSTED PRESS BOOK

John Wiley and Sons

New York and Toronto

Blackie & Son Limited
Bishopbriggs
Glasgow G64 2NZ

Furnival House
14–18 High Holborn
London WC1V 6BX

Published in the U.S.A. and Canada
by Halsted Press, a Division of
John Wiley & Sons, Inc., New York

First published 1981

Library of Congress Cataloging in Publication Data
Main entry under title:

Catalysis and chemical processes.

"A Halsted Press book."
Includes bibliographies and index.
1. Catalysis. 2. Chemical processes. I. Pearce, R. (Ronald) II. Patterson, W. R. (William R.)
TP156.C35C373 660.2'995 81-6694
ISBN 0–470–27211–2 AACR2

Filmset by Advanced Filmsetters (Glasgow) Ltd.
Printed by M. & A. Thomson Litho, East Kilbride, Scotland

Contributors

D. G. Bew	I.C.I. Petrochemicals and Plastics Division, Wilton, Middlesbrough
Dr J. P. Candlin	I.C.I. Petrochemicals and Plastics Division, Bessemer Road, Welwyn Garden City, Herts.
Dr T. Edmonds	B.P. Research Centre, Sunbury-on-Thames, Middlesex
Dr C. R. Harrison	I.C.I. Petrochemicals and Plastics Division, Wilton, Middlesbrough
Dr W. R. Patterson	I.C.I. Corporate Laboratory, Runcorn, Cheshire
Dr R. Pearce	I.C.I. Petrochemicals and Plastics Division, Wilton, Middlesbrough
Dr P. E. Starkey	I.C.I. Petrochemicals and Plastics Division, Wilton, Middlesbrough
Dr D. J. Thompson	I.C.I. Organics Division, Blackley, Manchester
Dr D. T. Thompson	Johnson Matthey Research Centre, Sonning Common, Reading
Dr M. V. Twigg	I.C.I. Agricultural Division, Billingham, Cleveland
Dr J. M. Winterbottom	Department of Chemical Engineering, University of Birmingham

The contributors gratefully acknowledge the cooperation of British Petroleum, I.C.I. Ltd, Johnson Matthey Ltd and the University of Birmingham in the preparation of this book.

Contents

PART 1 The Catalyst as Part of the Chemical Process

PART 2 From Raw Materials to Chemical Building Blocks

PART 3 From Chemical Building Blocks to End Products

Preface

General comments and scope of book

Catalysts are indispensible components of the majority of chemical reactions. As enzymes they regulate an array of transformations within natural organisms from single cells to man. As chlorophylls they play a key role in the photosynthetic chain in the plant kingdom. Man-made catalysts support our daily lives in the manufacture of, for example, fuels, clothing, plastic goods, fertilisers, and medicines.

Catalysis is well described as the cornerstone of all chemical processes. Around 80% of all chemicals rely on catalysts of one kind or another during their synthesis. In this book we are concerned with the field of catalysis and its application to the production of industrial organic chemicals.

There are many ways of treating the subject of catalysis. Perhaps the simplest and that most commonly adopted is classification according to carbon number (e.g. C_1, C_2, C_3 etc.), chemistry, or chemical type, e.g. alkanes, alkenes, aromatics. Encyclopaedic treatments of this kind are widely available and provide an essential frame of reference for all aspects of organic chemistry whether catalysis is involved or not. On the other hand, the field may be viewed at a molecular level, by consideration of the fundamental processes occurring at the catalytic centre, e.g. in terms of surface science or kinetics and mechanism.

Instead we take an alternative course and view it from two different standpoints. The first provides a general view that examines the roles played by catalysts in the whole processes of transformation of raw materials through to the end products. We take this broader view since it is important to see the operation of the catalyst in the context of the whole process and not in isolation. It is true that many processes could not exist in the absence of a suitable catalyst. However, the provision of one does not guarantee a successful operation in today's competitive chemical industry. Other factors such as process design, environmental impact, and the correct choice of raw material can be of equal or greater importance. The second takes a more detailed look at specific catalytic functions. Here we adopt a classification according to reaction type (e.g. oxidation, hydrogenation, polymerisation) since (a) there exist such strong similarities between different catalysts in performing closely related functions, and (b) the majority of research workers organise their thoughts about catalysis in this manner. These two treatments are closely interwoven throughout the book with the first occupying the earlier sections and the second dominating the later chapters.

In Part One we explore the first line of thought. Firstly, we examine the preparation and physical characteristics of the various types of catalyst, since there are greater similarities in synthesis and physical properties than in catalyst function. We then briefly examine the various ways of operating catalytic processes. A range of options is available depending on catalyst type and the reaction under consideration. It is here that correct chemical engineering can make for a successful process. We complete the section with an assessment of the contribution of the catalyst to the cost of the product and an evaluation of the incentives for either minor modifications or wholesale changes in production pathways.

In Part Two we take a closer view of applications and discuss the methods of transforming the various raw materials into fuels and the chemical building blocks from which we construct the end products. We consider the generation of both fuels and chemical intermediates in this section since the two are so intricately linked. Basic organic chemicals represent only a small fraction of the total consumption of fossil fuels. Thus, only some 5–6% of the crude oil extracted finds its way to the petrochemicals market. The remainder is used to provide energy for a range of applications, e.g. for heating, transport, electrical generation, or for driving industrial processes and machinery. In addition, it must be emphasised that these basic building blocks referred to are simply the fundamental units from which the end products are synthesised. They do not themselves reach the consumer to any significant extent. For example ethylene, the largest tonnage single chemical, finds no outlet as such but is widely available in other, more familiar forms, e.g. in plastic goods (polythene, PVC), surface active agents, antifreeze, solvents and speciality chemicals.

In Part Three, we open with a general view of the problems confronting the industrial chemist in managing the complex flow of raw materials through the various, often highly interdependent, process steps to the end products. We then develop a more detailed examination of catalytic processes according to reaction types, oxidation, hydrogenation and carbon–carbon bond formation. In the later chapters we deal with smaller scale processes, the synthesis of some speciality chemicals, and look at the interface between industrial chemistry and natural systems in the application of enzymes to the production of chemicals.

For all these areas we make a selective survey. We concentrate our attention on chemistry that is in current practice or shows considerable promise for future exploitation. There will almost always be more than one route to a particular chemical, but it is economy of operation and not chemical elegance and diversity that has guided our writing.

We try to assess present problems and future trends and opportunities that face the chemical process industries. These are areas in which catalysis will have a major role. The middle decades of the twentieth century have seen astonishing advances in catalysis, its applications and understanding. The petrochemicals industry has grown to its current dominance from virtually nothing in some thirty to forty years. Similarly, the range of physical methods of catalyst characterisation has seen something of an explosion during this period. The role of catalysis will not diminish in the last part of the century when a number of key problems must be faced and effectively resolved. These include: (a) the need to move away from our almost total dependence on oil as a raw material with increased contributions from coal and natural gas, and, in the longer term, biomass; (b) with the ever-increasing cost of raw materials in real terms, their more effective use via both increased selectivity in synthesis and the further upgrading of our currently less valuable streams; and (c) the more effective use of energy via conservation and improved efficiency in use.

Our survey cannot be comprehensive. To counteract this, we give a selected bibliography after each chapter, pointing the way for a more detailed study.

R.P.
W.R.P.

DECADES OF INNOVATION IN CATALYSIS IN THE PETROCHEMICAL INDUSTRY

(A selection of advances that are still relevant in the 1980's)

Decade*	Process	Catalyst
1900's	Butter substitutes via fat hydrogenation	Ni
	Methane from synthesis gas	Ni
1910's	Liquid hydrocarbons via coal hydrogenation	Fe
	Ammonia synthesis—Haber–Bosch process	Fe
1920's	Methanol from synthesis gas—BASF high pressure process	Zn, Cr oxides
	Hydrocarbons from synthesis gas—Fischer–Tropsch process	Promoted Fe or Co
1930's	Fixed-bed catalytic cracking—Houdry process	Aluminosilicates
	Ethylene oxide	Ag
	PVC	peroxides
	Polyethylene (low density)—ICI high pressure process	peroxides
1940's	Aldehydes (and alcohols) via hydroformylation of olefins	homogeneous Co
	Catalytic reforming of naphtha for production of gasoline and aromatics	Pt
	Cyclohexanol/cyclohexanone via cyclohexane oxidation (intermediates for nylon-6, 6)	homogeneous Co
	Cyclohexane via benzene hydrogenation	Ni or Pt
	Synthetic rubbers SBR (styrene-butadiene)	Li or peroxides
	NBR (acrylonitrile-butadiene)	peroxides
	Butyl rubber (isobutene)	Al
1950's	Polyethylene (high density)—Ziegler–Natta and Phillips catalysts	Ti or Cr
	Polypropylene—Ziegler–Natta catalysts	Ti
	Polybutadiene elastomer—Ziegler–Natta catalysts	Ti, Co or Ni
	Catalytic hydrodesulphurisation	Co, Mo sulphides
	Acetaldehyde via ethylene oxidation Hoechst–Wacker process	homogeneous Pd/Cu
	Terephthalic acid via *p*-xylene oxidation	homogeneous Co/Mn
	α-olefins via ethylene oligomerisation	homogeneous $AlEt_3$
	Hydrocracking	Pt
1960's	Maleic anhydride via butene oxidation	V, P oxides
	Acrylonitrile via propylene ammoxidation—Sohio process	Bi, Mo oxides
	Acrolein/acrylic acid via propylene oxidation	e.g. Bi, Mo oxides
	Hydroisomerisation of xylenes	Pt
	Ethylene (and but-2-ene) via propylene metathesis	W
	Adiponitrile via hydrocyanation of butadiene	homogeneous Ni
	Improved reforming catalysts	Pt, Re
	Improved cracking catalysts	zeolites
	Acetic acid via carbonylation of methanol	homogeneous Co
	Vinyl chloride via oxychlorination of ethylene	Cu
	Vinyl acetate via ethylene oxidation—modified Hoechst–Wacker process	Pd/Cu
	Phthalic anhydride via oxidation of *o*-xylene	V, Ti oxides
	Propylene oxide via oxidation of propylene with hydroperoxides	homogeneous Mo

Decades of innovation—*contd.*

Decade*		Process	Catalyst
1970's		Methanol from synthesis gas—ICI low pressure process	Cu · Zn · Al-oxides
		Acetic acid via carbonylation of methanol—Monsanto low pressure process	homogeneous Rh
		Improved catalysts for xylene isomerisation	zeolites
		Polyethylene (low density) via ethylene—α-olefin copolymerisation—Union Carbide "Unipol" process	Ti/Cr
		Polypropylene, improved catalysts	Ti
		α-Olefins via sequential ethylene oligomerisation and metathesis—Shell Higher Olefins process	homogeneous Ni and Mo/W
		Improved hydroformylation catalysts—Union Carbide–Johnson Matthey process	homogeneous Rh
		Chiral amino acids via hydrogenation of α-amino acrylic acids	homogeneous Rh
1980's		Gasoline from methanol—Mobil process	zeolites
		Acetic anhydride from synthesis gas via carbonylation of methyl acetate	homogeneous Rh
		Methyl methacrylate via oxidation of *t*-butanol	Mo oxide
		Improved coal liquefaction processes	Co, Mo sulphides
		Improved coal gasification processes	Main Gp Metal Salts
1990's	?	Vinyl acetate from synthesis gas via carbonylation of methyl acetate	homogeneous Rh
	?	Ethylene glycol from synthesis gas	homogeneous Rh
	?	Improved routes to propylene oxide	?
	?	Phenol from benzene directly	?
	?	Novel thermoplastics via biochemical routes	enzymes
	?	PVC via gas phase processes	?

* Dates relate either to first discovery or to large scale commercialisation.

CHAPTER ONE

INTRODUCTION

W. R. PATTERSON

1.1 The catalytic property and what it means

The discovery of catalytic action in the early part of the nineteenth century coincides with the emergence of chemistry as a rational science. True, applied chemistry of a kind existed before this period, but it flourished in the absence of any rational theory and could almost be regarded as a craft based on the traditions of the alchemists of the previous century.

By 1770, there were 15 known and recognised elements. In the 60-year period which followed, an additional 33 were isolated and a further 3 recognised as their oxides. This is a measure of the intense experimental activity following Boyle's recognition that chemistry was properly the study of matter. It was during this fruitful period that catalytic action was noted for the first time in a number of isolated and diverse observations. These included the following:

(a) the ability of heated platinum wires or foils to cause oxygen to combine with coal-gas, alcohol and ether (Davy, 1817) and to promote the union of hydrogen and chlorine (Turner, 1834);
(b) the oxidation of hydrogen (to water) at room temperature, caused by the presence of platinum sponge (Dobereiner, 1822);
(c) the decomposition of hydrogen peroxide by platinum, gold and silver (Thenard, 1818);
(d) the conversion of starch to sugar on the addition of mineral acid (Kirchoff, 1812);
(e) sulphuric acid remained unchanged when it caused the dehydrogenation of alcohol to ether—a reaction known since medieval times (Mitscherlich, 1834).

It was left to Berzelius to recognise the common theme. He concluded that a new phenomenon had been discovered and, in 1835, applied the Greek word *catalysis* (meaning decomposition or dissolution) to describe it. Erroneously, he believed that a catalytic "force" was responsible since the catalyst was unchanged by the reaction it precipitated.

During the latter half of the nineteenth century, the theory of chemistry progressed considerably, and by the end of the century the kinetic theory of gases and basic chemical kinetics had been formulated. This paved the way for a more rigorous view of catalysis, and Ostwald in 1911 redefined catalysts as substances which changed the *velocity* of a chemical reaction, thus dispelling the intangible "catalytic force" of Berzelius.

Ostwald's view has stood the test of time, so that the following is now accepted as a fairly complete definition: *a catalyst augments the rate of a chemical reaction without itself becoming consumed or altering the position of final thermodynamic equilibrium for that reaction.*

It should be realised that, adequate though the definition is, it applies to the ideal situation. In practice, catalysts are observed to change while acting upon chemical species. This is particularly so for heterogeneous catalysts whose surfaces are very sensitive to their environment and undergo changes due to sintering, coking, structural reorganisation of the surface, etc. In many cases these changes come about via processes not directly related to the main chemical reaction. Furthermore, in polymerisation reactions the catalyst is not recovered from the polymer produced. The catalytic cycle ends with the catalyst molecule or particle becoming trapped or deactivated in the polymer matrix. In the chapters that follow, the reader will become aware of catalyst deactivation, regeneration, lifetime, etc. as important features in the operation of chemical plant and he should realise that catalysts can be rather metastable materials susceptible to change. There is no such thing as a universal catalyst. Catalysts are merely chemicals which induce other chemicals to combine or fragment.

Specificity is an important issue in catalytic chemistry. In a situation where a number of chemical reactions are possible, a catalyst may affect the rate of all or just some of them. Different catalysts will have different relative effects on these rates. Nevertheless, for each of these separate reactions the final equilibrium position will be determined by the thermodynamics of the overall reaction and cannot be influenced by presence of the catalyst. This ability to direct reactions along certain paths is a property of catalysts which is as valuable as their ability to enhance rates of reaction.

An example of this is provided by some reactions of cyclohexene under different conditions:

$$\xrightarrow[\text{quartz tube, no catalyst}]{800^\circ\text{C}} \quad C_4H_6 \quad + \quad C_2H_4$$

$$\xrightarrow[\text{Pd catalyst}]{>300^\circ\text{C}} \quad + \quad 2H_2$$

$$3 \quad \xrightarrow[\text{Pd catalyst}]{\ll 300^\circ\text{C}} \quad + \quad 2$$

Cyclohexene $\xrightarrow[\text{Pd catalyst}]{\leqslant 300°C,\ O_2}$ benzene + $2H_2O$

Cyclohexene $\xrightarrow[\text{quartz tube}]{400°C,\ O_2}$ complex mixture of degradative oxidation products

1.2 Catalytic reactions

Space allows only the briefest sketch of the basic features of the catalytic act. However, the reader is referred to the many good text books which exist, and suitable general books on catalysis which are listed at the end of the chapter.

Catalysts fall into two classes—homogeneous and heterogeneous. The former are present in the same phase as the reactants. Normally, this is the liquid phase, although gas-phase homogeneous catalysis is not unknown. Heterogeneous catalysis applies to reactions where the catalyst is in a separate phase—these reactions may be gas/solid, liquid/solid and gas/liquid.

In either case the catalytic act may be represented by five essential steps:

(1) Diffusion to the catalytic site (reactant)
(2) Bond formation at the catalytic site (reactant)
(3) Reaction of the catalyst–reactant complex
(4) Bond rupture at the catalytic site (product)
(5) Diffusion away from the catalytic site (product).

In the case of homogeneous catalysis, steps 2–4 represent the formation and decay of the reactive intermediate; in heterogeneous catalysis they represent surface-adsorption and desorption with reaction of the surface intermediates.

In some cases of homogeneous catalysis, the general mechanisms are now well established—perhaps the simplest case is that of acid-catalysed rearrangements involving carbonium ions. In others, the identity of the reactive, intermediate complex is subject to debate; nevertheless, the fact that such a complex is a molecular entity often reduces the number of possibilities. This is not so with heterogeneous catalysts where the true nature of any surface species is still a matter of conjecture. Moreover, the surface of a heterogeneous catalyst is energetically non-uniform: that is to say, surface atoms are exposed with varying degrees of coordinative unsaturation. Therefore, it is possible that adsorbed reactants may be too strongly bonded to undergo further reaction. Equally, adsorption may be too weak to allow a reactive enough intermediate to form. This argument can be extended to compare the activity of different catalysts for the same reaction. There are, therefore, optimum conditions for adsorption/desorption in relation to any particular reaction. These conditions can affect not only the rates of the catalytic reaction, but also the nature of that reaction.

This can be more clearly seen by examining the effect of a catalyst on the activation energy for a given reaction. The process of adsorption on the surface of a heterogeneous catalyst is exothermic: conversely, desorption is endothermic. Since a chemical bond is formed with the surface, the process is known as chemisorption where heats of adsorption are usually greater than 80 kJ/mole. Physical or Van der Waal's adsorption is a usual prerequisite in the formation of a chemisorbed species. (Heats of physical adsorption are 20–40 kJ/mole). The formation of a chemisorbed species is shown diagrammatically in figure 1.1. If this chemisorbed species undergoes a reaction with activation energy E, the activated complex will have a potential energy which is decreased by the heat of chemisorption, ΔH_c (figure 1.2). In other words, chemisorption supplies some of the energy required to form the activated complex, energy which would otherwise only be available by raising the temperature of the system. Transformation of the activated complex gives the chemisorbed product which then absorbs energy from the system on desorption. Thus, it can be seen how in one way a catalytic surface lowers the activation energy of a chemical reaction. If the same reaction had taken place in the gas phase it would have energy of activation of E. On the catalyst surface, this has been lowered by an amount equal to the heat of chemisorption. Thus, in the appropriate rate equation, the

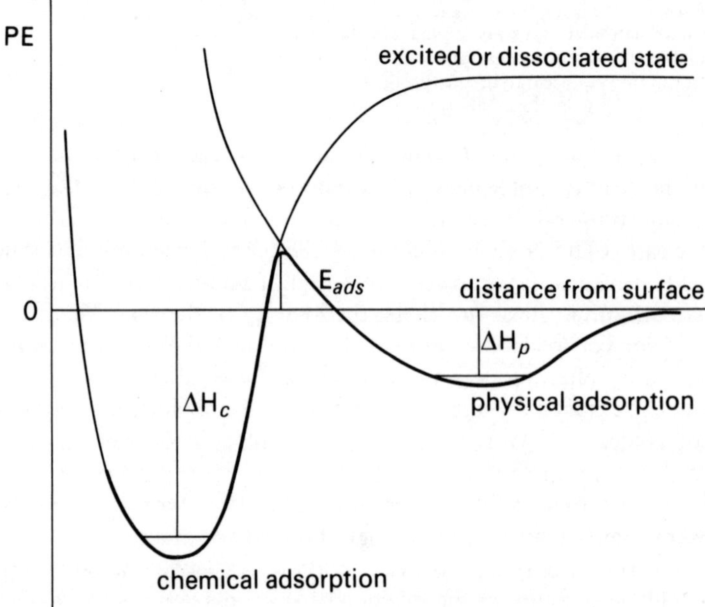

Figure 1.1 Formation of a chemisorbed species represented by the overlap of two potential energy curves. As an adsorbing molecule approaches the surface (thick line) it is initially physically adsorbed forming a species lying at a distance equal to the sum of the covalent radii plus the Van der Waal's envelopes (~0.3–0.4 nm). On closer approach to the surface, a chemisorbed species is formed at a distance approximating to that of a chemical bond (~0.1 nm). The energy of activation for the transition of a physisorbed to a chemisorbed species is usually low or negative.

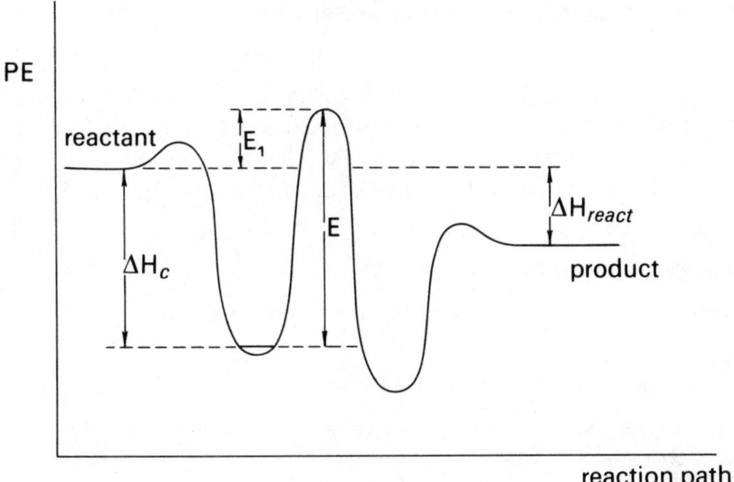

Figure 1.2 The energy profile for a typical catalytic reaction. Note that the activation energy for the same reaction in the gas phase is decreased by the heat of chemisorption, ΔH_c, of the reactant.

exponential term and hence the rate will have a higher value for the catalytic reaction than for the corresponding gas-phase reaction. Therein lies a simple explanation of catalytic *activity*.

The origin of catalytic *selectivity* is more complex. In the simplest case, the heat of desorption of product can determine how selective the reaction will be for that product. If the heat of desorption is low, the product can leave the surface easily and escape further reaction, which may be possible if the heat of desorption is high. More important, however, is the nature of the interaction between reactant and the active centres of the catalyst and, for that matter, the nature of the active centres. The type of activated complex which is formed will clearly be reflected in the products into which it decomposes.

It is not possible to calculate surface energetics of even the simplest catalytic reactions. It is not surprising, therefore, that catalysts are discovered by the experimental screening (often of large numbers) of candidates, rather than by calculated design. Many of the catalytic reactions and processes that are described in the following chapters have been discovered by this method, and it will become obvious to the reader that the implementation of a catalyst discovery by the chemical industry can occur long before any real scientific understanding of its mode of operation emerges. However, lest it should be felt that scientific endeavour in this area has little practical value, it should be emphasised that knowledge gained about the behaviour of a catalyst in a particular reaction can lead to improved (i.e. more selective or more active) catalysts for that reaction. In today's world of diminishing petrochemical feedstocks, and the resulting need to generate an alternative base for the industry, selectivity is the all-important requirement for future catalytic processes.

1.3 Available literature on catalysts and catalytic processes

The total literature on catalysis is vast. This section is intended to help the reader locate current progress in catalytic chemistry. This includes developments not only in the scientific sense, where selected journals are a prime source of information, but also developments in the chemical process area where the patent literature is often the only source of detailed knowledge.

1.3.1 *The periodic scientific literature*

(a) **Reviews** (the first three are multi-volume works that appear regularly)

Advances in Catalysis, Academic Press
Catalysis Reviews in Science and Engineering, Dekker
Advances in Organometallic Chemistry, Academic Press
Catalysis—Chemical Society (London), Specialist Periodical Reports
Aspects of Homogeneous Catalysis, D. Reidel
Fundamental Research in Homogeneous Catalysis, Plenum Press

(b) **Journals of prime interest—scientific**
Journal of Catalysis
Applied Catalysis
Journal of Molecular Catalysis
Kinetika i Katalitika (Academy of Sciences, USSR)—English translation, *Kinetics and Catalysis*, Consultants Bureau, New York
Reaction Kinetics and Catalysis Letters
Journal of Organometallic Chemistry
Journal of the American Chemical Society
Journal of the Chemical Society, Faraday Transactions I
Surface Science

(c) **Journals of prime interest—technical**
Chemtech
Chemical and Engineering News
Chemical Engineering
Chemical Engineer
Chemical Engineering Progress
Hydrocarbon Processing
Oil and Gas Journal
In addition, *European Chemical News*, *Chemical Age*, and *Chemical Week* provide topical items on new processes.

1.3.2 *The patent literature*

A patent document contains a description of the catalyst or catalytic process for which the inventor has been granted the protection to operate free from interference from competitors. It must contain exemplification of the invention which must be in sufficient detail to enable the reader to reproduce the basic chemistry of the invention claimed in the patent.

Patents therefore contain a considerable amount of experimental information in substantial detail. Details of catalyst preparation and the manner of conducting the reaction are, and indeed are required to be, given in full. The performance of the catalyst in promoting the described reaction is usually given fairly briefly—usually in terms of yields at a fixed, or at a series of temperatures. Occasionally, structural information on the catalyst is also given in the patent if it is felt that this strengthens it.

Unless an account of the catalyst or the process is published in the scientific journals, the patent is the only source of information on the preparation and reactivity of the catalyst; indeed it may be the only written record that such a catalyst has ever been prepared. Furthermore, patents, for obvious reasons, are filed and may be published some time before basic data appear in the scientific literature.

If a series of patents is granted to an inventor over a period of time, the later patents often give clues to some of the difficulties of earlier work, not necessarily revealed in the first patents. This is particularly valuable information, since it is not likely to be published elsewhere.

Patents are not required to contain explicit reference to potential disadvantages of the invention. Thus it does not invalidate a patent if, for example, a catalyst promotes a reaction to the extent described in the patent, but only for a commercially unacceptably short time. Deactivation of the catalyst through poisoning, sintering, etc. might not be described as such. It does not matter, for patent purposes, whether a catalytic reaction can be sustained for an hour or a year. However, later patents may reveal the invention of long-life catalysts in which the advantages of the new over the old are emphasised (as possibly the only reason for novelty).

As soon as possible after an invention is made, an account of it is placed with the Patent Examiner's Office. The inventor has a further year to alter or supplement the account of his invention, but by the end of this year he must submit the detailed description which goes forward for examination and is published by the Patent Office once granted. The time taken to examine and publish the patent varies from country to country but is rarely less than six months and may be as long as three years. Therefore $1\frac{1}{2}$–4 years may elapse between the filing of a patent application and its publication. Applications are usually filed in several countries and the patents are published in the national language. Because of the length of time taken to publish examined patents, some countries publish unexamined patents, e.g. Germany (as *Offenlegungsschrift*), Japan, and, more recently, Great Britain and the European Patent System.

The national patents which should be regarded as prime sources for catalysis are British, American, German, French, Japanese and the recently introduced European patents. In addition, Belgian patents are useful in that they are published more rapidly than any of the above countries.

Brief accounts of patents occur in chemical, trade and technical literature. More comprehensive surveys are provided by *Chemical Abstracts*, as part of its general scientific abstracting service, and by Derwent Publications Ltd (London), a specialist organisation that provides abstracts of patents classified by subject from every country.

The multiple filing of patents in several countries means it is frequently possible to obtain patents in one's own language. Thus, a patent of Japanese origin can be obtained in English via the U.S. or U.K. equivalent. Equivalent patents can be traced through the *Chemical Abstracts* Patent Concordance Index and through Derwent Abstracts.

How does one obtain a copy of a patent? In Britain, most major public libraries retain a complete collection of national patents. If such a source is not available, or non-national patents are required, application must be made to the Patent Office. In addition, abstracting organisations such as Derwent Publications provide a supply service for most countries' patents to their subscribers.

The book *Use of Chemical Literature* (ed. R. T. Bottle, Butterworth, 3rd edn., 1979) provides a useful guide to the main sources of chemical information.

1.3.3 *Bibliography*

At the end of each chapter we give a selected set of references and a bibliography. Here we list a number of the more important general texts.

(a) Encyclopedias

Encyclopedia of Chemical Technology, Kirk–Othmer, Interscience, New York, 2nd edn., 1962–71, 3rd edn., 1978—only partially complete, volumes appearing on a regular basis. This is the single most important reference work covering almost all aspects of industrial catalysis.

Encyclopedia of Polymer Science and Technology, Interscience, New York, 1964–71.

Encyclopedia of Chemical Processing and Design, Dekker, New York, 1977.

(b) General books on industrial chemistry

Industrial Organic Chemistry, K. Weissermel and H. J. Arpe, Verlag Chemie, Weinheim, 1978.

An Introduction to Industrial Organic Chemistry, P. Wiseman, Applied Science, 2nd edn., 1979.

Industrial Organic Chemicals in Perspective, H. A. Witcoff and B. G. Reuben, Wiley-Interscience, Vol. 1, Raw Materials and Manufacture, 1980; Vol. 2, Technology, Formulation and Use, 1980.

The Chemical Economy, B. G. Reuben and M. L. Burstall, Longman, London, 1974.

Faith, Keyes, and Clark's Industrial Chemicals, revised F. A. Lowenheim and M. K. Moran, Wiley, New York, 4th edn., 1975.

Basic Organic Chemistry V: Industrial Products, J. M. Tedder, A. Nechvatel, and A. H. Jubb, eds., Wiley, Chichester, 1975.

The Structure of the Chemical Processing Industries, J. Wei, T. W. F. Russell, and M. W. Swartzlander, McGraw-Hill, New York, 1979.

In addition, a 19-part series (completed in May 1980) on hydrocarbon processing, *From Hydrocarbons to Petrochemicals* by L. F. Hatch and S. Matar is to be published as a book by Gulf Publishing Co., Houston.

(c) General books on catalysis

Chemistry of Catalytic Processes, B. C. Gates, J. F. Katzer, and G. C. A. Schuit, McGraw-Hill, 1979.

Heterogeneous Catalysis, Principles and Applications, G. C. Bond, Oxford University Press, 1974.

Catalysis by Metals, G. C. Bond, Academic Press, 1962.

Introduction to the Principles of Heterogeneous Catalysis, J. M. Thomas and W. J. Thomas, Academic Press, 1967.

Heterogeneous Catalysis, S. J. Thompson and G. Webb, Oliver and Boyd, London, 1968.

Catalysis, P. H. Emmett, Reinhold, New York, 7 vols., 1954–60.

Structure of Metallic Catalysts, J. R. Anderson, Academic Press, 1975.

Contact Catalysis, Z. B. Szabó and D. Kalló, eds., Elsevier, vols. 1 and 2, 1976.

Chemical Physics of Surfaces, F. S. R. Morrison, Plenum Press, 1977.
Homogeneous Catalysis by Metal Complexes, M. M. Taqui Khan and A. E. Martell, Academic Press, vols. I and II, 1974.
Organometallic Mechanisms and Catalysis, K. J. Kochi, Academic Press, 1978.
Homogeneous Catalysis, G. W. Parshall, Wiley-Interscience, 1980.
Proceedings, International Congress on Catalysis: I (Philadelphia, 1956), *Adv. Catalysis*, vol. 9; II (Paris, 1960), Editions Technip. Paris; III (Amsterdam, 1964), North-Holland; IV (Moscow, 1968), Rice Univ. Press; V (Palm Beach, 1972), North-Holland; VI (London, 1976), Chemical Society; VII (Tokyo, 1980).
Homogeneous Transition—Metal Catalysis, C. Masters, Chapman and Hall, 1980.
Heterogeneous Catalysis in Practice, C. N. Satterfield, McGraw-Hill, 1980.
Design of Industrial Catalysts, D. L. Trimm, Elsevier, 1980.
Introduction to Chemical Process Technology, P. J. van den Berg and W. A. de Jong, Reidel, 1980.
For a broadly-based review on recent advances in catalysis, see D. A. Dowden in *Ann. Reports Progress Chem.*, Chem. Soc., London, 1979, **76**, 3.

1.4 Glossary of common terms

Conversion—the amount of a given reaction component transformed into products related to the initial quantity of that same component, usually expressed as %.

Selectivity—the amount of a given reaction produced related in common units (moles) to the amount of starting material converted, usually expressed as %. Where there are several reaction components and/or products, the basis of which selectivity is expressed must be stated. Selectivity can change with the degree of conversion when the product(s) can undergo further transformation. Primary reaction selectivities can only be obtained from measurements at low conversions.

Yield—the amount of a given reaction product related to the amount of a starting component, usually as %. When there are several reaction components or products the basis on which the yield is expressed must be stated. In the majority of cases, yield, conversion, and selectivity are related: yield = selectivity × conversion. Thus for conversions of less than 100%, yield is always less than selectivity. For example, if a catalytic reaction is operated under recycle conditions so that each pass through the reactor accomplishes 8% conversion at a selectivity of 95%, then the *single pass* yield is 7.6%.

Space-time yield—a term frequently used in industrial catalysis, giving the amount of reaction product formed per unit volume of catalyst (or reactor) per unit time.

Space velocity—the volume of gas (or liquid) passing through a given volume of catalyst (or reactor) per unit time (e.g. $m^3/m^3/h = h^{-1}$).

Catalyst activity—the ability of the catalyst to transform the reactant(s) to product(s). It can be expressed on several different bases: e.g. g (product) per g (catalyst) per hour, g (product) per cm^3 (catalyst) per hour, or mole (product) per mole (catalyst) per hour.

Specific activity—catalyst activity related to a particular catalyst property; e.g. activity per unit surface area or per active centre.

Turnover number—the number of molecules undergoing transformation per active site per unit time.

Poison, inhibitor—a substance that reduces catalyst activity. These terms are often used interchangeably in catalysis. Poisons (inhibitors) are typically species that bond strongly to the active catalyst sites, blocking adsorption by other species, or chemically transform the catalyst into an inactive state. The effect may be either reversible or irreversible. "Self-poisoning" occurs where the product is adsorbed too strongly to permit further reaction. Where a number of pathways are possible, poisons (inhibitors) may be used to block those less desirable, i.e. to improve catalyst selectivity.

Promoter—a substance that enhances the activity of the catalyst, usually with a simultaneous increase in selectivity.

Cocatalyst—a substance that augments catalyst activity, often dramatically. In the majority of cases, the cocatalyst is an essential component of the overall catalyst system.

Adsorption isotherms—the *Langmuir Isotherm* is one model of adsorption in which the amount of gas adsorbed onto a surface is defined in terms of its partial pressure in the gas phase. Since the rate of a heterogeneous catalytic reaction is a function of the surface concentrations of each reactant, rate equations based on their concentrations in the gas phase (readily measured) can be derived using the Langmuir isotherm. Such treatment is only approximately correct since, as it is normally used, the Langmuir isotherm assumes a heat of adsorption which is independent of surface coverage. This is usually incorrect. More refined, but mathematically more complex, isotherms have been derived by Freundlich, Tempkin, and Brunauer, Emmett, and Teller (BET Isotherm).

The simple Langmuir isotherm for first-order adsorption and desorption may be derived as follows.

If θ is the fraction of a surface covered by molecules of an adsorbed gas at pressure P, and the whole is at equilibrium, then

$$\text{rate of adsorption} \propto P, 1-\theta$$
$$\text{rate of desorption} \propto \theta$$

i.e.

$$\text{rate}_{\text{ads}} = k \times P \times (1-\theta)$$
$$\text{rate}_{\text{des}} = k' \times \theta$$

but as the system is at equilibrium,

$$kP(1-\theta) = k'\theta$$

or

$$\theta = \frac{bP}{1+bP} \qquad \text{where} \qquad b = k/k'$$

It can be shown that $b = a\exp(-\Delta H/RT)$ where ΔH is the heat of adsorption and a is a term containing the entropy of adsorption.

$$\therefore \qquad \theta = \frac{a\exp(-\Delta H/RT)P}{1+a\exp(-\Delta H/RT)P}$$

Thus at low heats of adsorption, where $bP \ll 1$,

$$\theta \propto P$$

i.e. θ is low and directly proportional to pressure. At high heats of adsorption, where $bP \gg 1$,

$$\theta \propto 1$$

i.e. coverage of the surface is complete and adsorption is independent of pressure.

For a diatomic gas, such as hydrogen, which may be dissociated on adsorption, it can easily be shown that

$$\theta = \frac{bP^{1/2}}{1+bP^{1/2}}$$

Langmuir–Hinschelwood mechanism—a mechanism in which reactants combine through being adsorbed on adjacent active sites on the catalyst surface. Most surface reactions are believed to proceed in this manner.

Eley–Rideal mechanism—an alternative mechanism for a surface reaction in which one reactant is adsorbed on active sites and the other combines with this adsorbed species directly from the gas phase and not from an adjacent site.

Dissociative adsorption—the adsorption of a molecule occurring with the breaking of an internal bond. This form of chemisorption is common with small molecules, e.g. H_2, O_2, N_2, and CO.

CHAPTER TWO

THE CATALYST: PREPARATION, PROPERTIES, AND BEHAVIOUR IN USE

M. V. TWIGG

2.1 Introduction

This chapter focuses attention on catalysts themselves, including how they are manufactured and what happens to them in use. Consideration is given to factors which limit their useful lives, but mechanistic details of how catalysts work are not covered.

2.2 Homogeneous and heterogeneous catalysts

Catalysts are conveniently divided into two groups, *homogeneous* and *heterogeneous*, according to whether or not the reactant and products share the same phase as the catalyst. The distinction between homogeneous and heterogeneous catalysis has been drawn in section 1.2. Examples of homogeneous catalysts include the oldest industrial catalyst, nitric oxide, which was used in the "lead chamber" process for the manufacture of sulphuric acid. Present applications of transition metal-based homogeneous catalysts are diverse,[1] and are discussed in Part 3. Their overall contribution to production of industrial organic chemicals is not large, but is increasing steadily. Their main attraction is high product selectivity, which is associated with uniformity of catalytic species that may be modified by changing the coordinating ligands. For example, the selectivity of rhodium-based propylene hydroformylation catalysts for *n*-butyraldehyde, rather than the *iso*-isomer, may be increased from 1:1 to 10:1 or more by incorporation of a suitable tertiary phosphine (cf. 8.4). The reaction paths of homogeneous catalysts are often well understood.[2] In the majority of cases all of the metal atoms form catalytically active centres which have a high degree of uniformity. This is in sharp contrast to the situation prevailing with heterogeneous catalysts where relatively few metal atoms are directly involved in catalysis and in which there generally exists a multiplicity of catalytically active environments. Moreover, some homogeneously catalysed reactions have no known, or no suitable heterogeneously catalysed counterparts.

Separation of catalyst from product can be difficult to achieve at moderate cost with some homogeneously catalysed reactions. This can be a serious disadvantage if the catalyst is expensive (e.g. precious metal compounds), when even small losses cannot be tolerated. When catalyst residues have undesirable properties such as intense colour or high toxicity, complete removal of catalyst from the product is also essential. Separation is straightforward where catalyst and product have very different volatility. Thus the palladium-catalysed oxidation of ethylene to acetaldehyde, the rhodium-catalysed carbonylation of methanol to acetic acid, and the rhodium-catalysed hydroformylation of lower olefins to aldehydes, in which the catalyst resides in a relatively high-boiling solvent, have been operated successfully.[3]

Problems arise when no simple means for separating the catalyst from the product are available. This difficulty has been a major obstacle preventing the more widespread adoption of processes using homogeneous catalysts. Nevertheless, these problems are being overcome, and during recent years several new processes have been introduced. It seems likely that demands on product selectivity created by escalating feedstock costs will create more opportunities for adoption of processes using homogeneous catalysts.

An interesting recent development is the "heterogenising" of soluble transition metal catalysts[4]—that is, bonding active species to the surface of an inert support such as alumina, or cross-linked organic polymer which has an appropriate high surface area, and using it as a heterogeneous catalyst. This aspect is considered in section 12.12.

2.3 Heterogeneous catalysts

Unlike homogeneous catalysts, which are dispersed at the molecular level, solid heterogeneous catalysts have to be fabricated into appropriate shapes with suitable physical properties, with the result that there may be many stages in their manufacture. They are, however, convenient to use on a large scale (there are few problems in separating catalyst from products), and they are used in almost all the major catalytic processes. Accordingly, the rest of this chapter is concerned with heterogeneous catalysts.

Most industrial catalysts fall into one of the four categories listed in Table 2.1. The general feature of heterogeneous catalysts is the presentation of a relatively high surface area of the catalytically active phase to the reactants. Surface areas of commercial catalysts cover a wide range: "low" surface area catalysts up to $10\,m^2\,g^{-1}$, "high" surface area $10–100\,m^2\,g^{-1}$, while "very high" surface area materials such as special aluminas have an area in excess of $200\,m^2\,g^{-1}$ In the absence of physical factors which may reduce the effectiveness of the catalyst, activity is proportional to the surface area of the active phase. High surface area is achieved in one of two ways: the active phase can be in a highly divided or porous form (such as Raney nickel or activated carbon), or (more commonly) an inert high surface area material can be used, which has dispersed on it the

Table 2.1 Types of heterogeneous catalysts

Type	*Examples*	*Applications*	*Comments*
Unsupported metals	Raney nickel	Organic hydrogenations	Generally unsupported metals rapidly lose activity by sintering, but the surface areas of gauzes used for ammonia oxidation increase during use
	Platinum gauzes	Ammonia oxidation to nitric oxide	
	Silver granules	Methanol oxidation/dehydrogenation to formaldehyde	
High surface area materials	Synthetic silica-aluminas including zeolites	Catalytic cracking of hydrocarbons	
	Activated carbons } Zinc oxide }	Absorption of impurities from process streams, e.g. $ZnO + H_2S \rightarrow ZnS + H_2O$	Absorption of impurities is not catalytic, but requires high surface area to be efficient
Supported dispersions			Supported dispersions make use of an inert high surface area support which often markedly retards sintering. In multicomponent catalysts interaction between the different phases can be very complex
Binary systems	Copper supported on zinc oxide	Hydrogenation of carbonyl compounds to alcohols	
	Nickel on refractory support	Steam reforming of natural gas and organic hydrogenations	
	Palladium on alumina	Selective acetylene hydrogenation	
Multicomponent systems	$Cu/ZnO/Al_2O_3$	Methanol synthesis	
	$Fe/Al_2O_3/CaO/K_2O$	Ammonia synthesis	
	Co + Mo sulphides on Al_2O_3	Hydrodesulphurisation	
Supported liquid phase	Phosphoric acid on porous support (e.g. Kieselguhr)	Ethylene hydration to CH_3CH_2OH	Supported liquid acids deactivated by alkaline impurities. Several catalysts may involve a liquid phase under operating conditions, e.g. V_2O_5-based sulphuric acid catalyst and $CuCl_2$-based ethylene oxychlorination catalyst
		Isopropylbenzene from propylene and benzene	

active phase in the form of many, very small crystallites. Such heterogeneous systems are called *supported catalysts*, and the inert material the support, or *carrier*.

2.4 Surface area and crystallite size

By assuming the crystallites in a supported catalyst are spherical (or some other simple shape such as cubic), and that all surface area is available (which of course is not actually the case), relationships between size, number, and the total surface area of the crystallites are easily derived.[5]

Spherical crystallites	*Cubic crystallites*	
$N = 6M/\rho\pi d^3$	$N = M/\rho d^3$	(1)
$SA = 3M/\rho d$	$SA = 6M/\rho d$	(2)

where N = number of crystallites in catalyst containing mass M of active phase
ρ = density of active phase
d = diameter or side of crystallites
SA = total surface area of crystallites.

Crystallites of the active phases in most industrial catalysts are in the range 50 Å to 500 Å. Thus, according to equations 1 and 2, one gram of catalyst containing 25% of a metal such as copper (i.e. 0.25 g) in the form of 50 Å crystallites contains almost 10^{18} particles of copper, having a surface area of about 30 m^2 (in fact contact regions reduce this by a factor of about two). Thus, some 10^{20} molecules each covering 10Å^2 would be required to cover fully the copper surface in the one gram of catalyst.

Figure 2.1 shows how the number of crystallites and their total surface area change with crystallite size for the example of one gram of catalyst containing 0.25 g of copper. Surface area rapidly decreases as crystallite size increases to about 200 Å, and then changes more slowly. Clearly for maximum activity, which requires maximum surface area, it is necessary to have the crystallites as small as possible. Manufacturing methods are available for producing such fine dispersions, but unfortunately this situation is not thermodynamically stable. High surface energy accompanies high surface area in very fine particles, and there is considerable driving force to reduce surface energy. This can be achieved only if small crystallites coalesce into larger ones. This happens during use, although operating conditions markedly affect the rate of the process, which is called *sintering*. Since sintering of the active phase reduces the activity of a catalyst, much research effort has been directed towards minimising it in commercial catalysts. Success is accomplished by the selection of suitable support, relative quantities of the catalyst's components, and most importantly by the actual method of manufacture.

Sintering involves transfer of material from small to larger crystallites, and

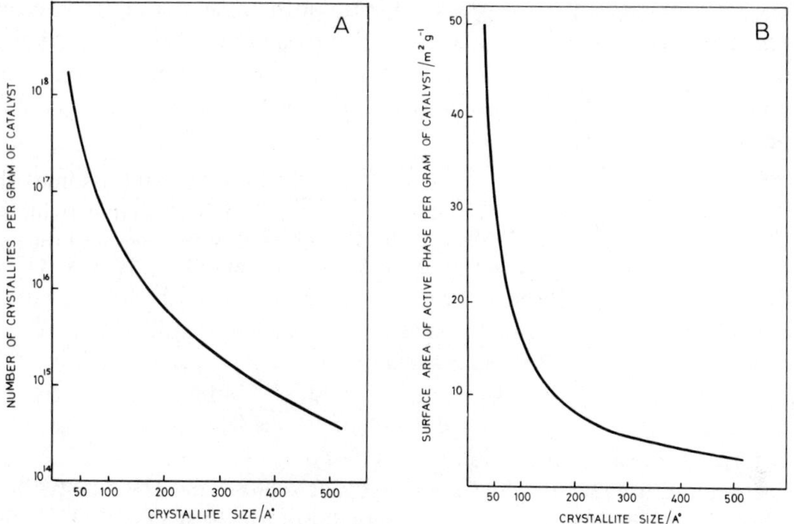

Figure 2.1 Influence of copper crystallite size on the number of copper crystallites (A) and theoretical copper surface area (B) per gram of catalyst containing 25% copper. These curves are obtained assuming spherical crystallites. In practice, surface areas approximately half those indicated are observed.

this occurs either through the gas phase, or via the surface.[6] Several surface transport mechanisms are available. Particles themselves can move across the surface and coalesce, or molecular fragments of the particles can migrate. Molecular transport can also take place via intermediate formation of mobile molecular complexes, and traces of impurities which form such species markedly increase the rate of sintering. Impurities which do this are irreversible poisons, for example the poisoning of copper catalysts by chloride ion. Operation at high temperatures and pressures increases sintering rates, as does the presence of large amounts of water through the formation of "mobile" species.

2.5 Role of the support

Most heterogeneous catalysts sinter to some extent in use, metal catalysts being the most susceptible—the activity of Raney nickel for instance soon decreases when used at higher temperatures. Supported metal catalysts show much higher resistance to sintering. This effect has been interpreted in terms both of steric factors and simple energetics. To be effective the support should be at least as finely divided as the dispersed metal. The support then acts as a physical "spacer" between the metal crystallites, and if there is adequate support available it will inhibit sintering of the metal crystallites. This model has the appeal of simplicity, and it can be made semi-quantitative by application of geometric considerations. The other approach depends on an interaction between the metal and the support. This being the case, the highly dispersed state will be somewhat stabilised with respect to sintering. Whatever the cause, supporting metal crystallites on a refractory oxide (e.g. alumina or titania) can result in a catalyst with good resistance to sintering. Most industrial metal-based catalysts are of this type.

Alumina is the most widely-used support material. It is available in high purity at moderate cost, and various polymorphs and hydrated forms can be made with a range of surface area ($\sim 1\,m^2\,g^{-1}$ to more than $200\,m^2\,g^{-1}$), porosities, and pore structures. The two most important aluminas are α-Al_2O_3 and γ-Al_2O_3. α-Al_2O_3 is obtained by heating γ-Al_2O_3, or any of the hydrous oxides, above 1000°C. It has hexagonally close-packed oxide ions with aluminium ions in octahedral interstices (corundum structure). α-Al_2O_3 is chemically inert, has a low surface area ($< 15\,m^2\,g^{-1}$), and is used as a support for impregnated catalysts that operate under arduous conditions (e.g. natural gas steam reforming, or some selective oxidation catalysts).

There are a number of aluminium hydrous oxides, and their dehydration involves several intermediate phases (except diaspore, a form of $AlO \cdot OH$ which dehydrates directly to α-Al_2O_3) in which aluminium ions occupy tetrahedral and octahedral sites. Higher temperatures increase structural order until α-Al_2O_3 is formed. These aluminas, and in particular γ-Al_2O_3 (prepared by heating ($\sim$500°C) hydrous oxides) are widely used in catalyst manufacture as support material for both impregnated and precipitated catalysts. The particle size of γ-Al_2O_3 is very small (typically 50 Å). It has high surface area, and catalytic activity in its own right (e.g. dehydration/hydration reactions).

2.6 Promoters

The patent literature contains very many references to additives which are described as promoters, and their use in industrial catalysts is widespread. There is also evidence that some supports have a promoting effect, besides their main role described in the previous section. The correct choice of the support material can therefore be important.

A classic example of a "promoted" catalyst is the ammonia synthesis catalyst.[7] Basically, this is an iron catalyst manufactured by fusion of a mixture of oxides and reduction of the iron compounds to metal before use. Small amounts of alumina (support) improve performance by inhibiting sintering and so maintaining iron surface area, but addition of smaller quantities of alkaline material such as calcium oxide (typically 2 %), and potassium oxide (typically 1 %) brings about further improvements. Calcium oxide combines with traces of silica to form calcium silicate, which would otherwise react with the potassium promoter. Potassium compounds increase the intrinsic activity of the iron surface by a factor of two or more, it is thought by lowering its work function, although this is the subject of controversy. During manufacture and use there is interaction between the various catalyst components, so optimum quantities of promoters have to be carefully determined. Commercial catalysts which contain all three activity-improving components are described as being *triply promoted.*

2.7 Amorphous and crystalline silica-aluminas: strong acid catalysts

2.7.1 *Introduction*

Solid strong acids are required in the petroleum industry for reactions such as the catalytic cracking of hydrocarbons. The best known solid acid catalysts of this type are the silica-aluminas, whose properties are most pronounced in the crystalline zeolites. The annual usage of zeolites in catalyst applications is about 250 000 tonnes per year (1979).

The basic structure of silica-aluminas resembles that of silica with some of the Si^{4+} sites occupied by Al^{3+}. This creates a charge imbalance which is neutralised by the presence of free cations. If the cation is a proton (it will be located near several oxygen atoms) the silica-alumina will behave as a Brönsted acid. On heating to above ~450°C these Brönsted sites are converted to Lewis acid sites by a dehydration process, which can be represented as shown, although there is disagreement about the exact details of this reaction.

O H⁺ O O — O O O
2 Al⊖ Si $\xrightarrow{-H_2O}$ Al⊖ Si
O O O O — O O O O

+ O O
Al ⊕Si
O O O O

Acid strength increases with the silicon/aluminium ratio, and is comparable with strong mineral acid (e.g. concentrated sulphuric acid). It is therefore possible to generate a range of surface carbonium ions with these solid acids. These are the intermediates exploited in hydrocarbon processing reactions. Synthetic acid silica-aluminas were extensively used for the catalytic cracking of hydrocarbons, but these have now been displaced by supported synthetic zeolites.[8] These crystalline silica-aluminas can have high activity, and also high selectivity in cracking reactions, resulting from controlled access to the internal surface area by a well-defined pore structure. Zeolites containing small amounts of platinum or palladium are used as hydrocracking catalysts, and the properties of the catalysts also depend on the structure and composition of the zeolite.

2.7.2 *Structure of zeolites*[9]

The structure of zeolites is based on SiO_4 and AlO_4 tetrahedral units. These tetrahedra are then combined to form several families of regular assemblies—

rings, chains, or complex polyhedra. Common structural units are a feature of these different families. Thus, the truncated octahedron is the building unit for the commercially important A, X, and Y zeolites (figure 2.2).

The open framework structure of zeolites gives rise to well-defined pore systems extending in one, two, or three dimensions which can be intersecting or nonintersecting, linear or zigzag. The total pore volume can be up to about 50% of the crystal volume, and this is associated with a high internal surface area of crystalline material, which has properties unique to each particular zeolite. Access to the pores and internal surface area is controlled by the size of the apertures created by 4, 5, 6, 8, 10, or 12-membered rings of oxygen atoms. The well-defined pore structure of zeolites is in direct contrast to amorphous silica-aluminas, since the latter consist of assemblages of particles having a range of sizes.

The charge imbalance in zeolites is compensated by the presence of free cations. These can be exchanged for a range of cations, which occupy particular sites within the zeolite (e.g. octahedral, tetrahedral) depending on factors such as their charge, size, and coordination preferences, as well as their degree of hydration.

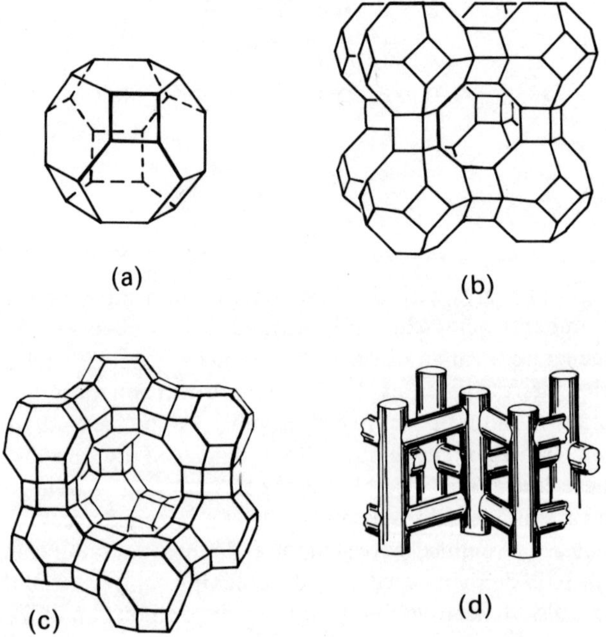

Figure 2.2 The structure of the truncated cubo-octahedron and some representative zeolites. (a) Truncated cubo-octahedron—lines represent oxygen atoms, and points of intersection, aluminium or silicon atoms. (b) Zeolite A—cubo-octahedra joined between square faces by oxygen bridges. (c) Zeolites X and Y—basic structure obtained by oxygen bridges between hexagon faces of cubo-octahedra. (d) Idealised pore structure in ZSM-5 represented by interconnecting tubes.

Hydrogen-exchanged zeolites are normally made by heating ammonium exchanged zeolite. They tend to undergo degradation during use as cracking catalysts. Rare earth-exchanged zeolites, containing for example La^{3+}, are more stable. Such polyvalent cations cannot occupy several charge imbalance sites, and those not directly associated with a cation become acid sites in association with hydrolysis of the hydrated rare earth cation.

$$La^{3+} + 2H_2O \rightleftharpoons La(OH)_2^+ + 2H^+$$

Stable rare earth-exchanged zeolites having shape selectivity are now used as cracking catalysts. Some applications of zeolites are shown in Table 2.2.

Table 2.2 Some applications of selected zeolite catalysts

Catalyst	*Applications*
RE-Y, H-Y	Catalytic cracking of heavy hydrocarbons
Pd, H-Y	Hydrocracking of heavy hydrocarbons
Pt, H-erionite H-ZSM-5 H-mordenite	Selective hydrocracking of linear alkanes in the presence of branched alkanes
Pt, H-mordenite	Linear to branched isomerisation of alkanes
H-ZSM-5	Ethylbenzene synthesis
Ni, H-ZSM-5	Isomerisation of xylenes
H-ZSM-5	Disproportionation of toluene
H-ZSM-5	Conversion of methanol to gasoline

Nomenclature. Although the problem of nomenclature is receiving attention, there is as yet no satisfactory definite system for naming zeolites. This has led to confusion in the past, but general practices are now recognised for synthetic zeolites. Previously, investigators referred to a synthetic zeolite by the name of a related mineral. Now synthetic zeolites are designated by a letter (or letters), e.g. A, X, ZSM-5. Prefixes are used to indicate the nature of exchangeable cation within a zeolite: the hydrogen form by H, alkylammonium cations by N, and other cations by their elemental abbreviation, e.g. Ca-A. The prefix RE is used for a rare earth-exchanged zeolite, e.g. RE-X.

Shape selectivity.[10] The restricted size of the pores in zeolites places a geometric constraint on the molecules which can be accommodated within them. This effect is commonly made use of in the application of zeolites as "molecular sieves". For example, straight chain hydrocarbons can be selectively adsorbed from mixtures containing branched isomers, and this effect can be used as a means of separation. Similarly, but at high temperatures, particular acidic zeolites can be used to selectively crack straight chain hydrocarbons, whereas in an unconstrained environment over amorphous silica-aluminas, preferential cracking of branched species takes place.[11] The effect discussed above may be referred to as *reactant selectivity*, since the geometric constraint is imposed on the reactants. The converse is *diffusional product selectivity*, where one (or more) of a number of product molecules is able to diffuse more rapidly through the pore system and escape. Less mobile molecules may fragment or rearrange before they are able to escape. In extreme cases selectivity may result from geometric constraint on some of the possible transition states. Only certain of these are accessible and so product selectivity results.

2.8 Catalyst size and shape

Industrial catalysts can be manufactured in a range of sizes and shapes, the size and shape used for a particular duty depending on the design requirements of the process and the type of reactor used.[12,13]

Under normal operating conditions the effective activity of many catalysts is to some extent controlled by mass-transfer effects (either pore diffusion, or film diffusion limitations). Consequently, activity is proportional to the total external surface area of the catalyst particles in the reactor. The total external surface area increases as the size of the individual particles decreases, and so, in terms of activity, it is desirable for them to be as small as possible. This however is not practicable. In large fixed-bed reactors, excessive power would be required to force the reactant gases (or liquid) through a bed of very fine particles. Thus a compromise has to be achieved between effective activity, pressure drop across the reactor, the amount (volume) of catalyst used, and the catalyst particle size.

Not only size, but also shape affects external surface area and pressure drop characteristics. A good example of this is the comparison of a cylindrical pellet with a ring of similar length and outside diameter. A ring containing $\sim 80\%$ of the volume of catalyst material contained in the pellet has $\sim 120\%$ of the pellet's external surface area, yet produces a significantly reduced pressure drop (e.g. 50% less) in a reactor.

The mechanical properties of catalyst particles are also important: in particular, the strength of particles used in large fixed catalyst beds, and resistance to attrition for particles used in fluidised beds. Here we will consider only strength. If the strength of particles is insufficient to carry the forces acting on them (which are considerable at the bottom of a large reactor), the particles break, and the pressure drop across the reactor increases. Maldistribution of gas through the bed may result, so making inefficient use of the catalyst charge, which is reflected in poor overall performance of the reactor system. In some instances (e.g. exothermic oxidations) excessive temperatures may result which can damage the reactor.

Catalyst size can in some situations markedly influence the life of a catalyst charge. When a catalyst charge is progressively poisoned (i.e. from inlet to exit) under conditions where progress of the poisoned zone is inversely proportional to the total external surface area of the catalyst particles, a reactor containing small catalyst particles has longer life than one containing larger particles.

Table 2.3 Some common catalyst shapes

Shape	*Typical catalysts**	*Comments*
Spheres 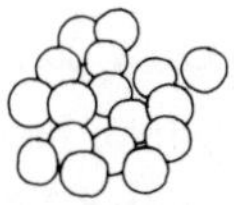	Some hydrodesulphurisation and some ammonia plant methanation catalysts. Certain impregnated precious metal catalysts (alumina spheres)	Low manufacturing costs. Some granulated material can be weak, and cementing agents are frequently used to increase strength. Packed beds of spheres produce relatively high pressure drop
Irregular granules 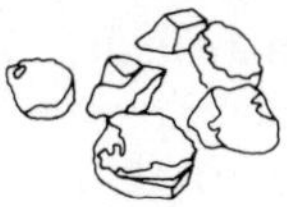	Ammonia synthesis catalyst (fused oxides), silver catalyst for formaldehyde production from methanol	Not a common catalyst shape. Consequence of restricted methods of manufacture. Used for low surface area catalysts such as bulk metal (cf. gauzes which in some instances are alternatives)
Pellets 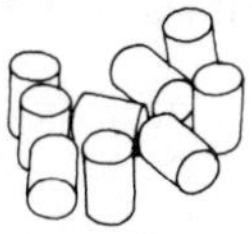	Low and high temperature carbon monoxide shift catalyst. Many hydrogenation catalysts (e.g. supported nickel) used in manufacture of organic fine chemicals	Highly regular shape, good strength. Although expensive to produce, the most common catalyst shape. Some materials are however difficult to pellet
Extrudate 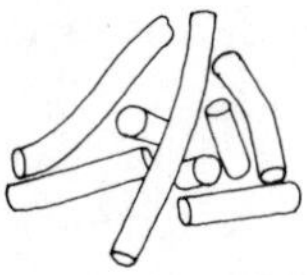	Very large quantities of hydrodesulphurisation catalyst are produced in the form of extrudate	Low bulk density and low pressure drop. Often poor strength. Some formulations can be extruded which are difficult to pellet. Extrusion aids (additives) and cementing agents frequently have to be used. Production costs are considerably less than those for pelleting
Rings 	Steam reforming of hydrocarbons (primary and secondary). Ethylene oxide catalyst, and other applications where pressure drop is an important consideration	Impregnation of preformed rings can produce very high strength catalyst. Low pressure drop, and so used in tubular reactors. Manufactured by pressing (like pellets) or extrusion. Pressed rings have superior physical properties, but costly to produce

* Frequently a particular catalyst is available in more than one shape. The size and shape used for a particular duty depends on reactor design, possible plant limitations, operating conditions, and cost considerations.

Listed in Table 2.3 are some of the more common shapes in which catalysts are fabricated. Typical dimensions of catalysts used in large fixed bed reactors are in the range 3 mm to 20 mm. As discussed above, the exact size and shape of catalyst used for a particular duty depends on a number of parameters. One of the most important of these is the requirements of the design of the plant concerned. It is not uncommon for different shapes or size of catalyst particle to be used for the same duty in different plants.

2.9 Catalyst manufacture

2.9.1 *Introduction*

The successful operation of catalytic processes depends on the properties of the catalyst itself. Usually deterioration of any of these properties will cause the process to be less efficient, perhaps to such an extent that it becomes uneconomical. Often the formulation and manufacturing procedure which produce the optimum catalyst are derived only through lengthy empirical processes which are time-consuming and costly. It is therefore not surprising that exact details about formulation and manufacture are considered proprietary information, and most published procedures are found only in the patent literature. Even routine catalyst manufacture can present difficulties. The complex way in which the components of catalysts interact causes small changes in production procedure (e.g. firing temperatures) to have large effects on how they perform in use. Some of the techniques used to manufacture heterogeneous catalysts are outlined below.

2.9.2 *Manufacturing methods*

(*a*) *Unsupported metals.* Few industrial catalysts are used in the form of unsupported metals, since they cannot present very high surface areas of active phase (the metal). The most notable example is platinum alloy gauze used for the selective oxidation of ammonia to nitric oxide, which requires short contact times.[14] This process accounts for virtually all nitric acid manufactured (40×10^6 tonnes during 1979), the major outlet for which is the production of fertilisers. Several layers of these platinum alloy gauzes are used, which may be up to eight feet in diameter, and are woven from wires only a few thousandths of an inch diameter. During use (they have a life of a few months) metal is lost from the gauze, equivalent to a few grams per tonne of nitric acid produced. Fabrication of metal catalyst gauzes is relatively straightforward, and installation in plant is not difficult. Another example of a massive metal catalyst is that of silver, used in the production of formaldehyde from methanol and air. The silver is used in the form of small granules or whiskers.

Raney nickel may also be considered as an unsupported metal catalyst, although it may contain some residual alumina which acts as a support. Raney nickel has the highest surface area of the unsupported metal catalysts discussed here, and is normally used as an aqueous suspension. In this form it has good activity for the mild hydrogenation of organic compounds. It is however particularly sensitive to poisons and loses activity fairly rapidly when used at higher temperatures.

(*b*) *Fusion methods.* This technique is not extensively used for manufacturing catalysts. However, the very important ammonia synthesis catalyst is made in this way.[15] Preparation involves melting the component catalyst oxides (including promoters) in an electric furnace (~1600°C). After cooling, the solid mixture is ground to give the desired particle size (e.g. 0.5 to 1.0 cm) for charging into a reactor. At this stage the catalyst has a very low porosity. Subsequent reduction in the reactor (or in some instances before installing in the reactor—such catalyst is called *prereduced*) produces iron crystallites (~200 Å), and a significant increase in porosity. Typically a weight loss approaching 20% results, and as much as twenty tonnes of water are formed during reduction of a charge of ammonia synthesis catalyst.

(*c*) *Impregnation of preformed supports.* Impregnation techniques are widely used in the manufacture of commercial catalysts, either when catalysts having low content of active phase are required (as with precious metals), or when a catalyst with particularly high strength is needed.[16] The preformed support is normally some refractory oxide, typically alumina, which is fabricated into small cylindrical pellets, rings, spheres, or extrudates. Various techniques are used to manufacture supports. By way of illustration, the preparation of alumina pellets involves precipitation of hydrated alumina from solution drying and processing the precipitate so it will readily form pellets when compressed. At this stage the pellets are not strong. Firing at high temperature causes partial sintering of the alumina which results in considerable increase in mechanical strength. The material so formed is very porous, and may have a surface area of up to $\sim 100\,m^2\,g^{-1}$ or more, depending on the firing temperature. It is often very important that the support material has defined physical properties, such as a particular surface area or distribution of pore sizes and these can be varied over a wide range depending on the method of manufacture.

The function of the support material is to give the final catalyst shape and mechanical strength, and to separate the crystallites of the active phase (deposited in the pores via impregnation), and so reduce their sintering rates. In some instances (e.g. cracking catalysts) supports are used to reduce the cost of the catalyst. The impregnation procedure involves filling the pores of the support with a solution of a suitable compound (e.g. a nitrate), which, when decomposed by heating, affords either the active phase itself or a compound which is easily converted to the active phase.

Industrially two types of impregnated catalyst are available. One has a uniform distribution of active material through the support pellet. The nickel-based catalysts used in steam reforming of natural gas to hydrogen and carbon oxides are of this type. The other type of impregnated catalyst does not contain a uniform distribution of active material through the catalyst—this is restricted to the outer regions of the catalyst particle. This type of "shell" catalyst is particularly appropriate for impregnated precious metal catalysts where it is important to make the most effective use of the metal and minimise overall catalyst cost. However, not all precious metal catalysts are of this type.

(*d*) *Precipitation methods.* Precipitated catalysts are generally prepared by rapid mixing of concentrated solutions of metal salts.[16] The product precipitates in a finely-divided high surface area form. Normally, mixed hydroxides or carbonates are prepared in this way. After filtering and washing out unwanted soluble salts, the precipitate is dried and heated to decompose the hydroxides/carbonates to the corresponding oxides, before adding a lubricant such as graphite which facilitates the compression of the powder into pellets. Binders may be added before pelleting to provide additional strength. Because of the difficulties in compacting low density material, pelleting is often done in two stages.

The final size of crystallites present in precipitated catalysts typically in the range 40 Å to 140 Å, while overall surface areas are in the range $75\,m^2\,g^{-1}$ to $150\,m^2\,g^{-1}$ or higher. As a result of this the activity of precipitated catalysts is usually high, but they tend to be mechanically weaker than their impregnated counterparts. Moreover, with impregnated catalysts it is easier to change the nature of the support, and to control or vary its pore structure.

Most precipitated catalysts are multicomponent systems. A typical example is the copper-based low-temperature carbon monoxide shift catalyst. The most widely used catalyst is made by precipitating basic copper and zinc carbonates together with alumina. However, as discussed below, there are several phases which may be present in the operating catalyst, e.g. metallic copper, zinc oxide, zinc spinel, and alumina.

(*e*) *Crystallisation* (*zeolite manufacture*). Zeolites are the only catalysts in which crystallinity is of overriding importance. They are manufactured by mixing solutions of silicon and aluminium compounds (e.g. silicates and aluminates), together with any additives, at a suitably high pH. A gel is formed, which in some instances must be aged at room temperature before heating it to induce crystallisation of the desired product.[17] This is freed from unwanted soluble salts by washing, and is often, though not always, a single pure zeolite with a fairly well-defined particle size distribution.

The exact zeolite which crystallises depends critically on the silicon/aluminium ratio, concentration, the presence of additives, and conditions such as temperature. Many aspects of zeolite preparation are empirical, and it is notable that in recent years relatively little academic work has been concerned with their preparation. Although the properties of zeolites are widely studied,[18] most of the published preparative details appear only in the patent literature. Indeed, many aspects of the manufacture of zeolites (like that of other catalysts) can be unpredictable. The order of adding reagents, rates of mixing, temperature, ageing times, and even the size of vessels used can influence the product obtained. In extreme cases the source, and prehistory of raw materials are important.

2.10 Catalyst characterisation

Newly-developed techniques (such as ESCA) for examining surfaces are now being applied to the study of solid heterogeneous catalysts, and this will lead to a better fundamental understanding of practical catalysts. Here the established techniques are considered first.

2.10.1 *Established methods*

The most important characteristics of a catalyst are its activity and selectivity. Considerable care is needed in designing activity measurements so that reliable and meaningful results are obtained. For example, reaction conditions must be chosen carefully so that the attainment of chemical equilibrium for the catalysed reaction is avoided. If equilibrium is reached it is clearly impossible to differentiate between catalysts having different activities. Similarly, conditions must be adjusted so the intrinsic activity of the catalyst is measured, rather than the rate of mass transfer. These mass transfer effects may be minimised by using a catalyst in the form of fine particles wherever practical.

When measuring catalyst activity, the behaviour of a catalyst is often compared with that of a standard sample operating under identical conditions, in order to overcome some of the difficulties in obtaining absolute measurements. Determination of catalytic activity is intimately associated with the development of catalysts for new processes, or improved formulations for existing duties.[19] It is usual for new catalysts to be tested on a relatively small scale before progressing to a larger scale. This approach is required because of the high costs involved in the larger scale tests. Table 2.4 lists most of the different kinds of test equipment used for determining activity. Even once a catalyst is in plant use, activity testing of modified formulations may continue in order to optimise the catalyst with respect to activity and other properties (see Table 2.5).

Table 2.4 Equipment for determining activity of catalysts

Equipment	*Sample size*	*Comments*
Laboratory tubular reactors		
(a) Microreactors	0.1–1.5 g	Rapid because of small scale, can operate at pressure, and results used to derive detailed kinetics. Useful for screening new catalysts
(b) Small scale	1–50 g	Useful for routine work, frequently operated at atmospheric pressure
(c) Semi-technical	50–1000 g	Costly to operate, time-consuming. Important when checking scale-up procedure (usually operated under plant conditions)
Laboratory recycle reactors		
(a) Internal recycle (e.g. spinning basket) (b) External recycle	5–50 g	Equipment complex, but particularly useful for obtaining kinetic data for full-scale plant design. Results free of diffusion limitations, effect of product inhibition and poisons can be evaluated
Plant side-stream	50 g–10 kg	Makes use of actual plant reactants containing levels of poison catalyst has to operate with
Full plant charge	Very large	Ultimate test, failure at this stage can be costly

Relatively straightforward techniques are widely used to characterise catalysts. Chemical analysis and mechanical measurements are two such examples. It is important before starting an extensive study on a catalyst to know its elemental composition, and in particular the level of possible poisons. Many automated analytical methods, such as atomic absorption spectroscopy, X-ray fluorescence, and neutron activation analysis (when available) are used routinely in catalyst characterisation. Mechanical properties such as strength of catalysts are important, as are the dimensions of the catalyst particles. In some circumstances these properties, which may appear trivial, can determine the performance of a charge of catalyst (see section 2.8).

Characterisation methods, considered important tools in heterogeneous catalysis, are concerned with the active phase (crystallite size and its total surface area), and the porous structure of practical supported catalysts. The determination of crystallite sizes in the range of 50 Å–500 Å is easily done by

Table 2.5 Established common laboratory methods used for studying heterogeneous catalysts

Information required	*Examples of methods used*
Activity and selectivity under possible operating conditions; their variation with time and in the presence of poisons	Small scale activity tests (see Table 2.4) usually tubular reactors but detailed kinetics require use of recycle reactors. Operation at pressure (plant conditions) is desirable
Elemental composition and poison levels	Chemical analysis. Electron probe analysis provides a means of obtaining poison distribution across pellets, and relative levels in micro regions
Surface area	Gas adsorption methods (B.E.T.)—total surface area using N_2, Ar or Kr. Metal surface areas in favourable cases using O_2, CO, H_2 or N_2O (e.g. $Cu + N_2O \rightarrow CuO + N_2$)
Porosity, pore size and pore size distribution	Density measurements with helium and mercury give porosity. This with total surface area enables a mean pore size to be deduced. Pore size distribution from mercury porosimetry
Type of catalyst (e.g. impregnated or precipitated)	Visible microscopy, scanning electron microscopy. Careful examination of phases present using X-ray diffraction
Phases present, and their crystallite size	X-ray diffraction gives crystalline phases, and line broadening their mean crystallite size. Careful line shape analysis gives the size distribution. Electron microscopy provides sizes and size distribution, while electron probe analysis gives elemental composition of crystallites. Magnetic susceptibility can give crystallite size of suitable phases (e.g. Ni, Fe)
Surface composition, and information about species in the surface	Electron and ion spectroscopies, e.g. ESCA and SIMS (rapidly becoming established)
Nature of adsorbed species	Infrared spectroscopy, thermal desorption, microcalorimetry

several techniques. These include transmission electron microscopy, measurement of broadened X-ray diffraction lines, more crudely via specific surface area measurements if the amount of material present is known, and in some instances magnetic methods.

Transmission electron microscopy[20] is a direct means of determining particle size, and the distribution of particle size can in some instances also be obtained. However, only a very small amount of material is examined and may not be truly representative of the bulk of the material. Moreover, changes may occur during sample preparation or during observation. Nonetheless, properly used, transmission and scanning electron microscopy are valuable tools in catalyst research.

Measurement of the extent of broadening of *X-ray diffraction* lines is frequently used.[21] Careful analysis of line shape can also produce information about size distribution, but this technique is normally used for obtaining mean crystallite sizes. X-ray line broadening has the advantage that the size measured is for a specific, known phase, and that it makes use of representative samples (a gram or two).

Surface areas are conventionally estimated by measuring the amount of a suitable gas (nitrogen, argon or krypton[22]), required to form a monolayer on the catalyst surface under particular conditions (e.g. at or about the boiling point of the gas). Given the area occupied by a molecule in the monolayer, the overall surface area can be calculated. Specific chemisorption of gases such as oxygen, hydrogen or carbon monoxide can be used to determine the surface areas of metals etc. dispersed on supports. Related methods make use of *surface reactions*, such as that between copper and nitrous oxide

$$Cu + N_2O \rightarrow CuO + N_2$$

In general the surface area of the active phase determines catalytic activity; the nature of the pore structure is also important. When the overall rate of reaction is pore diffusion-limited (i.e. when diffusion in the pores and reaction on the active phase have comparable rates) the reaction rate is proportional to the square root of the pore volume. In extreme cases pore size determines diffusion rates, and in some instances influences selectivity (compare with zeolites). Many selective oxidation catalysts require particular well-defined pore size distributions for optimum activity and selectivity.

The *porosity* of a catalyst is measured by the amount of free space, or pore volume it contains. This is normally expressed as cm^3 per gram of catalyst, and may be crudely estimated by filling the pores with a suitable liquid such as water and measuring the increase in weight. More accurate determinations are calculated from a knowledge of the true density of the catalyst material (measured in a helium atmosphere which fills all accessible pores), and the apparent density (measured in mercury which does not penetrate the pores). These measurements yield the pore volume. Knowing the total surface area and pore volume, a mean pore size can be calculated. However, real catalysts do not have single size pores, but rather a range of pore sizes. In some instances it is necessary to have knowledge of the pore size distribution. This is obtained by mercury porosimetry, in which mercury is forced under high pressure into the pores. The higher the applied pressure the smaller are the pores into which the mercury is forced, and by measuring the pressure and volume changes, the pore size distribution can be deduced.[22]

2.10.2 *Surface characterisation*

Much of our understanding of heterogeneous catalysis has been derived indirectly from the study of relationships between gas phase concentrations of reactants and products (overall kinetics), labelling experiments, measuring surface areas of active phases, determining the effects of different poisons and the like. This information has then been extrapolated back to the development of models of the molecular processes which take place on the catalyst's surface. There have been few, if any, techniques which provide direct information about the structure and nature of the catalyst's surface. During recent years techniques including scanning and transmission electron microscopy, have become available for the study of heterogeneous catalysts.[23] These can provide detailed information about the structure of the surface (and in some instances of adsorbed molecules), which can be of use in the manufacture of catalysts (quality control), and is beginning to provide the basis for an improved understanding of catalysis.

Material exposed to a relatively high energy electron beam emits primary and secondary X-rays characteristic of the elements present in the micro region investigated. This technique, electron probe analysis, is available on most electron microscopes (TEM and SEM), and is a powerful tool in catalyst research.[24] For example, the distribution of poison levels (e.g. sulphur, heavy metals) across the outer regions of a catalyst pellet can be determined by electron probe analysis. In some instances it is also possible to determine the phases with which the poison is associated.

During recent years a number of surface physics techniques have been developed for studying well-defined clean surfaces (i.e. single crystal faces) under ultra-high vacuum conditions. These techniques involve probing the surface by bombarding a sample with particles or radiation that result in surface selective emissions, and these have been the subject of numerous reviews.[23] Over the last decade such results have led to the development of models for surfaces and chemisorption, and they are now being applied to the investigation of practical catalysts. Considerable care however is needed when trying to elucidate catalytic phenomena. The working catalyst frequently operates in a high pressure/temperature environment contaminated with various impurities, rather than the ultra-high vacuum clean conditions used in most experiments. Nonetheless, it is clear that several of these techniques can be useful tools in catalyst research; so far XPS (ESCA) is the most widely used, from which valuable information can be obtained about the elements present in the surface and their approximate relative concentrations. However, for these techniques to become widely adopted and established as routine tools in catalyst research and catalyst evaluation they must still prove themselves cost-effective.

In the long term it is to be hoped that modern methods of surface characterisation will put practical catalysis on a firm scientific basis,[25] perhaps removing some of the ill-defined empiricism from current practice.

2.11 The catalyst in use

This section deals with some practical aspects of the use of catalysts in industrial processes. Attention is focused on what happens to the catalyst in use and the factors which restrict its useful life.

2.11.1 *Pretreatments*

Most industrial catalysts require some pretreatment to activate them. This is usually, but not always, done after the catalyst has been installed in the plant, and involves formation of the active phase from some precursor compound. Common examples of pretreatments include the reduction of transition metal oxides (e.g. copper oxide) to the corresponding finely-divided metal,[26] and sulphidation of cobalt–molybdenum hydrodesulphurisation catalysts.[27]

Pretreatments have to be carefully carried out because the activity of the catalyst in use depends on correct activation. This is particularly true for the formation of metals in reductive pretreatments, since they are very susceptible to sintering during reduction. Temperature, pressure, and steam concentration are the main factors which influence the extent of sintering during reduction, and it is important that each is kept to a minimum.

Pretreatments not only produce the active phase, they can also cause other considerable changes within the catalyst. For instance a large volume change may accompany reduction (e.g. more than 40% for bulk cupric oxide). Such transformations may therefore permit subsequent structural changes, including perhaps some relatively rapid sintering of the freshly reduced metal—which is associated with a corresponding loss of initial activity.

2.11.2 *Deactivation processes*[28]

(*a*) *Sintering.* After some fairly rapid loss of initial activity, most commercial supported metal catalysts sinter only very slowly during use. The actual rate and extent of sintering depends on a number of factors, including the metal concerned, the metal content, initial crystallite size, the nature of the support material, operating conditions, and most importantly the method of catalyst manufacture.[29] Many catalysts have useful lives extending over many years (see Table 2.6). Several theories of sintering predict activity/time curves of the type

Table 2.6 Typical lives of some industrial heterogeneous catalysts

Process	*Catalyst*	*Physical form*	*Typical life (years)*
Ammonia synthesis $N_2+3H_2 \rightarrow 2NH_3$	$Fe/Al_2O_3/CaO/K_2O$	Granules	5–10
Methanation (ammonia and hydrogen plants) $CO+3H_2 \rightarrow CH_4+H_2O$	$Ni/Al_2O_3/CaO$	Pellets	5–10
Low temperature carbon monoxide shift $CO+H_2O \rightleftharpoons CO_2+H_2$	$Cu/Al_2O_3/ZnO$	Pellets	2–6
Hydrodesulphurisation $R_2S+2H_2 \rightarrow 2RH+H_2S$	Co + Mo sulphides on Al_2O_3	Extrudate	2–4
Natural gas steam reforming $CH_4+H_2O \rightarrow 3H_2+CO$	Ni on ceramic support	Rings	2–4
Ethylene selective oxidation $C_2H_4+\frac{1}{2}O_2 \rightarrow C_2H_4O$	Ag/Al_2O_3	Rings	1–4
Partial oxidation of methanol to formaldehyde $CH_3OH+\frac{1}{2}O_2 \rightarrow CH_2O+H_2O$ $CH_3OH \rightarrow CH_2O+H_2$	Unsupported Ag	Granules/ crystals	0.3–1
Ammonia oxidation $2NH_3+\frac{5}{2}O_2 \rightarrow 2NO+3H_2O$	Platinum alloy	Gauze	0.1–0.5
Catalytic hydrocarbon reforming	$Pt/Al_2O_3/Cl^-$	Spheres	0.01–0.5
Catalytic cracking	Synthetic zeolites	Fine particles (fluidised bed)	Very short ("continuous") regeneration requires catalyst removal from main reactor to burn off coke

observed in practice (see figure 2.3). However, it is usually not easy to differentiate between the two general mechanisms of mass transport involved in sintering—surface migration and vapour phase transport.[6]

The rate of sintering of some supported metal catalysts during use is markedly accelerated by the presence of traces of particular impurities in the process stream (see below).

(*b*) *Poisoning.* The activity of almost all catalysts is much reduced by the presence of small amounts of chemical species referred to as poisons. Atoms or molecules having electron lone pairs available for bonding to the surface tend to be strong poisons (e.g. sulphur). Poisons are adsorbed on the catalyst much more strongly than the reactants, so effectively preventing their access to reaction centres. Only small amounts of poison are needed to deactivate catalysts. For instance, nickel hydrogenation catalysts are completely deactivated if sulphur compounds are adsorbed on them to an extent corresponding

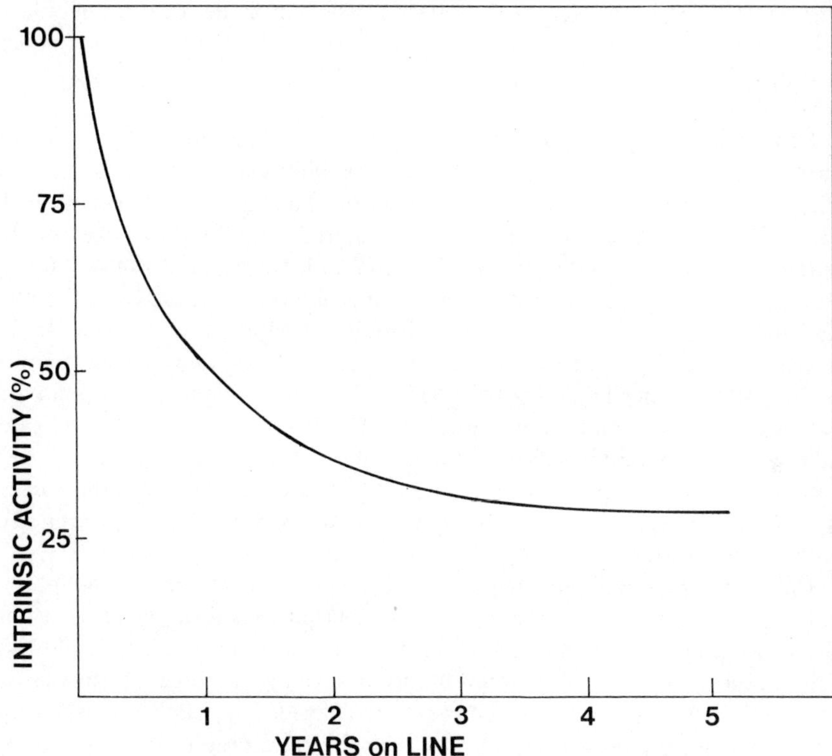

Figure 2.3 Typical activity–time curve for a long-lived catalyst subject to slow sintering.

to about 0.1 % of their weight. It is therefore important that reactant streams are carefully freed of poison. Sulphur compounds (one of the most common poisons) are removed from feedstocks by hydrodesulphurisation over cobalt/molybdenum sulphide catalysts followed by removal of the hydrogen sulphide (cf. 5.5.1).

The effect of poisons can be permanent or temporary, depending on whether the poison is removable from the catalyst surface. Hydrogen sulphide is a powerful poison for nickel catalysts used in the steam reforming of natural gas, but if after being exposed to relatively high levels of hydrogen sulphide (parts per million), the poison level in the gas is reduced, the equilibrium amount on the catalyst decreases, and full activity is restored. Here hydrogen sulphide is a temporary or reversible poison.

Often the effect of poisons (such as arsenic and lead compounds) is permanent because they are so strongly adsorbed on the catalyst that the process is irreversible under normal operating conditions. Further, the mode of action of some poisons leads to structural changes within the catalyst, and this causes truly permanent deactivation. A good example of this involves the poisoning of copper catalysts by hydrogen chloride. Relatively mobile chloro-copper species are formed on the surface of the catalyst, which provide an extremely facile mechanism for the sintering of the copper crystallites, resulting in permanent loss of activity.

(*c*) *Catalyst fouling*. In a number of processes catalysts are deactivated, not through sintering of the active phase or poisoning, but because reactants are physically prevented from reaching the catalyst. The best-known example is the formation of by-product carbonaceous material (often called coke) on the external surface and pores of silica-alumina hydrocarbon cracking catalysts.[8] This takes place very rapidly, but is not a major limitation on their use. Catalytic activity is fully restored by burning off the coke deposits in air. The catalyst structure is unaffected (under controlled conditions), and multiple regeneration is possible with negligible loss of activity. A fluidised catalyst bed is used in catalytic cracking, and "continuous" oxidative regeneration of deactivated catalyst in a separate vessel is an integral part of the process. Indeed, heat is recovered from the regeneration stage, and made use of elsewhere in the process. Careful temperature control during regeneration is needed to prevent the catalyst from sintering.

Catalysts which have sintered are more difficult to regenerate, although this can be done if the active phase can be converted to a form which is readily dispersed, and then reconverted to the active phase. A good example of this is the regeneration of hydrocarbon reforming catalysts (precious metals on alumina). Here sintered metal is converted to chlorides by reaction with chlorine compounds in the presence of oxygen. The chlorides become dispersed across the surface of the support, and subsequent reduction generates a fine dispersion of metal crystallites.

In general, methods used for regeneration include oxidation and reduction, exposure to high temperatures to desorb and decompose poisons, and chemical treatments, for example with dilute acids. The effects on plant output of the related aspects of catalyst degradation and regeneration are further considered in section 3.7. Excessively high temperatures can adversely affect the physical properties of catalysts, and so careful control of conditions during regeneration is most important.

Although there are a few exceptions, of which catalytic cracking is the most important, in general it is only economical to regenerate precious metal-based catalysts. However, with increasing catalyst costs, incentives for regenerating discharged catalysts which do not contain precious metals are becoming higher.[30]

2.12 Catalyst economics

In this section some basic economic aspects of catalysts are considered. Catalysts are premium high-value products, and in absolute terms the cost of catalyst charges for large scale plants is very high. For instance the total replacement cost of catalyst for a 1000 tonne/day ammonia plant is more than £0.5×10^6 (1980). However, modern catalysts are efficient and highly cost-effective, and rapidly repay their purchase price. Thus the annual replacement cost of the catalysts in a large ammonia plant is equivalent to about 3.5 days of product.[31] This is achieved because modern catalysts produce very many times their own weight of product—ammonia synthesis catalyst during its operating life synthesises 25 000 times its own weight of ammonia. Even considerably shorter-lived catalysts have impressive productivity ratios: vanadium-based maleic anhydride catalyst for instance produces 1200 times its own weight of product during its useful life.[32]

REFERENCES

1. *Homogeneous Catalysis*, by G. W. Parshall, J. Wiley and Sons, 1980; *Principles and Applications of Homogeneous Catalysis*, by A. Nakamura and M. Tsutsui, J. Wiley and Sons, 1980; G. W. Parshall, *J. Molecular Catalysis*, 1978, **4**, 243; *Homogeneous Transition—Metal Catalysis*, by C. Masters, Chapman and Hall, 1980.
2. For example, see discussion on the mechanism of the rhodium and iridium catalysed carbonylation of methanol: D. Forster, *Advances in Organometallic Chemistry*, 1979, **17**, 255.
3. For a description of the rhodium-catalysed low-pressure propylene hydroformylation process see: R. Fowler, H. Connor, and R. A. Baehe, *Chemtech*, 1976, 772; *Hydrocarbon Processing*, 1976 (September), 247; *Chemical Engineering*, 1977 (December), 110.
4. L. L. Murrell in *Advanced Materials in Catalysis*, edited by J. J. Burton and R. L. Garten, Academic Press, 1977, 235; M. S. Scurrell, *Catalysis*, Chemical Society Specialist Periodical Report, 1978, **2**, 215; R. H. Grubbs, *Chemtech*, 1977, **7**, 512; D. D. Whitehurst, *Chemtech*, 1980, **10**, 44; N. L. Holy, *Chemtech*, 1980, **10**, 366.
5. For discussion on pore geometry see *Introduction to the Principles of Heterogeneous Catalysis*, by J. M. Thomas and W. J. Thomas, Academic Press, 1967, 206–212.
6. *Sintering and Related Phenomena*, Materials Science Research, Volumes 6 and 10, edited by G. C. Kuczynski, Plenum Press, 1973 and 1975. For general theories and applications of sintering see *Sintering*, by M. B. Waldron and B. L. Daniell, Heydon, 1978.
7. S. Brunauer and P. H. Emmett, *J. Amer. Chem. Soc.*, 1940, **62**, 1732; M. E. Dry, J. A. K. du Plessis, and G. M. Leuteritz, *J. Catalysis*, 1966, **6**, 194; G. Ertl, *Applied Surface Science*, 1979, **3**, 99; *Chem. Phys. Letters*, 1979, **60**, 391; for general aspects of ammonia synthesis see: *Synthesis of Ammonia*, by C. A. Vancini, Macmillan, 1971; *High Pressure Ammonia Synthesis*, by G. W. Bridger, R. E. Gadsby, and D. E. Ridler, in *A Treatise on Dinitrogen Fixation*, edited by R. W. F. Hardy, F. Bottomely, and R. C. Burns, J. Wiley and Sons, 1979.
8. *Fluid Catalytic Cracking with Zeolite Catalysts*, by P. B. Venuto and E. T. Habib, Marcel Dekker, New York, 1979; *Properties and Applications of Zeolites*, edited by R. P. Townsend, Chemical Society Special Publication (No. 33), 1980; J. A. Rabo, R. D. Bezman, and M. L. Poutsma, *Acta Phys. Chem.*, 1978, **24**, 39; for other recent surveys of zeolites as catalysts see *Molecular Sieves II*, ed. J. Katzer, Amer. Chem. Soc. Symposium Series, No. 40, 1977; *Carboniogenic Activity of Zeolites*, by P. A. Jacobs, Elsevier, 1977; M. S. Spencer and T. V. Whiltam, Chem. Soc. Spec. Per. Rep. Catalysis, 1980, **3**, 189; B. Imelik *et al.*, *Catalysis by Zeolites*, Elsevier, 1980.
9. *Zeolite Molecular Sieves*, by D. W. Breck, John Wiley and Sons, 1974.
10. P. B. Weisz, *Pure Applied Chem.*, 1980, **52**, 2091.
11. Cf. V. J. Frillette, W. O. Haag, and R. M. Lago, *J. Catalysis*, 1981, **67**, 218.
12. *Chemical Engineers' Handbook*, 5th Edition, edited by R. H. Perry and C. H. Chilton, McGraw-Hill, 1973.
13. *Catalytic Processes and Proven Catalysts*, by C. L. Thomas, Academic Press, 1970; *Handbook of Catalyst Manufacture*, by M. Sittig, Noyes Data Corporation, 1978.

14. J. A. Busby, A. G. Knapton, and A. E. R. Budd, *Proceedings Fertiliser Society*, 1978, number 169.
15. "Ammonia synthesis catalysts," by G. W. Bridger and C. B. Snowdon, in *Catalyst Handbook*, Wolfe Scientific Books, 1970, 126.
16. S. P. S. Andrew, *Chemtech.*, 1979, 180.
17. J. S. Magee and J. J. Blazek, in *Zeolite Chemistry and Catalysis*, edited by J. A. Rabo, *American Chemical Society Monograph*, **171**, 1976, 615.
18. *Atlas of Zeolite Structure Types*, by W. M. Meier and D. H. Olson, The Structure Commission of the International Zeolite Association, 1978.
19. *Chemical Reactor Design for Process Plants*, Volume 1, by H. F. Rase, J. Wiley and Sons, 1977, Chapter 5. For examples of the testing and development of new catalysts see: *Catalyse de Contact—Conception, Preparation et mise en Oeuvre des Catalyseurs Industriels*, by J. F. Le Page, Edition Technip (Paris), 1978.
20. "Transmission and scanning electron microscopy", by C. M. Sargent and F. J. D. Emburn, in *Experimental Methods in Catalytic Research*, Volume II, edited by R. B. Anderson and P. T. Dawson, Academic Press, 1976.
21. *X-ray Diffraction Procedures for Polycrystalline and Amorphous Materials*, by H. P. Klug and L. E. Alexander, 2nd edition, J. Wiley and Son, 1974.
22. "Determination of surface area and pore structure of catalysts", by W. B. Innes, in *Experimental Methods in Catalytic Research*, Volume 1, edited by R. B. Anderson, Academic Press, 1968; *Adsorption, Surface Area, and Porosity*, by S. J. Gregg and K. S. W. Sing, Academic Press, 1967, 27–33; S. Parkash, *Chemtech*, 1980, 572.
23. *Electron Spectroscopy: Theory, Techniques and Applications*, Volume 1, edited by C. R. Brundle and A. D. Baker, Academic Press, 1977; S. J. Thomson, in *Catalysis*, Volume 1, Chemical Society Specialist Periodical Reports, 1977, Chapter 1; *Handbook of X-ray and Ultraviolet Photoelectron Spectroscopy*, edited by D. Briggs, Heyden, 1977.
24. "The use of the electron probe microanalyzer in catalysis", by G. R. Purdy and R. B. Anderson, in *Experimental Methods in Catalytic Research*, Volume II, edited by R. B. Anderson and P. T. Dawson, Academic Press, 1976.
25. *Characterisation of Catalysts*, J. M. Thomas and R. M. Lambert (eds.), John Wiley, 1981; G. J. K. Acres, *Platinum Metals Rev.*, 1980, **24**, 14.
26. D. R. Goodman, in *Catalyst Handbook*, Wolfe Scientific Books, 1970, 168–174.
27. *Sulphide Catalysts, Their Properties and Applications*, by O. Weisser and S. Landa, Pergamon Press, 1973, Chapter 3.
28. *Catalyst Deactivation*, International Symposium, Antwerp, 1980, eds. B. Delman and G. F. Fromert, Elsevier.
29. *Preparation of Catalysts I*, edited by B. Delmon, P. Jacobs, and G. Poncelet, Elsevier, 1976; *Preparation of Catalysts II*, edited by B. Delmon, P. Grange, P. Jacobs, and G. Poncelet, Elsevier, 1979. See also ref. 13.
30. For example a plant has been commissioned for the large scale regeneration of discharged hydrodesulphurisation catalyst, see *Chemical Age*, 1980, **19** (7 March).
31. D. R. Goodman, *The Chemical Engineer*, 1980 (February), 91.
32. For a recent comprehensive survey on the use and value of industrial catalysts, see D. P. Burke, *Chem. Week*, 1978, March 28th, 42, and April 4th, 46.

CHAPTER THREE

METHODS OF OPERATING CATALYTIC PROCESSES

P. E. STARKEY

3.1 Introduction

To find use in chemical manufacture, a catalyst must become part of a process with sufficiently attractive economics to justify major expenditure required in establishing a working process. The more novel the ideas and further the departure from known technology the greater the risks, and so the rewards must increase in proportion. Modern plant is expensive, partly to attain the economics of scale, and partly due to the increasing sophistication and control applied, but also to meet the safety standards required by today's social and political awareness of the dangers in modern plant.

For the chemist and research worker, the chemistry and catalyst can become the total statement of the problem. The place the reactor has to play in the process should not be underestimated, but the capital cost and physical size of the reactor itself are usually only a small part of the total. Many hurdles have to be overcome before the employment of a new catalyst in a process. Decisions have to be made which seem divorced from the catalyst's requirements or capabilities, but it is the interaction and balancing of the many conflicting factors which ultimately make a successful and profitable process.

An understanding of the basic philosophy of process design and the interactions of systems, both physical and economic, is vital in transforming a novel catalytic reaction into a viable process. The complete design is a major task split between many skills and disciplines. The initial predictions can only be tentative as the process can undergo many metamorphoses and emphasis changes during the conception and creation stages. The starting point is, however, crucial as early decisions tend to influence, restrict and blinker efforts further downstream. As the development of an idea can be undertaken within one to three months, the time period and work load compare favourably with experimental work programmes and the reward from early evaluation cannot be over-emphasised.

This chapter will attempt briefly to introduce and exemplify some of the choices available to the process designer. The basis of process design is one of the topics in the curriculum of chemical engineers and as such cannot justly be treated in this chapter. However, many excellent text books and articles exist which cover the required knowledge and logic, both fundamental and specific.[1]

3.2 Implications of catalyst development for process operation and design

In the evaluation of a catalyst in a process, some understanding of the catalyst and process interactions is necessary before advantage can be taken of any new discovery. Not all evaluations are of equal difficulty. The benefits may be obvious after a few moments' thought or only after many man-years of work and much expenditure. The initial idea for a new catalyst may come from a group primarily interested in the chemistry or physical nature of the catalyst, and which may not be aware of many other features of the total process. It is prudent to maintain an awareness of the total problem, and what the process needs of a new catalyst as well as what is practicable to provide. The divergence between these requirements is sometimes extreme but careful work to meet the desired combination can often be far more fruitful than attempts made in isolation from the real problems.

The catalyst discoveries requiring evaluation fall broadly into three categories.

3.2.1 *Minor change to an existing process*

A new or improved catalyst has been discovered for an existing process with a significant change in one of the properties which affect the economics of the process. The operating conditions of the new catalyst are sufficiently similar to those of the old that no major capital expenditure will be required on the plant. The plant has considerable operating and trouble-shooting experience, and has many years of productive life remaining. The market for the product is known, prices stable, quality and end-uses clearly defined. The benefits of the new catalyst could be in one of many areas, such as:

(1) Increased selectivity gives reduced raw material costs
(2) Improved activity allows higher plant outputs within the constraints of existing equipment
(3) Greater conversion through the reactor results in smaller recycle streams
(4) Reduction in catalyst temperature or pressure requirements reduces heating and compression costs
(5) Reduction in the rates of sintering, mechanical breakdown or poisoning lengthens catalyst life
(6) Wider tolerances to composition variations of feed and recycles puts less load on the purification systems.

Often combinations of these improvements can have a synergistic effect which outweighs any of the individual benefits. This is often the case with marginal catalyst improvements which give a total package of sufficient worth to risk a change. Such an example would be an activity increase beyond the existing plant limit but coupled with an increase in impurity tolerance. The feed and recycle units can now provide more feed at a lower quality and the advantage of the activity increase can be realised.

The scrutiny of the new option is relatively easy in this case. There is real experience to be drawn upon and often the target and properties of the new catalyst can be identified in advance and work appropriately initiated and directed. The economics of the change can be precisely quantified and tight confidence limits placed upon the expected benefits and costs. It must be borne in mind when initiating research on a new catalyst that the existing catalysts have a range of operating conditions and are not fixed to a specific set of conditions. They may have been optimised when the economic environment was different (e.g. cheap fuel or fewer pollution restrictions) and are still capable of change to meet the current situation. A new catalyst can turn out to be the old one viewed in a different light. Processes can be revitalised by a re-evaluation of the existing system in the light of current technological capabilities, e.g. equipment design, corrosion-resistant materials, control and instrument design—factors which may have constrained the original options. The redesign of a process, with the same catalyst, to meet today's needs can give marked improvements. Many a "next-generation" plant has taken a step forward without changing the catalyst. Process design opportunities abound for many of the existing process routes to established chemicals and this should not be ignored in any wide-ranging research programme.

Once the benefit of a new catalyst has been identified and agreed upon by all interested parties there are often few problems with the implementation of the change. The catalyst is often the easiest part of the process to change; there will nearly always be specially designed pieces of equipment and features within the plant to allow catalyst changes. This is because the catalyst is usually a consumable, with a limited life in any plant. The catalyst might almost be classed as a reactant, actually passing through the plant with the product, but even if the catalyst is fixed within the reactor there comes a time when its properties have deteriorated to such an extent that replacement is economically justified. The substitution of a new catalyst of different formulation is then an easy matter.

3.2.2 *Major change to an existing process*

A new catalyst has been discovered with a radical difference in one or more of the properties which affect the process. Major items in the process will have to be replaced or modified, or even a new plant built to take full advantage of the new catalyst.

Evaluation is now more critical as many factors will change and the need to make judgements on possible outcomes is introduced into the problem. The uncertainties have increased and hence the risks of failure are greater. The economic climate and rewards must justify the venture. Many processes successfully pass through this phase in the course of their maturity, especially at the stage when new plants are being built either to replace those which are obsolete or to expand output. Very few new plants are identical to the old in all respects, they change as the process evolves, building on past knowledge and experience by stages. Catalytic research is aimed at improving the economics of operation or responding to changing market forces on the raw material and product. Often increasing scale of production demands new solutions to problems.

It is often at this point that new producers are attracted to the business as it can give their product an economic advantage over existing manufacturers who, although able to use the same technology and catalysts, still have most of their output based on the old process. The basic technology is known, most of the problems have been solved, the product has a proved market and price structure and the initial risks of a totally novel process and product have been overcome.

Evaluation of the proposed changes will involve the design of the new features and their comparison with the original sections. Careful study is necessary to confirm the effect of these changes on the remainder of the plant. As there is a large and secure core of the process which has known costs, the study can be performed with a high degree of confidence in the validity of the conclusions.

3.2.3 *Radically new process or chemical*

A novel route to an existing or new product, with its associated catalyst, has been discovered. A total process has to be designed around the catalyst and proposed operating conditions to allow the evaluation of the new ideas. Comparison with existing or other projected process will indicate the benefits or drawbacks of such a radical change. The evaluation and design is a difficult task, subject to many pitfalls which can affect the final outcome of the analysis.

(1) Known technology is being used as a comparison to a conceptual, scanty and often ill-defined proposal. The level of model credibility can be low
(2) There exists a danger of loading the study with too many problems, as well as ignoring them
(3) Any technology has a learning curve. The extent of future improvements is always difficult to predict (cf. chapter 4)
(4) The catalyst and process concept may still be some way from a final definition; some indication of present worth is needed even to sanction the further work of development. If this is not forthcoming the ideas might be abandoned without their full promise being realised.

The risks involved in entering a new market or starting such a project are great. The financial outlay before a return is obtained can be a major strain on the resources of any organisation and cannot always be tolerated. Market development, environmental standards and plant safety all have to be successfully tackled before ultimate acceptance of the new process is possible. Clearly, the evaluation of a new catalyst and process is a difficult and large-scale exercise spread over some considerable time. Often pilot-plant studies and trial production are required before all the external constraints can be assumed to have been satisfied. The cost of such a venture can be justified only if the product is of very high value, e.g. special-effect or pharmaceutical, or the potential scale of output is large enough to warrant the risks and investment of time, money and effort (cf. 4.4).

3.3 Process mass and heat balances

An essential part of any process design is to perform a mass and heat balance to determine the duties of the various units and the overall material and energy usages. The balances can have varying degrees of numerical accuracy and physical reality, but they are such a useful guide and source of information that it should be policy to carry out an approximate balance at the earliest opportunity. Sections of the plant for which information is inadequate can be modelled by analogy to similar existing systems and more precise data sought as the project progresses. Further attention must be focused on these areas since unexpected features may emerge if grey areas are ignored until the project is well advanced. Any system with recycles, for example, *must* have an early mass balance performed as the results often pose questions relating to catalyst behaviour under slight variations in feed conditions made to benefit the remainder of the process. The procedure can become cyclic as the interaction between catalyst and process is considered.

Computer programs exist which can help solve the larger process models, and at various levels of design sophistication, detailed modelling of individual units can be used to obtain a very close approach to a realistic final design.[2] Texts and worked examples are also available to help the designer.[3]

3.3.1 *Mechanism of performing a mass balance*

The calculation of a mass balance follows certain general rules and requires a knowledge of the needs and performance of the catalyst and process units. Often the most difficult factor is picking a start point in a cyclic system, and being able to work back to that start point.

The reactor feed conditions are usually well-defined and normally provide a suitable point. From the reaction chemistry, with real stoichiometry and by-products, the composition of the product stream can be calculated. The remaining stream compositions and flow can be calculated from the model of the process, using the projected sequence of operations and the performance characteristics of the individual units. The calculation should be based on molar flows and use convenient round numbers for ease of calculation. Once the balance has been completed it can be scaled to give the desired plant output. Minor components of similar property can be lumped into one group, but the final balance must consider them in individual detail, and feed impurities must be included in the balance to ensure they do not build up in recycles. Several interactive cycles might be required to close the balance to the required mathematical accuracy. The balance can then be manipulated to investigate the effect of changes in reactor performance, feed compositions and separation efficiencies on the final scale, the recycle requirements and process complexity.

The mass balance will almost certainly indicate that the process raw material usages are higher than the reactor experimental results would indicate. The situation is aggravated by increasing numbers of recycles, purges and purification systems in the process. Each separation stage can cause the loss of 1% to 2% of the product flowing through the unit, which in recycle systems with low conversion reactions can be much larger than the output rate from the plant.

Once losses and mass flows have been determined, the features that cause loss or process problems can be concentrated upon in an attempt to improve the process. In an extreme case of processing loss it could be more advantageous to reduce these losses rather than enhance the reactor selectivity, and experimental work could be so directed. The separation stages in the process can have more potential for improvement than the catalyst system, often another possible cause of the steady improvement in productivity in each new generation of plant.

3.3.2 *Mechanism of performing a heat balance*

Once a mass balance has been performed it is possible to consider the heat balance. The aim is to be able to specify the heat to be added or removed at each stage of the process, and to balance the heat sources with the heat sinks to minimise the energy requirements. Sizing of heat exchanges and reactors can be started and the suitability of the thermal design checked. The heat balance can produce details which could be changed to advantage, and some stream heat load matching could require modification of the process temperatures or routing. Occasionally some feature of the thermal design requires recycling back to the mass balance before a solution is possible.

The heat balance requires a substantial body of available data, often including an estimation of the thermal properties of the more unusual chemicals.[4] The basic requirements are:

(a) The total flow at each point in the process
(b) The composition of each flow
(c) The heat of reaction, and its variation with selectivity
(d) The heat capacity of each component over the temperature and phase ranges of interest
(e) The latent heats of melting and evaporation of all components which undergo phase changes
(f) The vapour pressure data of all components, especially if distilled.

The heat to be added to or removed from a stream which undergoes a temperature change and/or phase change can be found by use of the following type of equation:

$$\text{Heat load} = Q = \sum_{i=1}^{n} q_i = \sum_{i=1}^{n} (M_i \overline{Cp_i} \Delta T + M_i \lambda_i)$$

where M = molar flow rate
$\overline{Cp}$ = specific heat
λ = latent heat
ΔT = temperature rise or fall

If no phase change occurs the second term is not required. The specific heat must be the value which covers the temperature range of the stream. This can be performed for each component in the stream and then summed.

Heat loads in reactors can be calculated from the known reaction rate and heats of reactor.

$$\text{Heat load on reactor} = (\text{mass or molar reaction rate}) \times (\Delta H)$$

where ΔH = heat of reaction

Some heat loads, such as in distillation, will require unit designs to be performed in detail before they can be assessed. After stream matching, accounting where necessary for approach temperature limitations,* the outstanding heat loads can be supplied by steam, burning fuel or cooling water. Summation of these services gives the energy inputs that the process requires.

3.4 Equipment design for process evaluation

No attempt can be made in this text to teach the equipment design principles required to carry out a process capital cost estimation. However, design rules, both basic and highly detailed, are available in the literature. Perry's *Chemical Engineering Handbook*[5] is an excellent general text which covers nearly all aspects of short-cut design techniques—no better source for initial data and design can be consulted. The aim of the preliminary process design is to establish approximate capital cost and operating expenses. An extensive and precise design can wait until the project has shown sufficient promise to warrant the expense.

Bearing this in mind, the information from the mass and energy balance can be used with the short-cut design techniques to:

(1) give a size to each unit (calculated from suitable first principles) sufficient to describe the scale of the equipment. Heat exchanges are costed on surface area, pumps on power and flow requirements, etc. No detailed mechanical design is necessary unless it is required for arriving at an estimated cost. Charts and tables are readily available relating cost to size, materials of construction, etc.;
(2) show up likely critical equipment design points or constraints. Attention can then be concentrated on these few items to resolve the problems;
(3) specify materials of construction, possible corrosion restrictions, temperature and pressure limitations;
(4) show how the individual units would be affected by some of the possible options which can be considered.

* Temperature approach in heat exchangers relates to the closeness of matching of the temperatures of the two streams. In practice, differences of 20 to 30°C are the best that can safely be assessed in an initial design without detailed consideration of the individual exchanger designs.

The precise order of designing the individual items in a process is somewhat arbitrary and is usually tackled in a sequential manner, as the units have most interaction on those immediately downstream. Units of similar duty, e.g. heat exchangers or pumps, can often be designed *en masse* but it is often more convenient to deal with all the units in a "box" at each stage. The typical process outline as given in the literature does not usually attempt to show the detail of the actual equipment or its arrangement. An outline process diagram shows boxes for the individual transformations that are performed on the streams, for example the box for an absorber could be more fully displayed as in figure 3.1, and on a final design drawing might cover several large sheets with all the control details included.

This explains how a process description consisting of several boxes loosely interconnected with lines and a few words such as "purification" or "reactor" can have estimated costs of many tens of millions of £ (sterling). The work and detail required in arriving at such an estimate is not usually displayed in the articles freely available.

A list of the typical process sections is given to indicate the extent of the design load, and the total number of options and decisions to be made. Each item of equipment requires decisions as to type, design criteria and process suitability. The units for which several options exist must be carefully considered, and an evaluation of the benefits and drawbacks of alternatives made whilst the design is progressing. The multitude of options is too great to leave the choice totally open, and some narrowing of the process definition has to be made at an early stage. The list is not exhaustive and each process must be considered carefully to ensure that no equipment items or processing requirements are overlooked. Units should be included in an initial assessment if any possibility exists that they will be required, and deleted later when the process definition has been

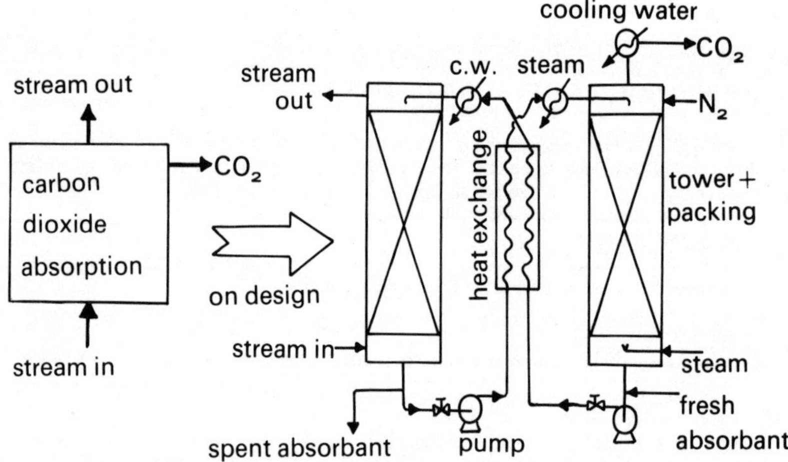

Figure 3.1 Process boxes encompassing a section of plant equipment.

confirmed and simplifications are carried out. Inclusion of such items can be problematic in the later stages of a process evaluation, causing reworking of the design to accommodate them.

(1) *Feed system.* This consists of the delivery and handling systems for the raw materials, catalyst and other consumables required in the process. Storage facilities and safe handling procedures must be considered and suitable equipment provided. Equipment for pretreatment, mixing and heating is required to give feeds at the desired temperatures, pressures and flow rates for the process needs.

(2) *Reactors.* A wide range of reactor types is available, each best suited to a particular catalyst system and set of operating conditions (cf. 3.5). Factors which affect the choice are the catalyst's physical form, the reactor volume, the level of agitation or mixing between catalyst and reactants, the heat released in reaction and the temperature control required. Additional features depend on the catalyst charging technique, activation and regeneration requirements. Reactor control mechanisms must also be provided, especially for oxidation reactions.

(3) *Heat exchangers.* Heat exchangers are used in nearly every section of a process and can be the most expensive single group of equipment after the piping. They are sized using the mass and heat balances, which define the heat loads and process fluids which each must handle. The cost of heat exchangers is a function of their surface area, design pressure and materials of construction. Problems of fouling and deposition must be considered as the performance of exchangers can be very adversely affected; efficiencies can drop to only fractions of the original design on fouling. The requirements of plant start-up must also be considered, as extra exchangers can be required to heat up the system.

(4) *Pumps and compressors.* These provide the means of moving the material around the plant and their power requirements can be calculated from the mass balance and initial unit designs, which provide the system pressures. Particular problems which can require costly solutions are dirty fluids, slurry handling, high temperatures, low-pressure gases and high differential pressures across individual pumps.

(5) *Piping.* The pipe and valve costs in a plant are usually about 1/3 of the total capital, being a function of materials construction and design pressure. It is difficult to cost piping without detailed design and so it is usually estimated as a percentage of the cost of each item of equipment, typically 50%. This is the biggest unknown in a preliminary evaluation and hence partly explains the tentative nature of the design of other units. To expend excessive effort on a design of a small unit is not sensible if the cost of the biggest item is to be wholly estimated.

(6) *Separation units.* A very wide range of equipment is available which can make use of any of the differences in physical and thermodynamic properties of the components to be separated. The choice will depend on the particular systems employed, and design will require extensive and accurate physical data. Units can be very important energy users (e.g. distillation columns), or may involve chemistry to effect the separations, e.g. carbon dioxide absorption in amines.

(7) *Product handling.* Provision for product storage, dispatch and packaging must be made. Plants which supply other users direct might pipe straight from the final unit to the consumer, others might require small packages to be filled and dispatched to the market-place. The capital burden of holding large stocks of partly finished and end product must be considered and material held up in the process minimised.

(8) *Services and effluent disposal.* These are the items external to the plant. Services are fuel, water, electricity and steam purchased from beyond the plant limit; their availability and cost depend on the particular location of the plant. Effluent disposal must be considered as most plants produce significant volumes, some with toxicity and B.O.D. problems attached.

Even such a wide range can only skim the options and decisions put before a designer. The particular part the reactor has to play in catalyst utilisation is however of prime interest to the catalyst researcher and this area warrants a clear understanding of the various types with their capabilities and restrictions. This is covered in more detail in subsequent sections.

3.5 Chemical reactor types and design constraints

The reactor is the heart of any catalytic process and upon its choice and design rests the success of the process. Reactor size and cost belittle the part it plays—its requirements and performance have a major influence on the capital and operating costs of the plant. Many decisions will be taken early in the course of any project and the process can take on a clearly-defined shape, i.e. become "frozen", well before the full chemical and catalytic investigations have been completed. The initial assumptions and projections made for catalyst behaviour are therefore critical, as any significant alterations can nullify the conclusion of the process evaluation. The confidence limits that can be placed upon the evaluation are a function of the degree of definition of the catalyst system—scouting evaluations could accept a very loose definition, but projects going forward to major capital expenditure will require firm and accurate data obtained from pilot trials on the catalyst system.

The basic aim in designing a reactor is to produce a given product from certain raw materials of defined quality and quantity at the desired rate in an economical manner. Once the particular conditions of reactor feed, temperature, pressure and expected product compositions are known, work can start on process design and evaluation. Much of the early assessment can take place by treating the units of operations as "black boxes" within which the desired stream transformations occur. The values of the known constraints can be used to produce the mass and heat balances. The design must, however, quickly move into a consideration of the detail of unit design. The reactor being central to the process, it is prudent for the catalyst designer to be aware of some of the fundamental differences between reactor types and the effect of the reaction conditions on the process design and costs. The following sections will aim to highlight some of these differences and cost factors.

3.5.1 *Homogeneous and heterogeneous reactors*

In a *homogeneous* reactor only one phase is present, usually a liquid or a gas, but not inconceivably a solid. The reaction is initiated by mixing the reactants to give a uniform mixture either at the desired reaction temperature, or, if below, with subsequent heating to raise the temperature into the desired range. Further mixing or agitation is not necessarily required, but exothermic reactions which need temperature control require agitation to facilitate heat exchange to some external source.

In a *heterogeneous* reactor more than one phase is present. The reactants and catalyst are discrete components and will tend to segregate. Not only is initial mixing required but some technique of maintaining the mixture in the reactor is necessary. Reaction ceases if the phases disengage, and becomes transport limited if the necessary degree of contact is not maintained. The wide choice of heterogeneous reactors is indicated in Table 3.1.

Not all the reactions occur at the phase boundaries. In most gas–liquid systems the reaction occurs in the liquid phase with a dissolved catalyst present and might therefore be more correctly classed as homogeneous. However, the multiphase nature of the system is dictated because the reactants are different phases, gas and liquid, at the conditions of reaction. The reactor is here essentially fulfilling a dual role, that of effecting both the solution of gas in the

Table 3.1 Heterogeneous reactors

Phases	*Reactor type*	*Method of surface area generation*
gas–liquid liquid–liquid	Agitated vessel Bubble column Spray reactors Tray column	Impeller dispersion Bubble formation Droplet formation Bubbles/drops at orifices
gas–solid liquid–solid	metal gauze beds, fixed bed, tubular fluidised bed	Physical structure of catalyst mass, with gas–solid velocity difference
gas–liquid–solid	Trickle bed Agitated vessel	Physical structure of catalyst Impeller dispersion

liquid and the chemical reaction. It is possible to separate these actions in some cases by predissolving the gas in the liquid before feeding to the reactor. This would then result in a homogeneous reaction system.

Heterogeneous reactors require special design features to maximise phase contact and prevent separation with the minimum of power consumption. The technique employed will depend on the phases of the reactants and their relative volume ratio. For example, high ratios of gas to liquids use a liquid dispersion technique, whereas high ratios of liquid to gas best employ a gas dispersion technique, the aim being to disperse the phase present in the smallest volume.

3.5.2 *Plug-flow and back-mixed reactors*

An additional classification, which can include both homogeneous and heterogeneous systems, is that of a plug-flow or back-mixed reactor. A plug-flow reactor is one in which all the reactants spend exactly the same time in the system, not mixing with feeds or products of any other time period. A batch reactor, in which a charge is fed, reacted and then withdrawn is an *intermittent* plug-flow reactor. A tubular reactor is a *continuous* plug-flow reactor, feeds and products being added and removed continuously. The reactants pass through without mixing with feeds of any earlier or later period (figure 3.2).

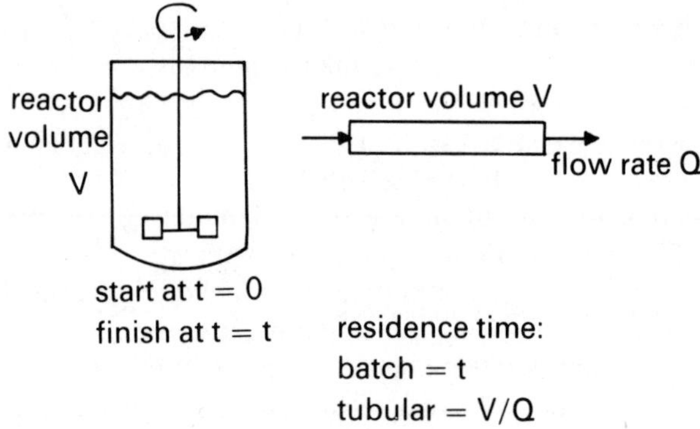

Figure 3.2 Plug-flow reactors.

A back-mixed reactor (figure 3.3) is a continuous reactor in which mixing is occurring between the fresh feeds and bulk of the system. Some reactants leave the system with the products immediately on entering. In the ideal case the mixing is sufficiently rapid for the composition to be uniform. The reaction occurs at the composition of the product stream.

The residence time distributions of the two types of reactor are shown in figure 3.4.

No real reactors can fully match these ideals. In practice, if the mixing is an

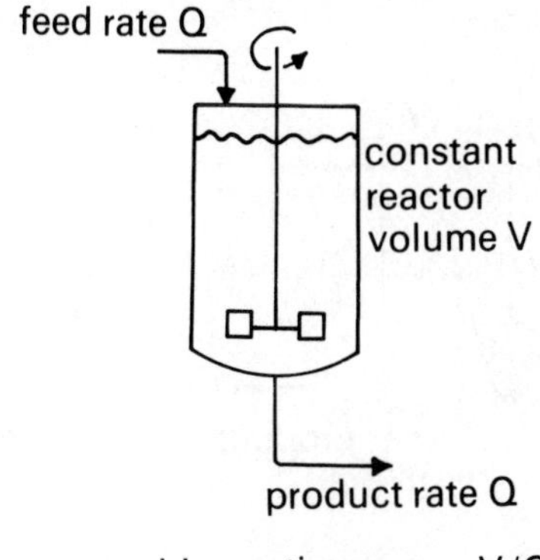

Figure 3.3 Back-mixed reactor, constant-stirred tank reactor (CSTR) type.

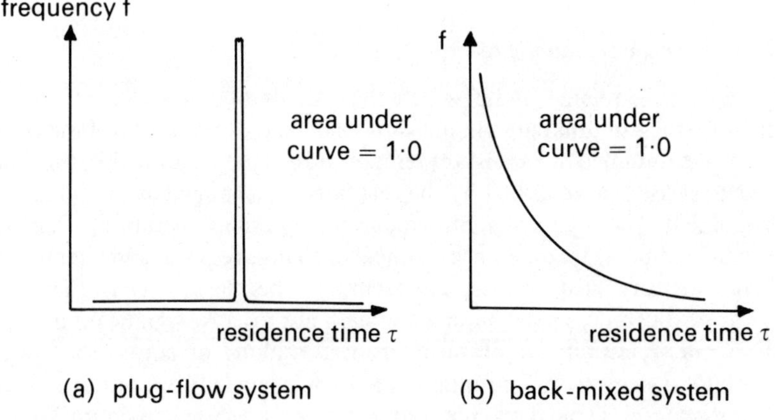

Figure 3.4 Residence time distributions of plug-flow and back-mixed reactors.

order of magnitude faster than the reaction the system can be assumed to be back-mixed, and vice-versa. Many reactor designs seek deliberately to obtain some special advantages by suitable combinations of plug-flow and back-mixed stages to enhance the effects of the chemistry and selectivity in the desired direction.[6] Multiple constant stirred tank reactors (CSTR's), each individually back-mixed, are often used for liquid systems which require agitation for mixing and temperature control but require a plug-flow reactor system to maximise outputs. The overall effect is to provide plug-flow behaviour as the number of stages increases, albeit at a capital cost penalty (figure 3.5). Practical tubular plug-flow reactors are in fact equivalent to 10 to 40 CSTR's in series; experimentally more than 25 CSTR's in series are indistinguishable from the ideal case.

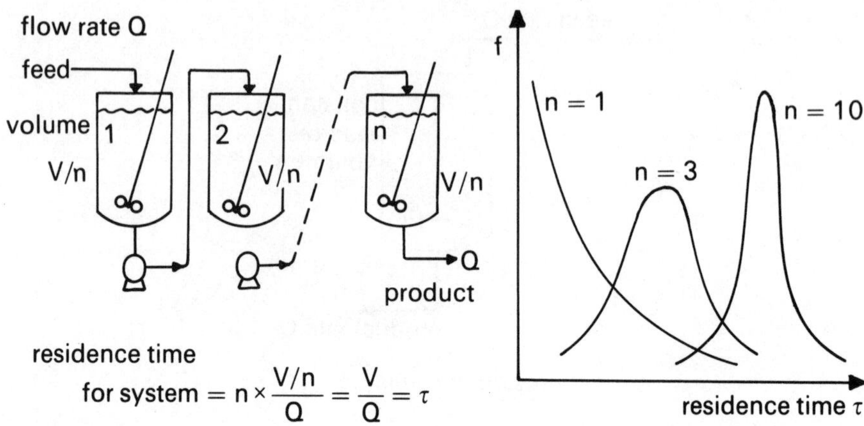

Figure 3.5 Multiple CSTR in series. As $n \rightarrow \infty$, back-mixed behaviour approaches plug-flow behaviour.

3.5.3 *Isothermal or adiabatic reactors*

The temperature regime under which the reactor operates can be classed as either isothermal or adiabatic. In an *isothermal* reactor the whole of the reactant medium is maintained at a constant temperature at all places within the reactor. Any heat released or absorbed by the reaction is exchanged at the same rate as its production. The practical achievement of isothermal operation will result in some small temperature differences, as heat exchange requires some temperature gradients for any heat transfer to occur. To be classed as isothermal the temperature differences must be small enough not to cause significant changes in reaction rate, selectivity or physical properties along or across the system—liquid phase systems with their inherent high heat capacity approach isothermal conditions closely. Typical reactor configurations are shown in figure 3.6.

An *adiabatic* reactor is one in which the heat of reaction is not removed from the reacting mass, but is allowed to change the sensible enthalpy of the system. The catalyst mass is uniform in temperature across its mass, but can vary along its mass as the reactants release heat which increases the stream temperature. Practical systems can approach closely the adiabatic ideal, heat being lost from the extremities of the bed, so calling for compact reactors with a minimum of wall area. Reactions with selectivity-reducing side reactions are not usually performed in adiabatic beds as temperature-dependent selectivity losses are severe. Endothermic reactions which cool on reaction can be used this way, the reaction being extinguished by the drop in temperature.

Figure 3.6 Heat removal from reactors. (a) Fixed tube, boiling liquid on shell side; (b) fixed tube, heat transfer medium, non-boiling liquid on shell side; (c) CSTR with jacket cooling; (d) CSTR, heat removal by boiling bed; (e), fluidised catalyst bed, coils in bed to cool; (f) fixed tube, autothermal operation.

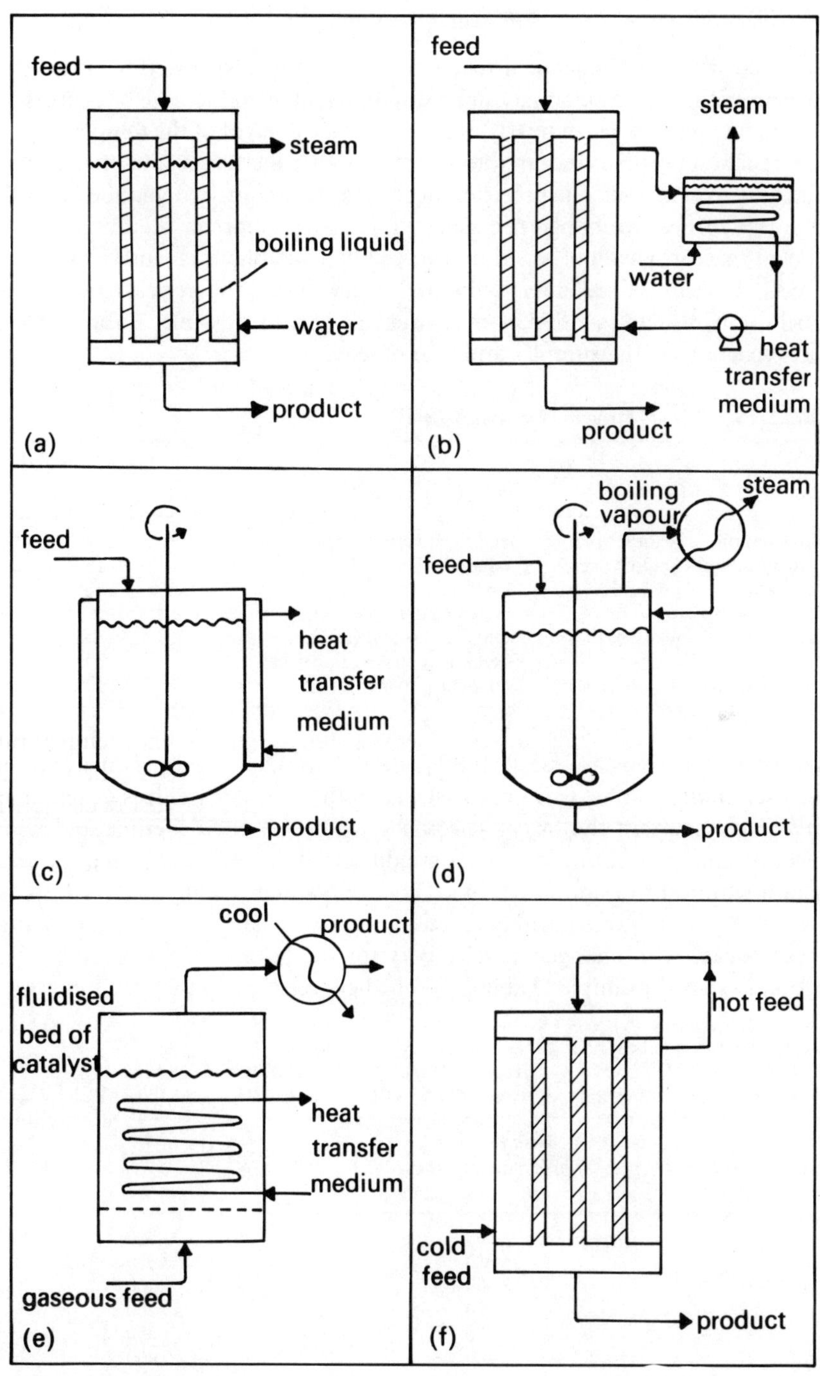
feed
steam
boiling liquid
water
product
(a)
feed
steam
water
heat transfer medium
product
(b)
feed
heat transfer medium
product
(c)
boiling vapour
steam
feed
product
(d)
cool
product
fluidised bed of catalyst
heat transfer medium
gaseous feed
(e)
hot feed
cold feed
product
(f)

3.5.4 *Effect of heat of reaction on reactor design*

Associated with every chemical reaction there is a heat of reaction. Viewed in abstraction, this energy change can be seen to result in an increase or decrease in the system temperature. In nearly every case some control of the temperature at which reaction occurs is required to maintain either the target rate or selectivity of the desired reaction. The precision of temperature control demanded and the magnitude of heat exchange necessary play a major part in the design of the chemical reactor. Heats of reaction are readily available in extensive tabulated sources[7] or can be reliably estimated from basic theoretical and group contribution principles.[8] Direct measurement is also possible. Table 3.2 lists typical reactions with examples and heat of reactions.

Table 3.2 Typical heat of reaction for ideal systems

Type		*Example reaction*	Heat of reaction, ΔH (kJ/mole)
Hydrogenation	(exothermic)	benzene to cyclohexane	−49
Dehydrogenation	(endothermic)	ethane to ethylene	+137
Isomerisation	(neutral)	*o*-xylene to *p*-xylene	−1.0
		m-xylene to *p*-xylene	+0.7
Oxidation	(exothermic)	ethylene to ethylene oxide	−105
		benzene to maleic anhydride	−290
Polymerisation	(exothermic)	ethylene polymerisation	−106

Industrial reactions are rarely 100% selective to the desired product and this loss of selectivity can have dramatic effects on the heat release. It is essential for catalysts to maintain the target selectivity, both over their lifetime and when subject to small variations in process conditions, if the reactor is to continue to function within the design intentions. Oxidation systems which have selectivity losses due to undesirable burning reactions are particularly affected. The change in heat release with changes in selectivity for the butane of maleic anhydride reaction is a vivid example (Table 3.3). The heat released in the reaction is used

Table 3.3 Variation of heat released with selectivity of reaction
Target reaction: $C_4H_{10} + 3.5O_2 \rightarrow C_4H_2O_3 + 4H_2O$ ($\Delta H = -23\,000$ kJ/kg butane)
Side reaction (total combustion): $C_4H_{10} + 6.5O_2 \rightarrow 4CO_2 + 5H_2O$ ($\Delta H = -49\,400$ kJ/kg butane)

Selectivity to maleic anhydride (%)	*Heat released, kJ/kg of butane consumed*	*Heat released, kJ/kg of maleic anhydride produced*
100	23 000	23 000
90	25 650	28 500
80	28 300	35 400
70	30 950	44 200
60	33 570	51 750
50	36 200	72 400

to generate steam to provide the energy for the other process steps. Figure 3.7 indicates the degree of integration typically achievable with exothermic reactions. The energy demand of the remaining process steps is largely a function of the product output and not of the feed input. Some heat loads in purification systems decrease as by-product formation drops, but the reduction is not pro rata. Increasing selectivity is desirable from the point of view of raw material utilisation, but as selectivity increases there comes a point at which heat released in the reaction no longer satisfies these requirements.

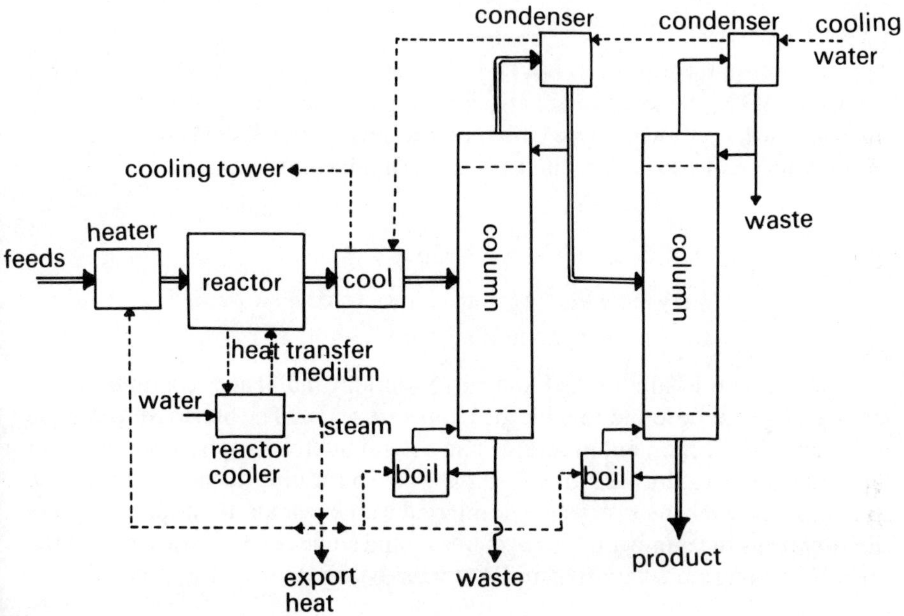

Figure 3.7 Typical interactions of process heat sources and sinks.

Large selectivity improvements in exothermic reaction systems might ultimately require additional sources of heat to be provided to compensate for the reduction in available waste heat. Any process in which the raw material is also a preferred fuel source (e.g. natural gas), might show no total decrease in the total feeds (i.e. of feedstock and fuel) with increasing selectivity. This point should be investigated in any studies on upgrading our current low-value fuel materials to chemical feedstocks; the total import of such items as coal or heavy tars to a works must be considered as well as the breakdown of its use as feedstock or fuel. However most preferred feedstocks have been pre-processed and have greater value than the fuels which would generate the additional heat required, and so improvements in feedstock utilisation will more than compensate for additional fuel import.

Table 3.4 Adiabatic temperature rise for typical reaction conditions

Reaction type	*Example*	*Conditions*	*ATR (°C)*
Flame	butane in air	no excess air	2060
Oxidation	butane to maleic anhydride	$1\frac{1}{2}$% butane in air, 80% selectivity	720
Polymerisation	ethylene to polyethylene	gaseous monomer to solid polymer	950
Isomerisation	*m*-xylene to *p*-xylene	gas phase	$\simeq$ zero
Hydrogenation	benzene to cyclohexane	liquid phase	290
Dehydrogenation	ethane to ethylene	gas phase	−1600

3.5.5 *Adiabatic temperature rise*

The adiabatic temperature rise (ATR) is a useful concept that indicates roughly the degree of heat removal and control necessary in the reactor. The temperature rise which will occur when a feed mixture is reacted to the desired end conditions without heat removal is calculated, using thermodynamic data, as:

$$\mathrm{ATR} \simeq \frac{(-\Delta \mathrm{H})}{\overline{\mathrm{Cp}}} \cdot \mathrm{Y}$$

Y = mole fraction of reactant in feed mixture

$\overline{\mathrm{Cp}}$ = heat capacity of mixture

Many reactions are carried out in essentially adiabatic conditions and methods have to be found to limit the temperature rise. The preferred technique is to carry the reaction out in several stages, with intercooling between stages to limit the temperature rise (figure 3.8). The controlling device is usually a standard heat exchanger, external or internal to the reactor. Costs of the system rise rapidly as the number of stages increase, and some compromise between the tolerable temperature variation and cost must be made.

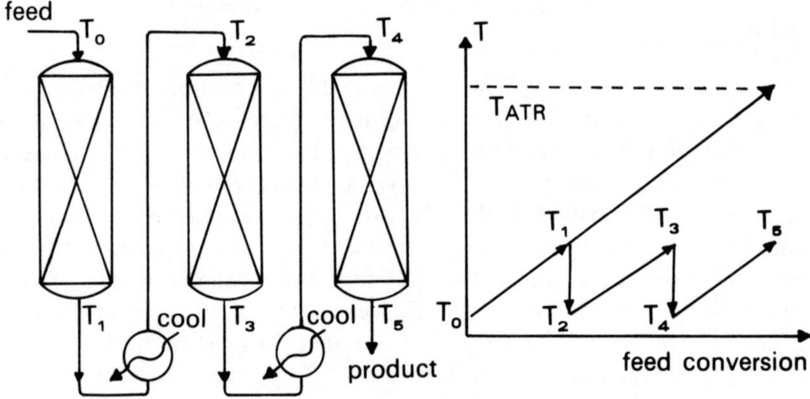

Figure 3.8 Temperature profile in multiple adiabatic bed reactors.

3.6 The effect of process temperature and pressure on the overall design

The reactor and process units will cover a range of operating temperatures and pressures and systems to provide and control them must be provided. The temperatures and pressures will normally be set by the chemistry and separation requirements. A heat transfer medium is often used to control temperatures and maintain the heat balance within a process. The total heat exchange between the process streams and the media can be many times the heat load of even the largest unit, as many heat cycles may need to be carried out on the feed before it is finally transformed to the desired product (cf. figure 3.7).

The cost of the heat transfer media and associated systems is often a major part of the process costs. The media can initially be expensive and later degrade, while the capital costs are affected by the system pressure, corrosion problems, media safety and process compatibility. An optimisation of heat loads can have dramatic effects on the total energy inputs and maximum economic exchange within the operability constraints should be attempted. Rising fuel costs make this an overriding aim, of equal importance to selectivity improvements. Ultimately a balance must be struck between the decreasing energy costs and increasing capital costs associated with extra heat exchangers used to extract all the available energy from the various process streams.

The choice of available media is wide, and suits different temperature ranges, although there are individual problems and constraints in use (Table 3.5).[9] The maximum temperature in a process is usually that in the reactor; any heat recovered from the reactor should be obtained at the highest temperature possible to maximise use in the rest of the process.

Table 3.5 Heat transfer media

Medium	*Temperature range °C (approximate)*	*Problems/constraints*
Refrigerants	−200°– 30°	Mechanical energy required. Low boiling liquids, often flammable
Water, ambient pressure	30°– 100°	Cheapest, minimal problems. Good transfer properties
Water, pressurised	100°– 240°	Cheap, restricted at high temperatures by pressure
Steam	100°– 240°	
Organic, ambient pressure	150°– 300°	Restricted by high pressure and decomposition. High media cost
Organic, pressurised	200°– 350°	Leaks cause process contamination
Molten salts, ambient pressure	350°– 500°	Difficult technology. Safety of oxidising salts problematic
Combustion gases	300°– 900°	Low heat capacity. Expensive recirculation and high losses
Electricity	700°–1500°	Only way of exceeding flame temperatures. Materials of construction limiting factor

The pressure employed in a process can have two major effects on the capital and operating costs.

(1) The capital cost of equipment and control systems increase with rising pressure, for a fixed size. For gas systems, the decreasing volume of individual units initially reduces costs and is compensating for pressures up to about 30 bar. Extreme pressures cause radical alterations to the design and construction philosophy, and safety and process operability problems increase with pressure.

(2) The cost of pumping reactants into the plant increases with pressure. For liquids the work involved is purely a function of the flow rate and pressure difference between the source and process unit. The work of compression for gases is a more complicated function of flowrate, pressure and pressure difference.

$$\text{work} = \frac{n}{n-1} \cdot P_1 V_1 \left[\frac{(P_2)^{n-1/n}}{P_1} - 1 \right]$$

where P = pressure, V = volume, 1 = initial, 2 = final

$n = \overline{Cp}/\overline{Cv}$

$\overline{Cp}$ = specific heat at constant pressure

$\overline{Cv}$ = specific heat at constant volume

Hence the work of initial compression to process pressure will increase with pressure, but the work of recycling against a fixed pressure drop is less at higher pressures. Also, medium pressure gas compressors are more efficient than low

Table 3.6 Effect of pressure range on system parameters

	Pressure (bar)					
Parameter	0	1	10	30	200	500
Volume—gas		decreasing →				liquid-like
—liquids		← no change →				compression problems →
Pumping costs, feed		← easy →		expensive →	special →	
recirculation	← expensive			easier for fixed Δp →		
Vessel wall thickness		← constant value →			increasing thickness →	
Heat exchange with gas	poor →		acceptable →		good →	
Liquid bpts.			increasing →			
Physical properties		← well known →				unknown →
Ancillary equipment		← some problems	← readily available →			special designs →
Safety				problems increase →		
Leakage	← Inwards		← minimal →		increasing →	

pressure machines. Some optimisation is therefore required. The catalyst bed structure should be arranged to give minimum pressure drops commensurate with the degree of mass and heat transport required. No machine is 100% efficient and the ideal energy requirements must be adjusted to account for pumping, gearbox and motor losses. Liquid pumps have total efficiencies in the range 50 to 80%, and gas compressors in the range 40 to 60%.

The cost and process design problems fall into broad pressure ranges, each with its own solutions and technology. The ranges merge, and can be shifted if unusual process requirements of temperature or materials of construction exist. Table 3.6 shows how some process factors change with pressure. Processes should avoid wide fluctuations in pressure between consecutive units; alternative techniques should be investigated to avoid any excessive pressure energy losses. Design carried out with estimated or extrapolated data at the higher pressures is potentially at risk, both from under- and over-design which can give false evaluation conclusions.

3.7 Effect of catalyst change and separation techniques on process design

The properties of a catalyst can change with time until it becomes unsuitable in some aspect for further use in the process (section 2.11). The change may be either chemical or physical and can make the process either economically or mechanically restricted. The rate of change of the key catalyst property is all-important in assessing the economic and process implications of such decay. Often the rate is a function of the operating conditions employed in the reactor, high reactor temperatures increasing the effects of sintering and phase change, poor purification systems giving rise to more rapid poisoning, and mechanical and thermal stresses causing catalyst structural breakdown.

3.7.1 *Effect of catalyst degradation on process*

Process design to accommodate deactivation requires information on several important factors.

(1) Is the catalyst deactivation reversible? Coke laydown on catalysts for instance can be removed to regenerate the performance.
(2) Is it practical to consider the reactivation of the catalyst within the reactor or process? Some regeneration requirements make it economical to use a specifically-designed catalyst regeneration plant, remote from the main plant.[10] Carbon removal from catalysts for instance demands the supply of oxygen to a process for burn-off. This can be a major safety hazard in what would otherwise be an oxygen-free system.
(3) Over what time scale does the property loss occur? Catalysts with lifetimes of the order of minutes can almost be considered to be reactants and must be recycled after regeneration to justify their employment. Catalysts with lifetimes

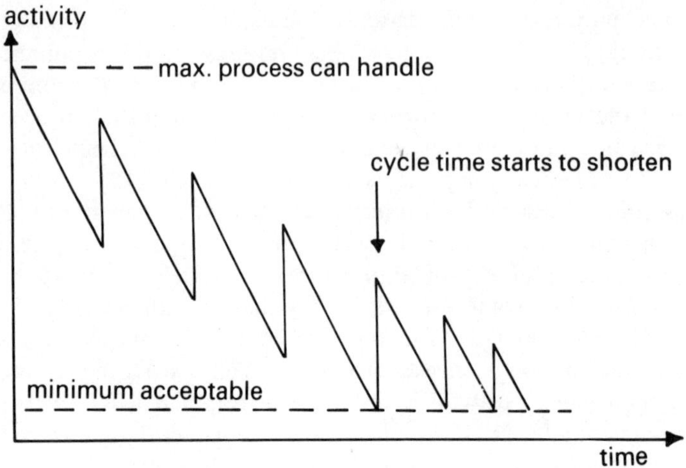

Figure 3.9 Decay of catalyst property with regeneration cycles.

of weeks are usually regenerated *in situ*. Typically a period of a day can be allocated to the task, and the equipment to provide the regeneration conditions will be both regularly used and moderately sized. Once catalyst lifetimes start to approach the half-year or more the policy of removal and replacement can be contemplated as a viable alternative to regeneration within the process. Changing the catalyst can often be scheduled to coincide with the normal maintenance programmes required on the plant.

Catalysts may have a dual decay behaviour—they may suffer rapid deactivation which is only partially reversible on regeneration. Thus a slow but progressive loss of activity occurs with each cycle. The total lifetime of the catalyst will be dependent on both the level below which the catalyst performance becomes unacceptable and the shortening period between regenerations. As activity decays, regeneration will be required more frequently to prevent the performance entering the unacceptable region. This can be seen diagrammatically in figure 3.9. Catalysts with excessive deterioration of activity or of other properties can cause significant problems for the process designer and in the economics of operation. The initial productivity of the catalyst is high, and equipment sized accordingly will become progressively under-utilised as the catalyst decays. This has significant capital cost implications, as the plant has an instantaneous capacity which is greater than the achievable average. Equipment has to be designed to cover a range of loadings, which can result in efficiency losses and require continuous adjustments in operating rates.

One method commonly used to circumvent the loss of the desired catalyst properties is to make progressive changes in the operating conditions. This is most easily understood in the response of activity to changes in temperature and is demonstrated in figure 3.10. Drawbacks to such a method of operation are:

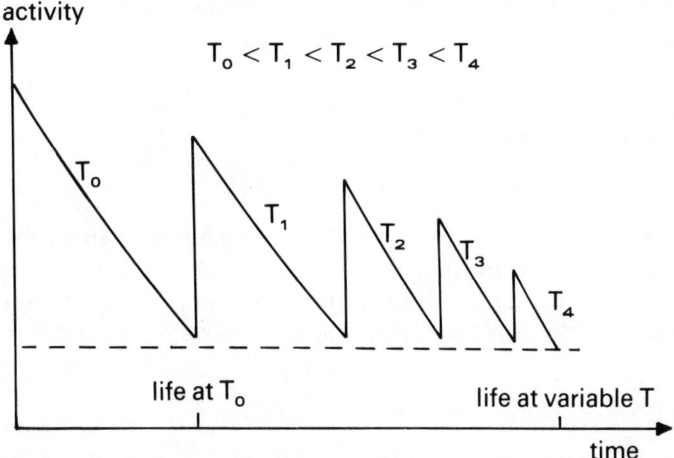

Figure 3.10 Increasing reaction temperature to maintain catalyst activity.

(1) the selectivity of reaction often falls with increasing temperature, so raw material usage deteriorates as the catalyst ages, compounding any intrinsic selectivity losses with time;
(2) the rate of activity loss often accelerates with rising temperature, giving progressively shorter periods of operation at each rise of temperature;
(3) the reactor and temperature control systems must be designed for a range of operating conditions. The cost of this additional feature can markedly increase construction and operating costs, depending on the range required and whether the design is already at the limits of its possible temperature or construction constraints.

The optimum catalyst system from a process design point of view is one with a steady long term performance, so that equipment can be designed for fixed loads, and costs of operation optimised for that condition. If catalysts which decay must be employed it is more convenient to have a long life at some intermediate performance level than short lifetimes with marked changes. The value of the catalyst depends on the total output over its lifetime and this can be taken (approximately) as the area under the activity vs. time curve. Because less variation in plant conditions is required, it can be cheaper overall to accept a larger catalyst charge of a more stable low-activity catalyst even at the expense of some increase in reactor volume. The regeneration and process changes are avoided and the design simplified. The basic strategy to minimise activity loss is often achieved by accepting operation at the lower end of the severity scale, i.e. by under-running the catalyst. This is a partial explanation for the "improvements" in many existing catalysts found when they are investigated by researchers in the short term in the laboratory environment.

Almost any of the properties of the catalyst can become the limiting factor in the process, not only the aspects of activity or temperature as exemplified. Mechanical breakdown can also be significant and is caused by:

(a) thermal stresses due to fluctuating temperatures in regeneration;
(b) grinding and crushing due to bed movements in fluidisation or sudden pressure changes;
(c) sublimation of catalyst constituents;
(d) changes in the structure of catalyst mass giving weakness as the catalyst sinters or breaks up by deposition of material in the pore structure.

All catalyst decay imposes extra costs and economic losses on a process. Additional equipment is required to cope with changes in temperature, changes in composition of streams, and the various products of catalyst breakdown. The systems must therefore operate under conditions of varying load and efficiency.

3.7.2 *The effect of catalyst poisoning on process design*

Poisoning of catalysts usually indicates that some component in the feed or recycle streams, or in the bulk of the catalyst, has destroyed the active sites which carry out the desired chemistry. If the poisoning is reversible then regeneration may be considered, but often the effects are permanent and the catalyst has to be changed. Catalysts which are particularly prone to poisoning, or are so valuable as to require extensive protection, can place major constraints upon the process design. Almost all aspects of plant operation have to be examined with a view to limiting poison ingress or production. Attention must be closely focused on the following three factors.

(a) *Raw material sources.* Special sources or pretreatment might be required since the natural variation in batches could introduce contamination, increasing raw material costs.
(b) *Feed storage and pretreatment.* Maintaining pure feeds, especially against water contamination, can place special demands on storage facilities and require guard beds to reduce poison ingress. Absorption materials are available for many common poisons, but as a last resort the catalyst itself can be used as the guard material in some suitable unit.
(c) *Recycles.* Processes with incomplete conversion over the reactor and having a recycle must be designed to provide the catalyst with recycles of the correct composition and purity. By-product separation, and ingress materials used to effect the product recovery, e.g. water or wash oils, must be investigated for catalyst poisoning characteristics.

Any catalyst modification which enhances resistance to poisoning is likely to provide process and cost benefit by reducing purification and recycle requirements. An increase of tolerance level to that present in a typical fresh feed will eliminate pretreatment requirements. Poison tolerance of about ten times the typical fresh feed levels could similarly reduce or eliminate the need for recycle treatment, poison concentration due to recycle of this order being fairly easy to limit by cheap process options.

3.7.3 *Catalyst/product separation*

By definition, the ideal catalyst has the potential for infinite operation, but in practice there will be a limit to its productivity and so the catalyst can effectively be considered a reactant, albeit of low usage. The high unit cost of most catalysts dictates their use as only a small proportion of the feed, and steps must be taken to avoid catalyst decay, recycle unused catalyst, or design the catalyst and reactor system to allow use of very low catalyst concentrations. The process design to meet these demands can be guided by considering three styles of catalyst-product separation:

(1) *The catalyst is a reactant*, used only once in the process. Fresh catalyst is mixed with the feed in the required proportion. Low catalyst concentrations and high catalyst activity are required. Spent catalyst is either removed from the product in a purification step, or (if compatible with the end-use) is left in the product. Recovery of active catalyst is often more costly than replacement (cf. polymerisation catalysts, chapter 10).

(2) *The catalyst passes through the reactor* with the feed but is recovered for recycle.

(a) No physical or chemical treatment is performed; often a small portion is purged and replaced with fresh catalyst to maintain properties and allow a simple recycle to be operated, e.g. the Co/Mn/Br catalyst used in *p*-xylene oxidation to terephthalic acid. The acid precipitates from the mother liquor and the homogeneous catalyst can be recycled.

(b) Treatment is required before recycle to restore the catalyst properties, e.g. coke removal external to the reactor for the hydrocarbon cracking catalyst.

Highly efficient catalyst recovery techniques are required for noble metal catalysts as the cost and availability demand minimal losses. Catalyst residue streams can be returned to manufacturers for recovery of active metals.

(3) *The catalyst remains in the reactor* and separation from products takes place within the reactor system. Problems of dust formation and sublimation of catalyst components can occur. The catalyst is usually heterogeneous (gauze, fixed or fluid bed) or can be homogeneous if the products are removed by evaporation.

The catalyst can have

(a) long-term stability with replacement when spent.

(b) short-term stability with frequent regenerative cycles in the reactor. Catalyst removal for regeneration outside the reactor can sometimes be justified.

Homogeneous catalysts have to be recovered from product streams by costly chemical or physical process steps, which do not necessarily retain the catalyst properties required for recycle. Heterogeneous catalysts either remain within the reactor or can be removed from product streams by mechanical separation techniques at minimal expense without loss of the desirable properties. Efforts directed at heterogenising homogeneous catalysts hope to reduce the costs inherent in the use of homogeneous catalysts while still retaining their chemical benefits of good selectivity and activity.

3.8 Process energy and service requirements

The economic evaluation of any process requires an estimation of the energy and services usage. Evaluation of options requires knowledge of the variation of usage with the scale of operation, selectivity of reaction and separation efficiencies, and good estimates can only be achieved by use of the mass and heat balances coupled with the basic unit designs, albeit elementary. The energy needs of individual unit processes can then be calculated and appropriate heat loads matched. Recent theoretical work on heat load matching in a process is providing a basis of comparison for actual designs with an objective ideal.[11] Comparison of competing models should be based on similar levels of integration if the correct conclusions are to be drawn.

The energy and service inputs to a typical process fall into six main groups: fuel, electricity, steam, cooling water, demineralised water and inert gas. The supply levels often have fixed maximum values set by an existing site system. The designer should use these but be alert to any real benefits in the use of other values for his process. The heat available from reactor cooling systems often provides the bulk of steam requirements; steam pressures should be chosen to maximise this use, as the fuel saving can be considerable. Any excess which is available for export must however meet the standards and constraints of the customer. The value (or credit) of such export is difficult to establish as the demand will be beyond the direct control of the manager.

The basic design will give the ideal inputs to the process streams and account must be taken of the efficiencies of usage in calculating the totals. Usages at start-up and at periods of plant hold must be estimated and included. Energy used in regeneration cycles must be assessed so that usage per ton of output can be calculated. The final figure can be significantly greater than the steady value in processes with frequent cycles of operations and can escalate rapidly under conditions of partial load.

Services, and various factors to be considered in the estimation of service usage, are given below, as well as details of the process design which call for their use.

1. *Fuel.* Gas, oil, coal or wood is used in specific equipment such as furnaces, boilers or engines; efficiencies of usage and loads will be well known from the design. Local supplies often have cost advantages, and high calorific waste streams suitable for burning can often be utilised to reduce imports.
2. *Electricity* is used for prime movers and pump drives, heating, lighting, instrument operation and to supply the general site facilities. Local conditions dictate the voltages and supply available. Off-peak or self-generated supplies can significantly reduce costs. Many electricity loads are not reduced significantly during periods of low plant output, total usage requiring some knowledge of plant occupacity (see 4.2.9).
3. *Steam* is used for heating process streams, vessels and pipework, for turbines and vacuum ejectors; and in the process as a reactant, diluent or to purge air

from the system at start-up. Steam is usually supplied at high pressure (40 bar), medium pressure (15 bar) and low pressure (3 bar) on large sites with centralised supply systems.

4. *Cooling water* is normally low-quality water, extracted from a river or lake (sea water is not usually acceptable). Raw water requires some chemical doping before use in the cooling system. The system can be open, returned at a higher temperature to source, or closed, recycled after cooling in towers. Closed systems using evaporative cooling have a water loss proportional to the heat load; water losses are about 0·2% of flow for each 1°C of cooling required in the tower. The water is also used for firefighting, area cleaning and other, miscellaneous uses.

5. *Process water* is high-quality water suitable for use in the process to maintain the water balance, for steam generation, critical vessel cleaning and personal uses. It is either potable or demineralised water; trace elements can be a problem in some locations or if the catalyst is sensitive to impurities.

6. *Inert gases* are used in the process as diluents, for regeneration and as a purge to eliminate oxygen from process vessels and storages. The availability of external sources can influence prices greatly. Piped supplies are cheapest, cryogenic liquid supplies reasonable and bottled supplies exorbitant if used in any quantity. Waste off-gas streams from the process can sometimes provide an inert supply after suitable treatment.

3.9 Environmental aspects to consider in process design

The operation of plants cannot be divorced from the pressure of environmental, social and regulatory forces which exist today. An evaluation must consider the suitability of the design and chemistry of a process in the light of current and future practice and restrictions. A detailed safety and environmental impact audit can accurately be done only once the process is fully designed, or even constructed and working, so that estimation techniques must be used for an initial assessment of the risks.

A very large body of knowledge concerning good designs and operating techniques is available in the construction and design guides published by Governments and professional bodies. Proposals within the current boundaries of technology and chemistry can adequately be guided by these codes, However, novel processes which step beyond the current limits of knowledge must utilise fundamental principles in assessing the risks.

Consideration of risks can be loosely grouped to cover three areas: the immediate environment, the surrounding environment and relevant long-term risks.

1. *The immediate environment*

Safety on the plant has been the main historical concern and only in recent times has significant emphasis been placed on factors beyond the confines of the factory wall. Areas of concern can be best summed up by identifying the factors which could affect the profitability of operating the plant, and listing action to be taken to avoid or minimise the risk. Typical factors are

(a) *Equipment design.* High pressures and temperatures can result in catastrophic failures if design and construction is not adequate, or safety margins are pared down to reduce costs. Corrosion margins must be generous.

(b) *Process safety.* Processes which involve reactions capable of combustion and explosion require careful control of mixture compositions. Reactor runaways must be avoided.

(c) *Process complexity.* The larger the number of stages and the greater the degree of interaction the more likely is danger to occur. Process simplicity in operation and control is desirable, combined with as many self-regulating and compensating systems as possible.

(d) *Process scale.* Large processes are not inherently less safe but the potential exists for accidents that can totally devastate a plant. Large inventories of liquids at temperatures above their boiling point should be avoided, as should excess intermediate and product storage.

(e) *Chemical toxicity.* Highly toxic materials will require expensive design features, and the operability of the process can be jeopardised. In the final analysis contact with all materials is hazardous and this should be avoided at all times. Manhandling of chemicals should be restricted. Special precautions for maintenance and vessel entry will be required.

2. *Environmental safety*

This involves the wider subject of the plant's effect on the surrounding area and population. To ignore the genuine concern and resistance which can occur would be foolish, but often problems arise because of the lack of perspective to the total problem which can blind opponents to reason or logic. Typical aspects which must be considered are

(a) *Plant location.* Risks to neighbouring population and facilities must be avoided or minimised. This is not always possible if new building is placed near to existing sites. (The reciprocity of this factor is not always accepted by objectors.)

(b) The *transport of feed and products* can be of significant concern to the neighbourhood. The problem of traffic increase alone, ignoring the chemical risks, can be severe. Suitable transportation might be required as part of the total design concept and costs.

(c) *Waste disposal.* Suitable methods of management are required for gaseous, liquid and solid wastes. Detoxification of highly active compounds could be required if local ecosystems are unable to handle the material.

3. *Long-term risks*

Many hitherto unsuspected problems are only now coming to light, often to haunt the original organisations who may have acted in good faith, within limits of legal and social responsibility formerly current. The long-term health hazards of exposure to chemicals—cancer, genetic effects and low level chronic toxicity—are recent discoveries, although the results of the poor waste disposal methods of the past might have been foreseen. The potential for future catastrophes can unfortunately only be expanding as new compounds are discovered and more exotic elements employed in catalyst formations. The dangers of asbestos, mercury, vinyl chloride and dioxin are well-known; others such as selenium, new pesticides and mutated organisms might be the agents of a future possible disaster. Avoidance can only be achieved by a high level of social awareness and foresight, coupled with sufficiently great penalties to ensure that commercial survival requires attention to such detail.

REFERENCES

1. *Chemical Engineering*, vols. 1, 2 and 3, by J. M. Coulson and J. F. Richardson, Pergamon, London, 1977, 1978, 1979; *Chemical Process Development*, Parts 1 and 2, by D. G. Jordan, Wiley-Interscience, 1968.
2. *Process Flowsheeting*, by A. W. Westerberg, H. P. Hutchinson, R. L. Motard and P. Winter, Cambridge University Press, 1979.
3. *Introduction to Chemical Engineering*, by E. V. Thompson and W. Ceckler, McGraw-Hill, London, 1977; *The Manufacture of Methyl-Ethyl Ketone from 2-Butanol: a Worked Solution to a Problem in Chemical Engineering*, by D. G. Austin and G. V. Jeffreys, Inst. of Chemical Engineers, London, 1979.
4. *The Properties of Gases and Liquids*, by R. C. Reid, J. Prausnitz and T. K. Sherwood, McGraw-Hill, London, 1978.
5. *Chemical Engineering Handbook*, ed. R. H. Perry, McGraw-Hill, London, 1973.
6. *Chemical and Catalytic Reaction Engineering*, by J. J. Carberry, McGraw-Hill, London, 1976; *Design of Industrial Chemical Reactors from Laboratory Data*, by J. Horak and J. Pasek, Heyden, London, 1978.
7. *Handbook of Chemistry and Physics*, 60th edition, ed. R. L. Weast, Chemical Rubber Co., 1979.
8. Cf. reference 4.
9. *High Temperature Heat Carriers*, by A. V. Chechetkin, Pergamon, London, 1963.
10. S. Geschurndt, *Chem. Age*, 1980 (7 March), 19.
11. B. Boland and B. Linnhof, *Chem. Engineer*, 1979 (April), 222.

CHAPTER FOUR

ECONOMIC ASPECTS OF CATALYTIC PROCESSES

D. G. BEW

4.1 Introduction

The aim of this chapter is to consider some aspects of the economics of catalytic processes. Use of catalysts is an essential feature of the production of major commodity chemicals and also of the production of many smaller-tonnage speciality or effect chemicals. Commodity chemicals are usually large-tonnage products made by a number of producers, the product made by one manufacturer being virtually industinguishable from that of another. Plant capacities in the commodity chemical area are commonly in the range of 10^4–10^6 tpa. Examples are ammonia, methanol, ethylene, polyethylene and caprolactam. Speciality or effect chemicals are bought for the effect they produce rather than as a specific chemical entity.

Products from different manufacturers can have very different compositions whilst producing comparable effects; examples in this category are catalysts, detergent formulations, resins and surface coatings. Capacities in this area range from 10–10^2 tpa, and, for some pharmaceuticals, up to 10^4 tpa.

At present the major user of catalysts and catalytic processes is the petroleum industry rather than the chemical industry. This results from the petroleum industry's continuing efforts to maximise the yield of valuable products from the crude oil barrel. Many of these processes have been developed to suit the American market where demand for gasoline has always been greater than demand for heavier fuels. As a result of this "market pull", processes for converting the higher molecular weight gas oils and heavy fuel oils to gasoline range products were introduced, e.g. catalytic cracking and catalytic reforming (cf. 5.2).

Other processes were subsequently developed to convert gaseous by-products from these "breakdown" processes and other refinery operations back into liquid gasoline range products. Processes for alkylate manufacture (reaction of isobutane with an olefin—typically propylene—to give a C_7 hydrocarbon) and the so-called polymer gasoline or polygas proccess (reaction of C_3/C_4 streams containing olefins and paraffins with acidic catalysts to give $C_6/C_7/C_8$ hydrocarbon products) are examples of the rebuilding processes.

A further important market which has developed for catalysts in recent years—particularly for noble metal catalysts—is for exhaust gas clean-up on motor vehicles.* In 1978 the U.S. motor industry used about $230m of catalysts and supports for this purpose which, in money value, was more than was used for catalytic cracking. This development provides a situation where the catalyst itself is the end product of manufacture rather than an aid to producing a chemical or fuel product.

The use of noble metals for exhaust gas converters is rapidly becoming a significant factor in the world consumption of these metals. One disquieting feature is that the logistics of collecting the spent catalytic converters from millions of cars for reprocessing is going to prove a major problem. This is in contrast to the chemical industry, where (say) a spent charge of a platinum reforming catalyst is readily available for metal recovery. On the other hand, the consumption of noble metals for this purpose may be a relatively short-term demand and could be eliminated by improvements in engine design leading to cleaner, more efficient combustion and avoidance of polluting by-products in the exhaust gases.

The effects of using a catalyst on the costs of producing a chemical, such as ammonia or methanol, are shown not only in the overall cost of raw materials—through improved yields—but also in the energy costs and capital associated costs borne by the product. Improvement in these latter components of process costs can result from the use of a catalyst permitting operation at lower temperatures and pressures or by facilitating operation to higher conversion levels with a resultant reduction in separation and feed recycle problems. These points will be considered in the following sections.

4.2 Cost of producing a chemical

The cost of manufacturing a chemical product is made up of a number of factors which can be arranged in different combinations depending on what information is required and the purpose for which it is to be used. A simple economic model which is useful for indicating the relative importance of the cost components is the Standard Cost table. Such a table can be drawn up for an operating process based on achieved results and can then be used to monitor plant performance. For a conceptual process being developed in a research department, the cost table can be used to show the targets to be achieved to give economic viability. A typical Standard Cost sheet based on published data is shown in Table 4.1.

4.2.1 *Elements of cost*

The elements contributing to production costs are usually combined into groups as follows:

* For a recent survey, see L. L. Hegedus and J. J. Gumbleton, *Chemtech*, 1980, 630.

Variable cost elements —total sum (£000/yr) varies with plant output	Raw material costs Cost of energy inputs Royalty or licence payments
Fixed cost items —total sum (£000/yr) is constant	Direct charges (operating and maintenance, labour and supervision, maintenance materials) Indirect costs Capital related charges

The first major grouping, *variable costs*, covers items which are consumed and charged to the operation account only as the product is being made. Hence the total expenditure during an operating period (whether a month, quarter or year) will vary depending on the plant output during that period. Items included in *fixed costs* are the charges which have to be met whether the plant is operating at full rate, at half rate or indeed is shut down and not producing for a short period. Since these total costs have to be paid whatever the production from the plant the cost per tonne will increase significantly at rates of output below 100 %. This point will be discussed a little later in the chapter. The groups in the cost table will first be considered in somewhat more detail.

Table 4.1 Standard cost build-up—dimethyl terephthalate (DMT)

Production scale 150 000 te/yr—operation at full capacity

Operating costs	*£000/yr*	*£/te DMT*
Paraxylene @ £300/te		189
Methanol @ £108/te		41
Catalysts		3
		233
Energy inputs		29
Direct fixed costs	3750	25
Overhead charges and other indirect costs	3000	20
Depreciation (15-year)	4867	32
Ex works cost		339
Desired return on capital of 10 %	8000	53
Required sales income		392
Capital involved (mid-1980 basis)		
Plant		£73m
Working		£7m
		£80m

With current DMT prices in Europe around £350–380/te the return, even at full output, will be reduced to 2–7.5 %.

4.2.2 *Raw material costs*

In the case of an operating plant, the usages of the process feedstocks can be obtained from measurements of process flows during periods of steady operation. For a conceptual process, or a process under development, raw material usages can be estimated from the process stoichiometry, applying assumed yields or yields obtained during research experiments. It is important to remember that the final yield of recovered product will be lower than the analysed ex-reactor product from research experiments would indicate, since allowance must be made for losses during the recovery and separation stages of the process.

General information on raw material prices can be obtained from the techno-commercial literature (e.g. *European Chemical News*, *Chemical Age*) and, as is often the case, information on U.S. prices is available for a wider range of materials (e.g. *Chemical Marketing Reporter*) than for U.K. or European prices. Within industry, additional internal company data will usually be available on raw material prices and company strategy must be considered. The commercial organisation will choose processes and raw materials to maximise advantage to the company—for example a process to a new product which consumes an existing by-product available within the company could be more attractive than an alternative which shows a higher yield but requires purchase of the feedstock.

Catalyst usages and costs are included in overall raw material costs, based either on catalyst loss per cycle for homogeneous catalysts, or cost of catalyst charge vs. output over operating life for a heterogeneous catalyst. In both cases, the value of catalyst components recoverable from purge streams or by reprocessing a spent heterogeneous catalyst charge will reduce the net cost of catalyst per tonne of product.

4.2.3 *Energy inputs*

This component of the product cost covers consumption of steam at various pressures depending on the temperatures required (for heating reactors and distillation columns), electricity (for motor drives operating pumps, compressors and agitators), cooling water (removing reaction exotherm, cooling and condensing still overheads), fuel oil or gas (for higher-temperature heat inputs), refrigeration (for operation below cooling water temperatures) and inert gas, usually nitrogen (for purging gas spaces and providing an inert operating atmosphere). These items form the major inputs of services required for process operation. For an operational process, usages can again be obtained from measurements of consumption during steady operation. For a process under development an estimate of the major services required can be prepared from the process energy balance. This can be built up from physical property data for feeds and products, together with information on operating conditions in reactors and product separation trains.

Prices for services are less readily accessible than raw material prices but indicative values can be found in chemical engineering journals. The services which are available and the prices charged for them will clearly vary from company to company and indeed from site to site within a company. A large integrated complex could operate its own power station, producing both steam and electricity at high overall thermal efficiency. Steam could be made available at high, medium and low pressures with electricity produced in associated turbine generators. Part at least of the fuel for the power station could be provided by waste product streams. A small isolated site on the other hand would have to obtain electricity from the grid and raise steam in a small package boiler using purchased fuel oil. The cost of the services provided would then be very different from those on an integrated site.

Following the massive rises in oil prices in 1973 and again in 1979 the cost of hydrocarbon feedstocks and energy has become an even more important factor in process costs than before. Processes giving high yields and conversions and operating under mild conditions with low energy requirements are even more desirable in today's conditions—a clear target for new catalytic processes.

4.2.4 *Royalty or licence payments*

These must clearly be taken into account when considering costs of alternative processes available from different licensors or when considering the relative merits of purchased technology against "in house" developed process technology. Such factors are subject to detailed negotiations between licensor and potential licensee and the results are not usually widely disseminated. The fees required will depend on the nature of the technology offered and the advantage offered over alternative processes. As an indicative figure, a charge in the range 3–5% of the estimated cost plus return for the product should give a reasonable guide.

4.2.5 *Direct costs*

The direct "people costs" associated with a process cover operators controlling the process equipment and maintenance workers who keep the equipment functioning and replace items which fail. Direct costs also include the immediate line supervision grades whose work can be readily related to a particular plant or process. Provision must also be made for materials used in maintenance work (e.g. new pumps, valves, motors) or used to repair worn or damaged items.

Charges for a running plant will be derived via numbers of men involved and works records of the annual cost of maintenance supplies consumed. For a process being developed in the laboratory such charges can be estimated by analogy with known processes on the basis of complexity, number of steps involved, process operating conditions and scale of operation. For preliminary estimates of costs, or where a suitable analogy is not available, an annual charge of 3–6% of the fixed plant capital (see below) would be a reasonable figure to include in the cost build-up.

4.2.6 *Indirect charges*

This category covers charges which are incurred in operating a number of process plants on a single site. The total costs are then allocated to individual plants and products. Factors included in the indirect charges are analytical laboratory services, general background research work, charges for support facilities (such as office blocks, canteen facilities and medical services), charges for clerical staff and works management. In addition, for a site which is part of a larger organisation, there will be an allocation of costs of head office services such as purchasing, personnel, and information services.

Even for an operating process the allocation of such charges can be a somewhat arbitrary procedure. Whilst important in determining total costs of production, they are of less relevance in process comparisons than are direct charges. For a process under development a rolled-up indirect cost of 1–3 % of fixed plant capital would be a reasonable basis for a preliminary study of total product costs.

4.2.7 *Capital-dependent charges*

Charges for depreciation of fixed plant—usually over a 10 to 15 year operating life, depending on accounting conventions and/or the process life expected before obsolescence—together with a return on capital, make up this item in the cost table. In considering the return on capital, both fixed plant and working capital must be included.

Fixed plant capital includes the reactors, distillation columns, compressors, filters and pipelines directly involved in carrying out the reaction (the so called "battery limits" capital). Since the items within the battery limits are dictated by the process they will be largely the same wherever the plant is sited. However, the fixed plant capital also includes capital for other items such as storage of feed and products, site roads, control rooms and maintenance buildings, supply of services and effluent treatment (the "offsite" capital) which can represent a large fraction of the fixed plant capital. The "offsite" capital is much more dependent on the particular situation at the site where the process is to be operated. There may be possibilities for sharing storage facilities, control rooms and workshops, and offsite capital is usually excluded from published process comparisons or data offered by contractors. Nevertheless, when estimating production costs for a chemical product, charges for the offsite capital must be included to compare with prices of materials on the market. For a typical petrochemical product offsite capital can range from 15–30 % of the battery limits capital but this is an area where consideration of the specific site involved is most important.

Working capital represents money which is tied up in stocks of raw materials, catalysts and finished product in storage. Spare equipment is also included in working capital and a major item is the net balance of (product sold but not paid for, less materials bought but not paid for). This net balance will depend on the credit terms allowed and for export sales—where credit terms of up to 6 months may be involved—can represent significant amounts of tied-up capital. In a highly inflationary situation the need to hold stocks of raw materials and products represents an increasing demand and reduces liquidity.

The above factors build up the cost of producing a chemical product. Other factors—costs of packing and transport, costs of sales agents, possibly taxes and duties—must be added to give a price for delivery to a customer.

4.2.8 *Effect of scale*

In considering the build-up of production cost an initial split into variable and fixed elements was made. The variable cost items per tonne of product (4.2.1) are practically independent of scale since the yield in the reactor and the efficiencies of separation are not significantly affected by the scale of the operation. So in the production of, say, cyclohexane from benzene, the raw material costs and services costs will be the same (in £/te of cyclohexane) whether the scale of operation is 50 000 or 200 000 te/yr. A caveat should be made: in larger scale process operations it is sometimes economic to introduce additional heat exchange or energy recovery equipment which is not viable on a small scale. This could give some reduction in energy requirements on the larger scale but, in general, variable cost items are not significantly scale-dependent.

Fixed cost items do, however, vary considerably with plant scale. The numbers of process operators and maintenance workers are by no means directly proportional to scale though the relationship will depend on the type of process involved. A highly automated process involving largely fluid handling operations (typical of the petroleum and petrochemical industries) will show a very small increase in manning requirements as the scale of operation is increased. In contrast, a small-scale batch operation involving solids handling will show an increase in manning requirements more nearly proportional to scale.

The main effect of change of scale is, however, on the plant capital and consequently the capital dependent charges per tonne of product. Plant fixed capital varies with scale by a relationship of the form

$$\frac{C_1}{C_2} = \left(\frac{S_1}{S_2}\right)^n$$

where C_1 and C_2 are fixed capital
S_1 and S_2 are plant scales
n is a fractional power.

Some early studies in the petrochemical industry showed a relationship where $n = 0.6$, and this became known as "the six-tenths rule". It is now clear that the value of n can vary widely, depending on the nature of the process and the operating scale. If the increase in scale involves doubling of equipment rather than a single larger item, then the scale factor will be considerably greater than

0.6. An example of this is shown in naphtha cracking for ethylene production, where cracking furnaces have a current capacity limit of ~30 000 te/yr. As the plant scale is increased, more furnaces are added and the scale-up factor for the furnace section of the plant is 0.8–0.9.

For relatively small plants or small changes in scale only the cost of the main equipment items will change. Other factors, such as cost of foundations and supporting structures, cost of installation, etc. will not change greatly, and the overall scale factor will be below 0.6.

An example of the effect of change in scale on production cost is shown in Table 4.2.

Table 4.2 Phthalic anhydride from *o*-xylene

Operation at full capacity*				
Plant scale (000 te/yr)	25		50	
Capital (£m)				
Fixed plant capital	12.0		19.5	
Working capital	1.5		2.5	
Total	13.5		22.0	
Operating costs	*£000*	*£/te*	*£000*	*£/te*
o-xylene @ £240/te		233		233
Catalyst and chemicals		5		5
Materials cost		238		238
Service costs (=energy)		10		10
Variable cost		248		248
Direct fixed costs	720	29	1170	23
Overheads and indirect	300	12	488	10
Depreciation (15 yr life)	800	32	1300	26
Works cost		321		307
10% return on total capital	1350	54	2200	44
Production cost + return		375		351

* Cost basis £ (mid-1980).

4.2.9 *Effect of low plant occupacity*

Table 4.2 shows calculated cost plus return for production on the two plants at full output. The larger plant provides cheaper product whilst still giving the desired return on capital. However, if the operator of the larger plant cannot sell all his product and has to operate at low output, then his cost per tonne of product will increase since all his large fixed costs have to be carried by the smaller tonnage. The effect of this is shown in Table 4.3 using the same phthalic anhydride example with sales limited to 35 000 te and a sales income of £375/te of phthalic anhydride to simplify the picture.

Table 4.3 Phthalic anhydride from *o*-xylene: effect of operation at less than full output

Plant scale (000 te/yr)		25		50
Potential Market (000 te)			35 000 te	
Sales possible (000 te)		25		35
Sales income at £375/te (£000)		9 375		13 125
Operating costs	*£000*	*£/te*	*£000*	*£/te*
Materials	5 950	238	8 330	238
Service costs	250	10	350	10
Direct fixed	720	29	1 170	33
Overheads and indirect	300	12	448	14
Depreciation (15 yr life)	800	32	1 300	37
Works cost	8 020	321	11 638	332
Profit margin	1 355	54	1 487	43
Return on capital	10%		6.8%	

Thus the operator of the smaller plant could obviously not meet all the potential market but could operate at full output showing a profit flow of £1.36m and a 10% return on his capital. The operator with a larger plant could satisfy all the market demand but his costs per tonne would be higher than those of the small plant at full output and his return on capital would be lower although the total profit flow would be larger, at £1.49m.

4.2.10 *Contribution of catalyst to production cost*

The direct contribution of catalyst cost to the total manufacturing cost is usually very small and is illustrated in Table 4.4

Table 4.4 Catalyst costs

Product	*Price* (\$/te)	*Catalyst cost* (\$/te)
Cyclohexane	639	1
Ethyl benzene	515	2
Styrene	860	5
p-xylene	617	3
Isopropanol	468	2
Ammonia	143	1
Methanol	218	1
Formaldehyde (as 37% solution)	165	1

Product prices are U.S. prices published in *Chemical Marketing Reporter*, June 1980. Catalyst cost figures are derived from data in *Hydrocarbon Processing*, November 1979.

However, the nature of the catalyst can have a dramatic effect on the overall costs of producing a chemical as a result of the influence on service requirements and process capital requirements.

4.3 The catalytic process in an integrated system

The contribution of catalyst cost to overall process costs and, indeed, the catalytic reaction itself, cannot be considered in isolation. Catalytic behaviour and the resultant effect on reaction conditions, conversion of feed materials, yield of and selectivity to desired product imply that the catalytic process must be considered in the context of the total process system. The physical and chemical nature of the catalyst and the kinetics and thermodynamics of the reaction have a controlling influence on reactor design and have been discussed in the previous chapter.

4.3.1 *Effect of catalyst on plant equipment*

In general the cost of the reactor system is a relatively small part of the total fixed plant capital. Despite this, the events taking place in the reactor in terms of conversion, product yield and yield of by-products affect all parts of the plant.

Downstream. Events in the reactor will clearly have a significant effect on all parts of the plant which handle material leaving the reactor. Low conversions in the reactor imply large reactors and a low concentration of product in the reactor exit stream. This in turn necessitates the recovery of the large quantities of unreacted feed material which must be recycled to the reaction stage to attain economic consumption of feedstock. A low catalyst selectivity will result in significant production of byproducts which could result in difficulties in subsequent separation steps and the problem of profitable disposal of these unwanted materials. Thus the catalyst performance in the reactor will control to a large degree the services consumption and capital requirements for the plant downstream from the reactor.

Upstream. The nature of the catalyst can also have an influence on the plant prior to the reactor where considerable investment may be needed to produce a reactor feed which is suitable for the catalyst. A moisture-sensitive catalyst will require equipment to dry the feedstocks to the required level—examples are the well known Ziegler catalysts produced from aluminium alkyls, and transition metal halides which are used in the production of high density polyethylene. One of the largest scale catalytic processes operated in the petroleum industry is hydrodesulphurisation, in which sulphur-resistant catalysts are used to remove sulphur and nitrogen compounds from feed to catalytic reforming units and to provide low sulphur fuel oils (5.5.1).

4.3.2 *Catalyst life*

A feature of catalyst performance which has not yet been considered is catalyst life, i.e. the period over which an acceptable level of activity and selectivity is maintained. The influence of this on plant design has been considered in the previous chapter. From the point of view of process economics, the simplest approach would consider the total product made over the period of active catalyst life plus associated shutdown periods to charge and discharge catalyst. However, the rates at which activity and selectivity (and hence overall productivity) change during the catalyst life are also very important. Two contrasting patterns of catalyst behaviour are illustrated in figure 4.1. Catalyst 1

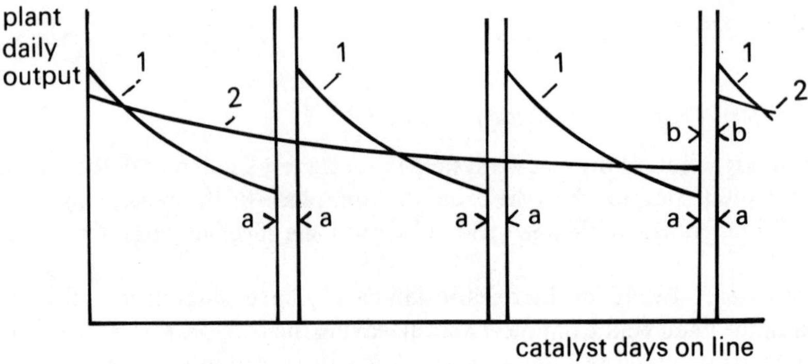

Figure 4.1 Contrasting patterns of catalyst decay. *a* >< *a*, catalyst 1 discharged and new catalyst charge in; *b* >< *b*, catalyst 2 discharged and new charge in.

shows a high initial productivity but a short life before the catalyst performance deteriorates to the point where a decision is made to shut down the reactor and install a fresh charge of catalyst. In contrast, catalyst 2 shows a lower initial productivity, but a slow decline and much longer life before a replacement is required. The total integrated production over an extended period (usually assessed on an annual basis) for the two cases will depend on the balance of initial activity, rate of decline, number of catalyst changes and duration of off-line time at each catalyst change.

From the point of view of the works personnel operating the process, catalyst 2 shows a preferred pattern of behaviour. A long on-line period without dramatic performance change allows operating conditions to be "fine-tuned" to achieve optimum production economics. With the pattern shown by catalyst 1, optimum conditions will differ considerably between the early and late stages of the catalyst life. This variation and the more frequent interruptions for shutdown and start-up will make optimum plant operation extremely difficult to achieve.

4.3.3. *Homogeneous catalysts*

So far consideration has been given largely to factors involved with heterogeneous catalysis which has the most widespread industrial application. However there are some industrially important processes which use homogeneous catalysts (e.g. hydroformylation) and during the last fifteen to twenty years the development of processes using homogeneous catalysis has been the objective of much catalyst research. The reasoning behind this approach was the belief that centres of activity could be defined and understood more completely in a known complex than in the traditional heterogeneous catalyst. This more precise knowledge should then lead to the development of more efficient and economic catalytic processes. Use of homogeneous catalysis does, however, lead to a number of problems which are not encountered in heterogeneous catalysis.

The first significant problem is the separation of the catalyst from the reaction product. Most homogeneous catalysts are based on expensive noble metals and, although used in small concentrations, these must be recovered and recycled if the system is to be economically viable. A further reason for the removal of catalyst residues from the product is that the residues, if not removed, would adversely affect the product properties. Examples of this are found in homogeneously catalysed polymerisation processes (chapter 10). Homogeneous catalytic processes which have found industrial application have, in general, shown ease of catalyst separation, usually because products of relatively low boiling point were involved – examples are the Hoechst–Wacker acetaldehyde process, the Monsanto acetic acid from methanol process and the Union Carbide–Davey–John Matthey *n*-butyraldehyde process.

A further difficulty which offsets some of the potential advantages of homogeneous catalysis is the question of catalyst regeneration. In some cases—such as the Monsanto and Union Carbide processes—the catalyst solution can be recycled, after product removal, without regeneration. The catalyst deactivation in each cycle is so small that a small (constant) purge taken for metal recovery, balanced by a small (constant) addition of fresh catalyst is sufficient to maintain activity. In other systems such as the Hoechst–Wacker acetaldehyde process, a complex bimetal redox system is involved (11.3.6). Initially a two-reactor process was used, with ethylene oxidised to acetaldehyde by the oxidised catalyst in one reactor and the deactivated catalyst reoxidised in a second reactor. This was subsequently simplified so that ethylene and oxygen were passed together into a single reactor where the two oxygen transfer reactions occurred simultaneously (figure 4.2).

A further disadvantage of many homogeneous systems is that corrosive solvents are used to maintain the catalyst in solution. This requires the use of expensive construction materials for the plant equipment and results in a considerable increase in capital requirements. In some cases the corrosive nature of the solvents can make an apparently attractive process virtually inoperable in plant terms[1] (see section 11.2.5).

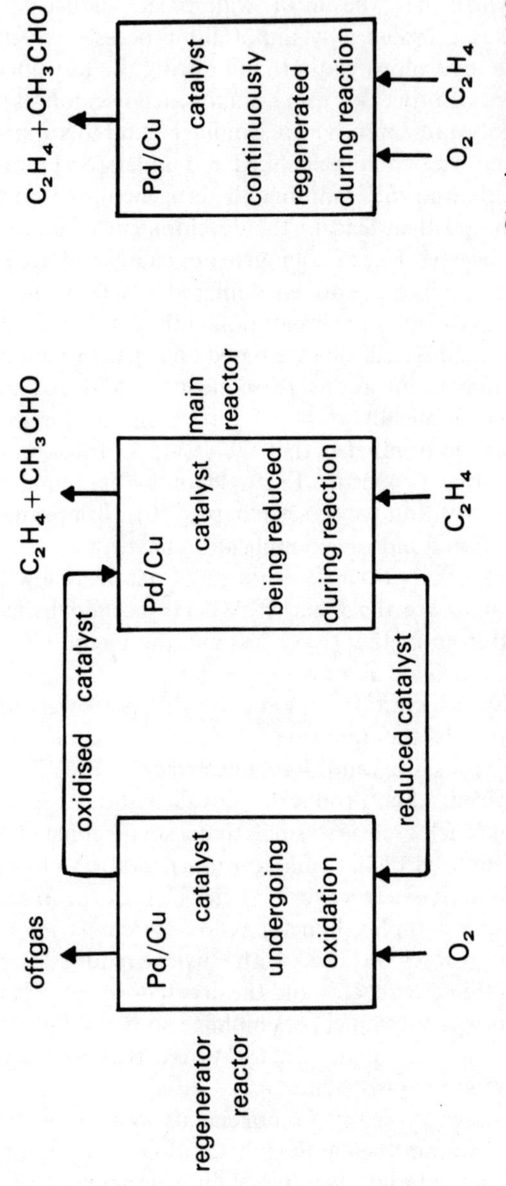

Figure 4.2 Catalyst regeneration in the Hoechst–Wacker process.

4.4 Effect of improvement in catalyst performance on process and economics

The petroleum and petrochemical industries (including in the latter ammonia production) are the major operators of catalytic processes and consumers of catalysts. In these industries the majority of processes are operated on a very large scale (10^4–10^6 te/yr) producing relatively low-value products—plastic and synthetic fibre intermediates, fertilisers and elastomers. The cost of introducing completely new technology or a new product in this area (the "entrance fee") is extremely high for several reasons. Existing products are produced efficiently, cheaply and in high volume with properties closely matched to the desired outlet. Because of the "economy of scale" effect a new product or process can only be introduced in a large plant comparable in size to units operating the existing process. The product must also develop a market share quickly to bring the plant up to capacity operation rapidly and minimise the period of low rate operation. This requires outstanding product or processing advantages and, with current plant scales and construction costs, a massive investment at risk. This is illustrated diagrammatically in figure 4.3. Two factors are significant, first the very large negative cash flow representing the high capital cost of a modern petrochemical plant, and second, the long period before a positive cash balance is achieved for the project. Sales tonnages or product prices lower than forecast will reduce the positive cash flow and delay the attainment of a positive cash balance.

As a result of the high risk and massive investment required to introduce completely new technology, the main theme of catalytic research in the commodity chemical industries is the improvement of existing catalytic processes to provide a cheaper, better product, preferably using existing production facilities and with the injection of a minimum amount of additional capital. Such

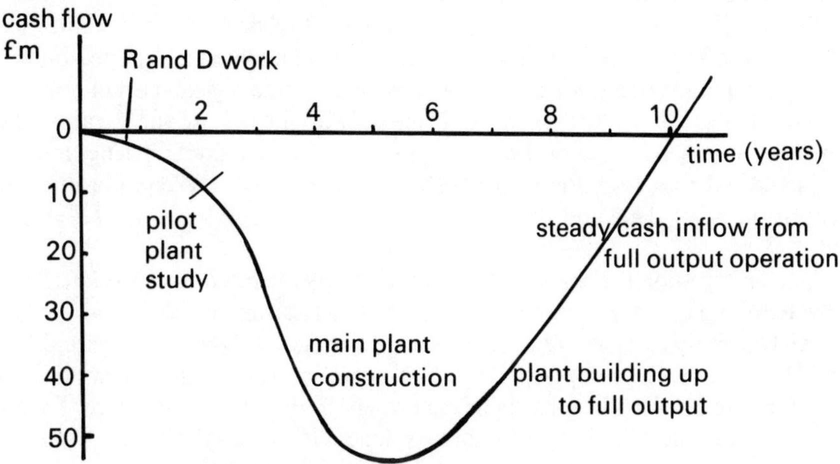

Figure 4.3 Typical "entrance fee" curve for a new process.

process improvement studies re-emphasise the importance of considering not only the catalyst *per se* but the whole process system. Indeed in processes involving a number of linked stages, such as methanol or ammonia manufacture, no change in any one stage can be implemented without detailed consideration of the effects on all the other process steps.

Improvements in the economics of chemical production can be made in several ways, for example:

1. *Improved existing process.*	(a) Use of zeolite catalysts in xylenes isomerisation.
	(b) Development of non de-ashing catalysts for high density polyethylene.
2. *Significant process or feedstock change.*	(a) Low pressure methanol process (ICI)
3. *New technology and feedstock.*	(a) Acrylonitrile from propylene and ammonia (Sohio).
	(b) Acetic acid from methanol (Monsanto).

There are numerous examples within the chemical industry of process and cost improvements resulting from catalyst development, some of which are indicated above, and these will be used to illustrate the changes which can result from successful catalyst development.

4.4.1 *Improved existing process: example 1*

An example of a technology in which a novel catalyst development allows existing plant equipment to be used but with a considerable improvement in performance is found in the xylenes isomerisation field.

The *o*-, *m*- and *p*-xylene isomers together with ethyl benzene are obtained as a C_8 aromatic fraction from the catalytic reforming of naphtha (5.2). The isomer in greatest demand is *p*-xylene for production of terephthalic acid used in polyester fibre manufacture. Unfortunately it forms only about 18% of the C_8 aromatics fraction. After separation of the *p*-xylene (cf. 9.4) the remaining xylenes mixture is passed over an isomerisation catalyst to re-establish the equilibrium concentration of *p*-xylene and the isomerised mixture is recycled to the separation stage.

An isomerisation process which has been widely operated since the late 1950's (the octafining process) carries out the isomerisation at 12–20 bar over a $Pt/Al_2O_3/SiO_2$ catalyst at temperatures in the range 440–480°C with circulation of 7–14 moles hydrogen per mole of hydrocarbon. The octafining catalyst has the ability to convert the ethyl benzene present in the isomerisation feed to low-boiling paraffinic materials and some xylenes. If the ethyl benzene were not reacted in this way an expensive fractionation would be required to prevent build-up of ethyl benzene concentration in the recycle stream. However, the production of low-value paraffinic materials represents a downgrading of feedstock value which increases as the ethyl benzene content of the isomerisation feed increases.

In the early 1970's the Mobil Oil Corporation developed a series of highly active zeolite catalysts (coded ZSM—Zeolite Socony Mobil) which were highly effective in xylenes isomerisation and also isomerised ethyl benzene. The new catalysts were of such high activity that they had to be diluted with an inert material such as alumina; they could then be used in an existing hydroisomerisation unit. Even after dilution the zeolite catalyst (ZSM-5 particularly) was effective in isomerisation at much higher space velocities than existing catalysts, thus permitting greatly increased quantities of xylenes to be processed in an existing unit. The new catalyst also gives an isomerised product containing slightly more *p*-xylene and less ethyl benzene and *m*-xylene. After separation of *p*-xylene, this means that a reduced quantity of xylenes have to be recycled to isomerisation, with a consequent reduction in energy consumption. Breakdown of ethyl benzene is more rapid over ZSM-5 than over alternative catalysts and occurs by disproportionation to benzene plus toluene, and C_9/C_{10} aromatics, both of which are high-value streams, rather than low-value paraffinic materials. As a further advantage, at a fixed hydrogen : hydrocarbon feed ratio, the ZSM-5 catalyst allows a longer run life between catalyst regenerations, again resulting in lower operating costs. Table 4.5 shows a comparison of isomerisation runs with $Pt/SiO_2/Al_2O_3$ and diluted ZSM-5 catalysts.

The potential benefits to be gained by installing the new ZSM-5 catalyst in place of an existing hydroisomeriser catalyst may not all be realisable owing to limitations in other parts of the system. However, benefits resulting from smaller quantity of xylenes to be circulated, higher value by-products and longer cycle life should be obtained even if the increased capacity offered by the higher space

Table 4.5 Comparison of platinum and zeolite catalysts in xylenes isomerisation*

		$Pt/SiO_2/Al_2O_3$	$ZSM\text{-}5/Al_2O_3$ (10% : 90% wt)
Temperature		430°C	430°C
Weight hourly space velocity—			
on total packing		1.6	10
on active catalyst		1.6	100
H_2 : hydrocarbon (molar)		6.5	6.5
Composition (wt %)	*Feed*	*Product*	*Product*
Ethyl benzene	17	10.8	8.2
m-xylene	65	46.0	44.7
p-xylene	11	18.0	18.9
o-xylene	7	16.7	18.8
C_5 paraffin and lighter	—	6.75	1.8
Benzene + toluene	—	0.75	6.3
C_{9+} aromatics	—	1.0	1.3
xylene loss	—	3.0	1.0

* Data from U.K. patent 1,484,481 (to Mobil).

velocity cannot be exploited. The savings to be made will vary from operator to operator but in the case of higher-value by-products could be 5–7 % of *p*-xylene value, i.e. £15–20/te. With *p*-xylene capacities of 100 000 te and greater, savings of £1.5–2 m/yr are involved. The process is reported now to be used in about 90 % of U.S. and 60 % of existing western world xylenes isomerisation capacity.[2]

4.4.2 *Improved existing process: example 2*

A second example where improved catalysts show significant cost savings in an existing process is in the production of high density polyethylene (HDPE) using Ziegler catalysts. As originally introduced in the late 1950's, the process involved five stages:

i. catalyst preparation—usually from aluminium tri-isobutyl and titanium tetrachloride
ii. reaction in dry hexane forming polymer slurry
iii. steam stripping of solvent, the wet solvent being subsequently dried for return to the reactor
iv. catalyst removal from the bulky polymer by sequential washes with alcohol, acid, alkali and water
v. drying, pelleting and packing the polymer.

The removal of the deactivated catalyst was necessary since (with the catalyst activities then attainable) the residual metal level would otherwise have an adverse effect on polymer properties.

By the mid-1960's several companies had developed high-activity variants of the basic Ziegler catalysts, typically by addition of magnesium compounds to the aluminium alkyl/titanium halide combination. With these modified catalysts the components can be fed directly to the reactor without the initial catalyst mixing and preparation step. Catalyst activity is so enhanced, and the resulting metal and halogen levels in the polymer slurry so low, that solvent can be removed by flashing from the reactor product and subsequently purified by simple distillation without drying. The resulting polymer melt can be pelleted without catalyst removal. For an operator planning a new plant the improved catalysts offer the prospect of considerable capital saving over a plant using a conventional Ziegler catalyst, and savings of the order of 20 % on capital have been quoted. Even with an existing plant the introduction of the new catalyst would allow the by-passing of catalyst preparation, the solvent drying step and the catalyst removal wash stages. A reduction of manning of ca. 40 % (4–5 men/shift) is thus possible on a plant producing 80 000–100 000 te/yr of polymer. Elimination of the solvent drying step and the polymer de-ashing washes also results in considerable energy savings. Savings of 15–20 % on services cost appear possible, which at current energy prices would represent ca. £18/te of polymer. More recent catalyst developments have resulted in vapour phase processes which offer even greater capital and energy savings.[3]

4.4.3 *Significant process change*

A process where development of an improved catalyst resulted in a more fundamental change in processing conditions, and hence a need for a significant change in process plant, is shown in the methanol synthesis reaction. Synthesis of methanol requires a process system in which "synthesis gas" (a mixture of hydrogen and carbon monoxide) is produced from a hydrocarbon feedstock and then reacted over a catalyst to give the desired product.

$$CO + 2H_2 \rightarrow CH_3OH \tag{1}$$

There are several possible sources of synthesis gas available, some of which depend in turn on catalytic processes. Synthesis gas mixtures can be produced by the catalytic steam reforming of natural gas (methane) or naphtha (a mixture of C_5–C_9 paraffins and cycloparaffins). If naphtha is used as the feedstock a desulphurisation catalyst is needed before the synthesis gas reactor to protect the sensitive steam reforming catalyst.

$$CH_4 + H_2O \rightarrow CO + 3H_2 \tag{2}$$

$$H(CH_2)_nH + nH_2O \rightarrow nCO + (2n+1)H_2$$

With the current escalation of oil (and hence naphtha) prices, natural gas is the preferred feedstock for steam reforming to produce synthesis gas. Synthesis gas can also be produced by partial oxidation of heavy fuel oil (Shell, Texaco processes) or from coal (Lurgi, Koppers–Totzek processes). In the present search for ways of reducing dependence on oil, interest in the production of synthesis gas from coal (which has been practised on a large scale in South Africa for about 25 years) is being revived throughout the western world. Manufacture of synthesis gas from heavy fuel or coal by partial oxidation produces significant quantities of carbon dioxide and also sulphur dioxide from sulphur-containing impurities in the feed. There are several physical or chemical adsorption processes which will remove the bulk of these acidic gases, but a final catalytic desulphurisation step is required to protect the methanol synthesis catalyst.

As can be seen from the simplified stoichiometry (eqns. 1 and 2) steam reforming of methane produces excess hydrogen over that required for methanol synthesis. The methanol reaction is normally carried out in the presence of excess hydrogen to aid heat removal from the reactor but, to avoid build-up of hydrogen concentration in the recycle gas, the excess must be purged from the reactor exit gases—possibly as feed hydrogen to an adjacent ammonia plant. If a cheap source of carbon dioxide is available from an adjacent ammonia synthesis plant (where it is produced by water gas shift in removing CO from the hydrogen feed), or from a partial oxidation process, then it can be added to the hydrogen-rich synthesis gas to produce additional methanol.

$$\underset{\text{syn gas}}{(CO + H_2)} + H_2O \rightarrow CO_2 + (2H_2 \text{ for ammonia})$$

$$CO_2 + 3H_2 \rightarrow CH_3OH + H_2O$$

A methanol synthesis process can be regarded as a synthesis gas production stage and a methanol synthesis loop with separation and recycle of unreacted gases. Before the 1960's the methanol synthesis loop was operated at temperatures of 300–400°C and pressures of 300–350 bar over zinc/chromium oxide catalysts. The high pressure required for the reaction necessitates the use of reciprocating compressors for bringing the synthesis gas to the reaction pressure. These are large, expensive machines whose uncertain reliability results in high maintenance costs. The combination of high temperature and pressure in the synthesis reactor means that this is a massive, costly unit with problems associated with gas leaks. Close control of operating conditions is needed with the high-pressure reaction since methane can be formed as a by-product. This latter reaction is even more exothermic than methanol synthesis and if not controlled could lead to temperature run-away and catalyst damage.

In the 1960's a highly active copper-based catalyst was developed for the synthesis loop which allowed the methanol synthesis stage to be carried out at ca. 250°C and pressures as low as 50–100 bar.[4] These much less severe operating conditions—particularly the lower pressure—meant that a much simplified reactor could be developed, with considerable saving in capital and greatly reduced problems from gas leakage. The new catalysts also showed better feedstock efficiency, giving reduced by-product formation, and hence lower feedstock cost per tonne of methanol. A further benefit of the lower operating pressure is that it permits the use of centrifugal rather than reciprocating compressors to raise the synthesis gas to the reaction pressure. The centrifugal compressor is of slightly lower efficiency than a reciprocating machine but is very much smaller, cheaper and much more reliable, allowing significant savings in operating costs. Use of the more reliable centrifugal compressors and the greater operability of the low pressure reaction loop also permitted the design and construction of large single stream plants, with the attendent benefits of scale in reducing capital cost per tonne of product.

In summary, the development of the new high-activity catalyst for the methanol synthesis loop offered the following advantages over the earlier catalyst systems:

Reduced byproduct formation resulting in lower feedstock cost/te of methanol
Lower operating pressure giving reduced costs for gas compression
Lower capital cost as a result of using centrifugal compressors, low pressure equipment and a simple reactor design.

Early published figures show an advantage over high-pressure technology of 4–5% or ca. £5/te at current prices. Although this may not seem very important, with plant capacities of 500–1000 te/day savings of £0.75–£1.5 m/yr are involved.

4.4.4 *New technology and feedstock: example 1*

A more radical change in technology can occur when development of a new catalyst allows development of an entirely new route to an existing product. A case of this type is the development of the Sohio process for the conversion of propylene to acrylonitrile by reaction with ammonia, originally over bismuth molybdate catalyst systems.

$$CH_2{=}CHCH_3 + NH_3 + 1\tfrac{1}{2}O_2 \rightarrow CH_2{=}CHCN + 3H_2O$$

Prior to the introduction of the propylene-based process in 1960, acrylonitrile was produced by a number of processes—in particular the addition of hydrogen cyanide to acetylene or ethylene oxide.

$$CH{\equiv}CH + HCN \rightarrow CH_2{=}CHCN$$

$$\underset{\diagdown O \diagup}{CH_2{-}CH_2} + HCN \rightarrow HOCH_2CH_2CN \rightarrow CH_2{=}CHCN$$

These raw materials are hazardous to handle and expensive and, although the addition reaction proceeds relatively easily and in high yield in both cases, the acrylonitrile produced by these processes is expensive.

Development of the Sohio process enabled two relatively cheap and readily available raw materials to be used and, although the yield was originally only ca. 60% on propylene, the process offered considerable advantages over the older competitors. The Sohio acrylonitrile process was introduced in the early 1960's and rapidly became the process of choice as demand for acrylonitrile-based fibres expanded. Sohio worked continuously to improve the process and also encouraged licensees to develop catalyst variants. The Sohio process and its variants now completely dominate acrylonitrile manufacture. Continuous development of process and catalysts which has been carried out over the years has improved efficiencies and lowered production costs, pushing the selling price of acrylonitrile along a "Boston Experience Curve" (see 4.5).

Published data on acrylonitrile production costs have been updated and compared with materials costs (based on reported yields) for the acetylene/HCN route). They show that, if acetylene and HCN are charged at cost + return, then the materials cost for the acetylene/HCN route is approximately equal to the full cost and return for the Sohio route (Table 4.6).

4.4.5 *New technology and feedstock: example 2*

A second example of a radical change in technology which is rapidly assuming a dominant role is the Monsanto methanol-based route to acetic acid. Before the development of this route, acetic acid was largely manufactured by oxidation of acetaldehyde (produced by Hoechst–Wacker oxidation of ethylene) or by hydrocarbon oxidation (*n*-butane, by Celanese Corp., or naphtha, by BP). A methanol-based process is operated by BASF using a cobalt catalyst, but this involves high pressures (700 bar) and temperatures of 250°C. The acetaldehyde route uses an expensive base material (ethylene) and intermediate (acetaldehyde) but gives a high yield in a relatively simple final process.

Hydrocarbon oxidation, by contrast, uses comparatively cheap feedstocks but produces a wide range of byproducts—including acetaldehyde, acetone, methyl ethyl ketone (MEK), formic acid, propionic acid—and considerable capital and energy costs are incurred in product separation. The raw material efficiency to

Table 4.6 Comparison of propylene- and acetylene-based routes to acrylonitrile (90 000 te acrylonitirle mid-1980)

	Sohio	*Acetylene/HCN*	
Fixed plant capital	£68m		
Working capital	9m		
Operating costs	*£/te*		*£/te*
Propylene @ £230/te	248	Acetylene @ £450/te*	275
Ammonia @ £100/te	46	HCN @ £350/te	210
Catalyst and chemicals	10	Catalyst and chemicals	10
Credit for by-products	(18)		
Materials	286		
Services	5		495
Fixed costs	30		
Capital dependent charges:			
21 % fixed†	159		
10 % working	10		
Cost + return	490		

* Estimated cost + return.
† This includes 10 % return on capital, 7 % depreciation, and 4 % overheads.

acetic acid is low and the final cost of acetic acid production is critically dependent on byproduct realisations. With recent rapid increases in hydrocarbon costs, such low-yield, high-energy-consuming processes become increasingly unattractive.

Table 4.7 Comparison of methanol- and ethylene-based routes to acetic acid (acetic acid 150 000 te mid-1980)

	Monsanto methanol/CO		*Ethylene via acetaldehyde*
Fixed plant capital	£36m		£51m
Working capital	6m		8m
Operating costs	*£/te acetic acid*		*£/te acetic acid*
Methanol @ £108/te	59	Ethylene @ £320/te	171
CO @ £130/te	69	Oxygen @ £20/te	6
Catalyst and chemicals	3	Catalyst and chemicals	4
Materials	131		181
Services	20		30
Fixed charges	25		25
			236
Capital-dependent charges:			
21 % fixed	51		71
10 % working	4		5
	231		312

Around 1970 Monsanto introduced their methanol-based process involving reaction of methanol with carbon monoxide in the liquid phase in the presence of a rhodium-based catalyst.[5] The process operates under mild conditions (30 bar, 150°C) and gives very high conversion and 98–99 % selectivity to acetic acid. Significant savings are offered in raw material, energy and capital costs compared with acetaldehyde oxidation, as shown in Table 4.7.

With present ethylene values an acetaldehyde-based producer will not be making any margin towards rewarding his capital when priced against product from a new plant. (Compare full cost plus return for £231 for Monsanto route with plant cost of £236/te for acetaldehyde route). In view of these figures it is not surprising that virtually every announcement of new acetic acid capacity since about 1973 has involved the Monsanto process.

4.5 Prospects for new catalytic processes

Development of new catalysts and catalytic processes is carried out with a moving target set by the economics of the existing technology. Producers are continuously working to improve efficiencies and reduce production costs in their existing operational units. In working on the development of a new catalyst or process development a wary eye must be kept on what the opposition's production costs are likely to be in three or four years' time when the new development could be operating.

One aid to considering future costs and prices is the "experience curve" which is a plot of log (cumulative production) against log (cost or price). The concept of experience curves was introduced in 1968 by the Boston Consulting Group in their report *Perspectives on Experience*.[6] This postulates that costs in constant money value decline by a characteristic amount for each doubling of experience. From a study of the then rapidly growing industrial sectors of electronics and the chemical industry, the Boston group stated that a characteristic decline in cost of 20–30 % occurs for each doubling of cumulative production. In a stable situation costs and prices decline at approximately the same rate, i.e. along parallel curves, and plots of more readily available price information can be used to illustrate the effects of cumulative production experience. When costs and prices decline at different rates an unstable situation will develop. If costs fall faster, then the most efficient producer will make a high return on capital and will be tempted to reduce prices to increase his share of the market to his capacity limit. If prices lead the way, the least efficient producer will find his profit reduced until there is no margin over direct out-of-pocket costs and he ceases manufacture. In either case the result is a shake-out period until stability is restored.

Progress along the experience curve does not just happen, but is the result of tight management, continuing process development and efficient operation of plant. As prices and costs decline with experience any producer who does not keep pace with the technological development in the industry will, in time, be eliminated. In the long run costs (and hence prices) will be determined by the

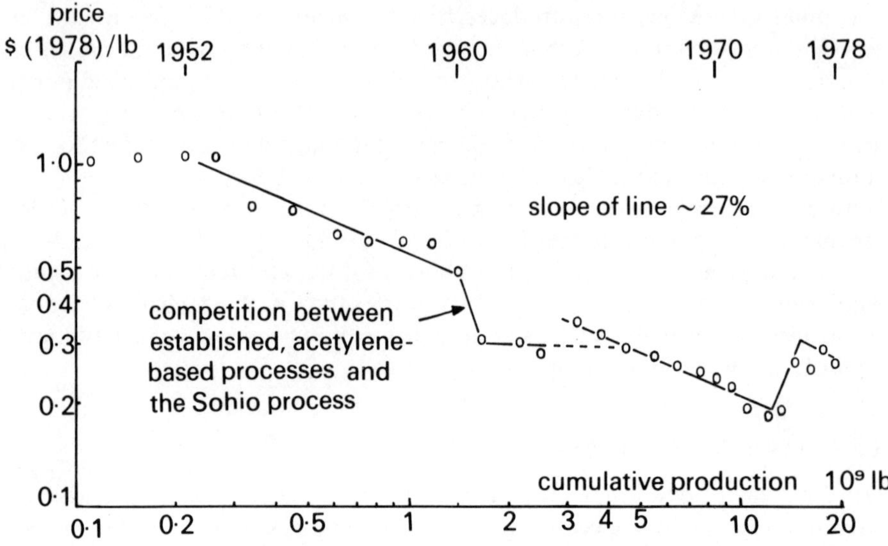

Figure 4.4 "Experience curve" for the production of acrylonitrile.

manufacturer with the best and most efficient technology. With his lower costs he will be able to gain market share and move down a steeper than average experience curve as his experience (production) builds up at a rate above the average for his competitors.

The price history for acrylonitrile—both before and after the introduction of the Sohio process—illustrates the principle of the experience curve (figure 4.4) U.S. data for prices and production are shown since they are more readily available for past years.

From the early 1950's the acrylonitrile price declined in a more or less stepwise fashion as demand for acrylic fibres grew. The average price decline was ca. 27% for each doubling of cumulative production. In the early 1960's the price dropped dramatically as the Sohio propylene/ammonia route was introduced, and established manufacturers reacted by reducing prices. Since, in most cases, the manufacturers of acrylonitrile from acetylene and hydrogen cyanide also produced these raw materials on the same site in old plants they could accept lower profits on these and charge lower prices for the raw materials to acrylonitrile production. This enabled the established producer to sell acrylonitrile at a price which gave a poor return to a propylene-based unit with new plant to pay for. The benefit of the new technology was thus effectively passed to the acrylonitrile consumer rather than the producer which stimulated growth in consumption. However, the low prices would not support the very large investment required for integrated acetylene/HCN/acrylonitrile units and as demand increased, large-scale propylene based plants were built which gave a reasonable return at prevailing prices due to scale economies, and eventually monopolised the market. By the late 1960's when virtually all acrylonitrile production was propylene-based a more moderate price decline was resumed, whose slope again represented about 27% price reduction for each doubling of cumulative production. The curve shows clearly the effect of the massive oil price rise in 1973 and this effectively introduces a new base line for the curve. Subsequent data (including preliminary 1978 data) suggests that the characteristic decline has been resumed but the oil price rises of 1979 and the present economic climate will add additional perturbations.

Similar curves are shown by many other products but a recent study[7] indicates that only products with high growth rates show the characteristic cost/price decline described by the Boston Group. For products with lower growth rates the relationship between price and cumulative production is much more erratic and shows a lower rate of decline than the 20–30% postulated by the Boston Group. In the low-growth future currently forecast for the industrialised nations it will be interesting to see whether the former experience curve relationships hold for chemical products or a more erratic picture develops. With a low growth future and escalating raw material and energy prices, the value of high yield, energy efficient catalytic processes will be even greater than in the past. The incentives for development of new, more efficient catalysts can rarely have been greater.

REFERENCES

1. *Hydrocarbon Processing*, 1972, **51** (May), 154.
2. P. B. Weisz, *Pure Appl. Chem.*, 1980, **52**, 2091.
3. *Chem. Engineering*, 1979 (Dec. 3rd), 80.
4. *Chem. Engineering*, 1969 (Sept. 22nd), 154.
5. J. F. Roth, J. H. Craddock, A. Hershman, and F. E. Paulik, *Chemtech.*, 1971, 600; D. Forster, *Adv. Organometallic Chem.*, 1979, **17**, 255.
6. *Perspectives on Experience*, The Boston Consulting Group Inc., U.S.A., 1968.
7. J. H. Taylor and P. Craven, *Process Economics Internat.*, 1979, **1**, 13.

Further Reading

F. A. Holland, F. A. Watson and J. H. Wilkinson, *Introduction to Process Economics*, Wiley, 1974.

M. S. Peters and K. D. Timmerhaus, *Plant Design and Economics for Chemical Engineers*, McGraw-Hill, 2nd edn., 1968.

B. Twiss, *Managing Technological Innovation*, Longman, 1974.

D. Davies and C. McCarthy, *Introduction to Technological Economics*, Wiley, 1967.

Assessing Projects: A Programme for Learning, ICI, Methuen, 1972.

D. H. Allen, *A Guide to the Economic Evaluation of Projects*, London, Institute of Chemical Engineers, 1972.

J. Happel and D. G. Jordan, *Chemical Process Economics*, Marcel Dekker Inc., 1975.

W. D. Baasel, *Preliminary Chemical Engineering Plant Design*, Elsevier, 1976.

G. M. Kaufmann and H. Thomas (editors), *Modern Decision Analysis*, Penguin Modern Management Readings, Penguin, 1977.

J. Dewing and D. S. Davies, "The economics of catalytic processes", *Adv. Catalysis*, 1974, **24**.

Two recent articles on capital cost estimation: J. Cran, *Chem. Engineering*, 1981 (April 6th), 65; J. L. Viola jr., *ibid.*, 80.

CHAPTER FIVE

OIL-BASED CHEMISTRY

T. EDMONDS

5.1 Introduction

Historically, the exploitation of crude oil as our main source of hydrocarbons covers a relatively short period of time, roughly from the 1940's to the present day. This short time-span belittles its contribution: in less than half a century oil has grown to dominate the fuels and, in particular, the chemical industries. Its use has contributed much to the major advances in the standard of material well-being of those nations that have reserves and the capability for their utilisation.

The primary use for oil has always been as fuel, particularly for transportation. The use of some of the crude oil fractions as feedstocks for the production of chemical building blocks arose as a secondary consideration. It was low price and availability of oil-based hydrocarbons that provided the stimulus for the changeover from coal and biochemical bases.

Oil is now the primary raw materials base for the production of chemicals ($\sim$90%). Even so, this accounts for only 5 to 10% of the oil barrel. Thus, any discussion of the chemistry of the transformation of oil will be largely concerned with the production of the various types of fuels. The chemical industry is also selective in its use of the various oil fractions and relies heavily on the lighter end of the barrel, e.g. naphtha and the coproduct gases, ethane and propane and butane (LPG). These are materials which have both the desired carbon: hydrogen ratio and the molecular weight range (C_2–C_9) for the production of the major chemical building blocks (cf. chapter 7).

The main aim of this chapter is to describe the major catalytic processes used to convert crude oil into the product spectrum required to fuel a developed economy: a second aim is to identify sources of chemical building blocks. Fuels derived from crude oil are used chiefly for transportation and heating. Transportation, provided by the internal combustion engine, sets the more stringent criterion in the form of the research octane number—the indication of anti-knock performance. To meet the range 85–95 necessary for modern high compression ratio engines, with their associated low fuel consumption, a number of processes have been developed. These are conversion processes such as catalytic reforming and isomerisation, where the overall molecular weight does not change greatly, alkylation, in which the molecular weight roughly doubles, and a cracking process, catalytic cracking, in which the molecular weight is reduced dramatically. Middle distillates used for heating, diesel engines, and jet fuel have considerably broader specifications. They should be paraffinic, rather than aromatic or naphthenic, to ensure good burning qualities, and meet the sulphur requirement. This is achieved by catalytic conversion in the presence of hydrogen. Two highly flexible processes lie within this category and are described at the end of the chapter. In *hydrodesulphurisation*, sulphur is removed from any crude oil fraction with little reduction in molecular weight. In *hydrocracking*, a high boiling range distillate, after sulphur removal, is converted to a product whose molecular weight is controlled by the nature of the catalyst employed and the severity of the conditions.

It is important to note at the outset that, with petroleum processes, feedstocks and products are discussed in terms of boiling point (and hence molecular weight) *ranges* and not in terms of single compounds. Thus, although we may write simple chemical equations to describe the reactions occurring in the process, it must always be remembered that this is an over-simplification and that it is the chemistry of compound types that is being discussed. To this end, the composition of the feedstock and products is usually represented by generic terms: paraffins, naphthenes, which are fully hydrogenated cyclic compounds, and aromatics, with olefins sometimes included. Only with isomerisation and alkylation does the chemistry of petroleum processes approach the relative simplicity of petrochemical processes.

5.2 Catalytic reforming

Catalytic reforming is the basic refinery process used to improve the (research) octane number of naphthas, or light distillates, for burning in modern high compression ratio internal combustion engines. Coincidentally, it is also *the major source of aromatics for chemicals.*

Research octane number (RON) needs to be defined before it can be related to the chemistry of light distillates. It is the volume percentage of iso-octane (2, 2, 4-trimethyl pentane) which, when blended with *n*-heptane, gives the same intensity of knock in a single cylinder engine of variable compression ratio as the sample fuel under the same operating conditions. Iso-octane is allocated a value of 100 on an arbitrary "octane number scale" while *n*-heptane has a value of 0.

The feed naphtha has a boiling range of 30°–190°C. Chemically it is composed of paraffins, naphthenes (fully hydrogenated 5- and 6-membered ring compounds) and aromatics; the level of sulphur is insufficient to cause catalyst poisoning. The research octane number is highest in the lightest fraction and decreases with increasing molecular weight (boiling point). The reason for this

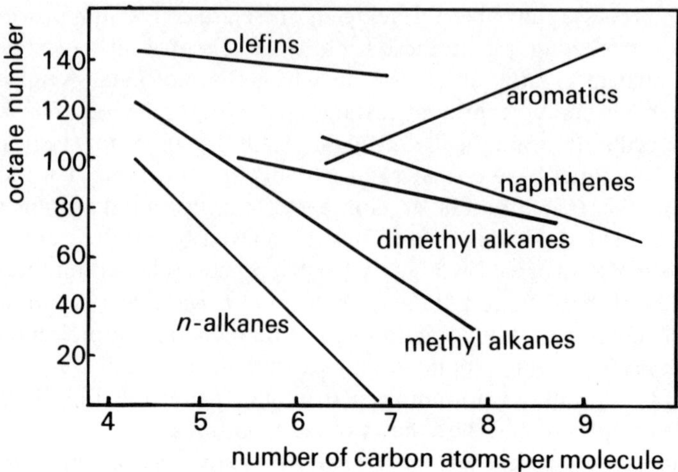

Figure 5.1 Octane numbers of various low molecular weight hydrocarbons showing the effect of charges in the number of carbon atoms.

is apparent from figure 5.1 which shows the relationship between RON and chemical composition. The rating of chemical compounds with respect to octane number falls as follows:

olefins > aromatics > naphthenes > iso-paraffins > *n*-paraffins

and, with the sole exception of aromatics, the octane number decreases with increasing molecular weight.

The specific aim in catalytic reforming is to produce a product whose RON lies between 85 and 95. There are four routes by which this aim can be met:

(a) increase the concentration of aromatics;
(b) increase the concentration of iso-paraffins and decrease their molecular weight;
(c) increase the concentration of olefins;
(d) decrease the overall molecular weight.

Because a high concentration of olefins in an engine causes fouling problems, the route chosen in catalytic reforming is to maximise the concentration of aromatics, while decreasing the overall molecular weight to some extent. To achieve these ends use is made of the following reactions which are known to occur with naphtha feedstock components (and are illustrated for C_6 molecules).

Dehydrogenation reactions

(a) naphthenes to aromatics

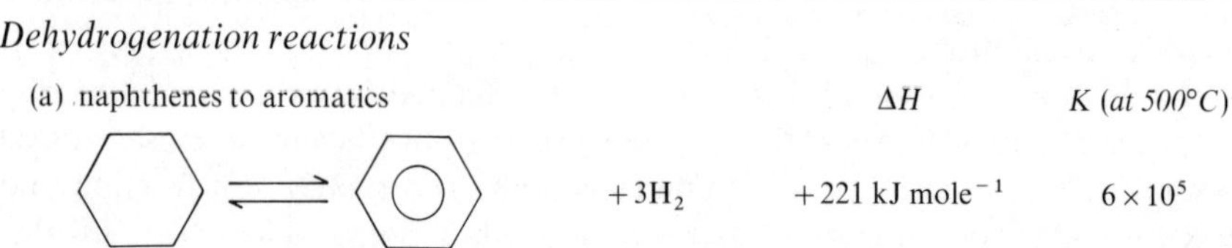

	ΔH	K (at 500°C)
(b) paraffin to olefins		
$C_6H_{14} \rightleftharpoons C_6H_{12} + H_2$	+130 kJ mole^{-1}	3.7×10^{-2}

Isomerisation reactions

(c) ring expansion

methylcyclopentane ⇌ cyclohexane	−15.9 kJ mole^{-1}	8.6×10^{-2}

(d) *n*-paraffins to iso-paraffins

$nC_6H_{14} \rightleftharpoons$ 3-methylpentane	−4.6 kJ mole^{-1}	7.6×10^{-1}

Reactions combining dehydrogenation and isomerisation

(e) Dehydrocyclisation of paraffins

$C_6H_{14} \rightleftharpoons$ benzene $+ 4H_2$	+266 kJ mole^{-1}	7.8×10^4

(f) Dehydroisomerisation of C_6 ring compounds

(c) + (a) above

Cracking reactions

(g) Hydrocracking of paraffins

$C_6H_{14} + H_2 \rightarrow 2C_3H_8$

(h) Ultimately carbon will be deposited on the catalyst.

In these equations ΔH is the heat of formation at 25°C and *K* the equilibrium constant at 500°C. For molecules with carbon numbers greater than 6 dehydrogenation of naphthenes and dehydrocyclisation of paraffins are more favourable. Overall a high yield of aromatics is favoured by high temperatures and low pressures.

The first catalytic reforming process using a noble metal catalyst—Platforming (platinum reforming)—became commercial in 1949. It was the prototype for all existing plants. Typical operating conditions are a temperature between 500° and 525°C, a pressure of 23.3 bar, a liquid hourly space velocity of 2 and a hydrocarbon to hydrogen ratio of 6:1. The catalyst is 0.3–0.5 %wt platinum on a high surface area γ-Al_2O_3 support promoted with 1 %wt chlorine. The two components of the catalyst perform the major reactions independently—platinum is the site for dehydrogenation and the promoted alumina the site for isomerisation. The loading of platinum is a compromise, balancing not only dehydrogenation activity but the maintenance of that activity and the basic cost of the metal. Ideally the platinum is atomically dispersed. The loading of chlorine on the alumina is set (and indeed adjusted during operation) to achieve the correct balance of isomerisation and cracking activity over the Lewis acid sites. Cracking to carbon can occur at both the platinum and the alumina surface. This controls the operation of the process for it not only limits the life of the catalyst but also necessitates the use of higher pressure and higher hydrogen: hydrocarbon ratios than are theoretically required. Improvements in catalyst life and/or a reduction in operating severity, allowing a closer approach to ideal thermodynamics, have been achieved within the last ten years by incorporating a second metal component, such as rhenium or germanium. The extension of catalyst life has also been assisted by the introduction of *rejuvenation* so that a single charge of catalyst can now be used for many years. The deactivating carbon is initially burnt off the catalyst for which the term *regeneration* is coined. However, this also causes the particle size of the platinum to grow considerably. Rejuvenation redistributes the platinum over the alumina and readjusts the chlorine level by treatment with water, chlorine and oxygen (see section 2.11.2).

The basic reactions taking place in catalytic reforming can be set out in a single reaction scheme shown below for C_6 hydrocarbons:

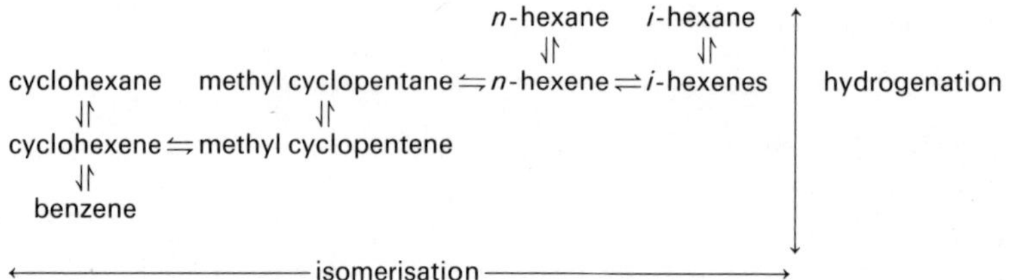

Dehydrogenation/hydrogenation reactions are shown vertically and isomerisation reactions horizontally. All the steps in this scheme have been proved by the detection of intermediates. Pilot plants were modified to operate at normal temperatures and pressures but at very high space velocities (up to 32 000). Under these conditions methylcyclopentene and cyclohexene can be isolated and their interconversion shown to be the rate-limiting step in ring expansion. Dehydrogenation reactions occur readily. A carbonium ion mechanism adequately accounts for the isomerisation steps in the above reaction plan. This will be described in detail under cracking, hydrocracking and isomerisation.

In a typical process scheme, preheated naphtha feed and hydrogen are passed through three reactor vessels in series. The more facile dehydrogenation reactions occur in the first two reactors and dehydro-isomerisation and cracking reactions take place in the third. Because the reactions are endothermic. heating is required between reactors. Product from the reactors is heat exchanged with the feed to conserve energy and then separated into a liquid and gas phase. The gas, which is mainly hydrogen, is split into two fractions; a portion is recycled internally and the remainder exported. The liquid is stripped of its lighter hydrocarbons which are taken overhead in a distillation column yielding high octane gasoline as the residue.

The net effect of catalytic reforming in converting paraffins and naphthenes to

Table 5.1 Catalytic reforming—process achievements
Hydrocarbon composition (%wt) of a reformer feed and product

No. of carbon atoms	*Paraffins*		*Naphthenes*		*Aromatics*	
	Feed	*Reformate*	*Feed*	*Reformate*	*Feed*	*Reformate*
C_3	0	trace				
C_4	0.1	2.0				
C_5	1.9	6.8	0.1	0.3		
C_6	6.4	9.3	2.5	0.6	0.4	4.4
C_7	12.1	8.2	6.6	0.5	3.6	17.6
C_8	15.6	4.4	6.8	0.3	5.6	23.0
C_9	13.3	1.3	5.6	0.1	4.3	16.4
C_{10}	8.8	0.2	3.0	trace	1.0	4.1
C_{11}	1.8	0	0.5	0	0	0.5
Total	60.0	32.2	25.1	1.8	14.9	66.0

Notes: Feed RON = 60
Reformate RON = 100

aromatics is shown in Table 5.1. This gives a detailed compositional breakdown of feed and product as a function of carbon number. Aromatics increase from 14.9 to 66 %wt; naphthenes are essentially eliminated and paraffins reduced from 60 to 32.2 %wt.

The high yield of aromatics in catalytic reformer product streams makes them key sources of aromatics for chemical feedstocks as well as major components of motor gasoline. The most widely adopted method of separation is via liquid–liquid extraction using polar species such as sulpholane, dimethylsulphoxide, *N*-methylpyrrolidone, and mixtures of mono-, di-, tri-, and tetra-ethylene glycols. This is followed by distillation into the key benzene, toluene, and mixed xylene (BTX) fractions. The aliphatic hydrocarbon fraction is a suitable feedstock for steam cracking (cf. 5.6).

5.3 Catalytic cracking

Catalytic cracking is the most extensively used petroleum process in the world. It aims to convert a wide range of distillates and heavier petroleum fractions to lower molecular weight products. Operating conditions and design can be varied to maximise the production of high octane gasoline, light olefins for either alkylate or petrochemical feedstock, and either heating or diesel oils. With limited crude availability, it is increasingly the key process for converting heavier petroleum fractions to high octane gasoline.

Catalytic cracking feedstocks normally have initial boiling points around 220°C with end points about 540°C or even higher, i.e. they fall within the range of straight-run and vacuum gas oils. As many as tens of thousands of individual molecular species may be present in such a wide boiling range material. Species are again subdivided into three general categories: paraffins, naphthenes and aromatics. In terms of modelling the process it will be seen that the aromatic fraction needs further subdivision into side chains and base rings.

Reactions in catalytic cracking take place in accordance with the general principles of carbonium ion reactions, and are rationalised by the carbonium ion stability sequence:

$$\text{tertiary} > \text{secondary} > \text{primary}$$

5.3.1 *Paraffin cracking*

There are two possibilities for the initial step in the catalytic cracking of paraffins. The first involves the simultaneous loss of a hydride ion from the paraffin molecule and of a proton from the acidic catalyst surface. This produces a carbonium ion in combination with acid anion and molecular hydrogen:

$$R_1{-}CH_2{-}CH_2{-}R_2 + H^+A^- \rightarrow R_1CH_2{-}\overset{+}{C}H{-}R_2 + H_2 + A^-$$

Alternatively a small amount of olefin, created by thermal cracking could initiate the reaction:

$$R_1{-}CH{=}CH{-}R_2 + H^+A^- \rightarrow R_1CH_2{-}\overset{+}{C}H{-}R_2 + A^-$$

The latter route is favoured since olefins crack at lower temperatures than paraffins.

Chain propagation involves an exchange reaction in which a carbonium ion reacts with a paraffin to give a new paraffin and a carbonium ion of the paraffin to be cracked (hydride transfer).

$$R_1{-}CH_2{-}\overset{+}{C}H{-}R_2 + R_3{-}\underset{}{\overset{\overset{CH_3}{|}}{C}}H{-}CH_2{-}CH_2{-}R_4 \rightarrow R_1{-}CH_2{-}CH_2{-}R_2 + R_3{-}\underset{+}{\overset{\overset{CH_3}{|}}{C}}{-}CH_2CH_2{-}R_4$$

The next step is the decomposition of the activated molecule. The primary rule involved is that the carbon-carbon cleavage occurs at the position one carbon atom away from the carbonium ion, i.e. β-scission:

$$R_3{-}\underset{+}{\overset{\overset{CH_3}{|}}{C}}{-}CH_2{-}CH_2{-}R_4 \rightarrow R_3\overset{\overset{CH_3}{|}}{C}{=}CH_2 + \overset{+}{C}H_2{-}R_4$$

A hydride shift then converts the primary carbonium ion formed into a secondary carbonium ion:

$$\overset{+}{C}H_2{-}R_4 \rightarrow CH_3{-}\overset{+}{C}H{-}R_5$$

Subsequent steps involve further β-scission and hydride transfer and proceed until the chain becomes so short that cracking at the β position is no longer a rapid reaction. The residual carbonium ion can then exchange with another larger paraffin molecule to produce a new carbonium ion and a small paraffin corresponding to the original carbonium ion thus propagating the chain.

Large amounts of iso-compounds are formed in catalytic cracking. This is readily explained by the rearrangement of the secondary carbonium ion:

$$CH_3{-}CH_2{-}\overset{+}{C}H{-}CH_2{-}CH_2{-}R \rightarrow \overset{+}{C}H_2{-}\underset{\displaystyle CH_3}{\underset{|}{C}H}{-}CH_2{-}CH_2{-}R$$

$$\rightarrow CH_3{-}\underset{\displaystyle CH_3}{\underset{|}{\overset{+}{C}}}{-}CH_2{-}CH_2{-}R \rightarrow CH_2{-}\underset{\displaystyle CH_3}{\underset{|}{C}H}{=}CH_2 + \overset{+}{C}H_2{-}R$$

[etc.]

5.3.2 *Naphthenic cracking*

Naphthenes can undergo both dehydrogenation and cracking. At 500°C with cyclohexane, dehydrogenation predominates; however, at a higher temperature, the rate of cracking to olefins and paraffins exceeds the rate of dehydrogenation. Bicyclic naphthenes show a greater tendency to cracking. This is undoubtedly connected with the presence of two tertiary carbon atoms in decalin as compared to only secondary carbon atoms in cyclohexane. This follows the general observation that cracking is accelerated by tertiary grouping relative to secondary and quaternary groupings.

There can also be hydride transfer between carbonium ions and polycyclic naphthenes to produce cycloalkyl carbonium ions. These crack by β-bond scission to open a ring and produce lower molecular weight naphthenes and olefins. In addition, naphthenes can isomerise by expansion or contraction of the ring. The cyclohexane-methyl cyclopentane equilibrium, cf. catalytic reforming, is a simple illustration but it undoubtedly occurs with polycyclic naphthenes as well.

5.3.3 *Aromatic hydrocarbon cracking*

Alkyl aromatics, with the exception of methyl or ethyl, crack to split off the side chain at the ring. Evidence with pure hydrocarbons suggest that the initial carbonium ion is formed on the aromatic ring rather than on the side chain. For example, isopropylbenzene cracks readily to benzene and propylene. If the carbonium ion were in the alkyl group, it should be on the tertiary carbon atom but here there is no β bond to sever. If the proton adds to one of the double bonds on the aromatic ring, one gets an intermediate that leads to cracking by β-bond scission with the carbonium ion going with the departing moiety while the remaining part is converted into an olefin (in this case, part of the aromatic ring).

$$\mathrm{C_6H_6^{+}\text{—}C(C)C} \longrightarrow \mathrm{C_6H_6} + \mathrm{{}^{+}C(C)C}$$

Catalytic cracking reactions are endothermic and hence require high reaction temperatures. The major side reaction is the production of coke in significant quantities, which must be continuously removed from the catalyst. The problem to be solved in reactor design is to operate a continuous process while alternatively using and regenerating the catalyst. At the same time, pollution legislation controlling the emission of CO must be satisfied.

The solution chosen utilises a fluidised bed. In its most modern form—the riser reactor—the process is illustrated in figure 5.2. It has three major sections: the riser reactor, the stripper and the regenerator. The oil feed meets the hot catalyst, substantially free of carbon, as it falls through a valve at the bottom of the regenerator and passes into the riser reactor. This is designed for short, but effective, contact time between the oil and the catalyst followed by rapid catalyst/vapour separation to permit the use of modern highly active zeolite catalysts. Separation is completed in the stripper which includes a cyclone to prevent catalyst particles passing overhead in the product. The carbon-covered catalyst finally falls through a valve at the bottom of the stripper into the regenerator where the carbon is burnt off leaving the catalyst regenerated and ready to be returned to the reactor.

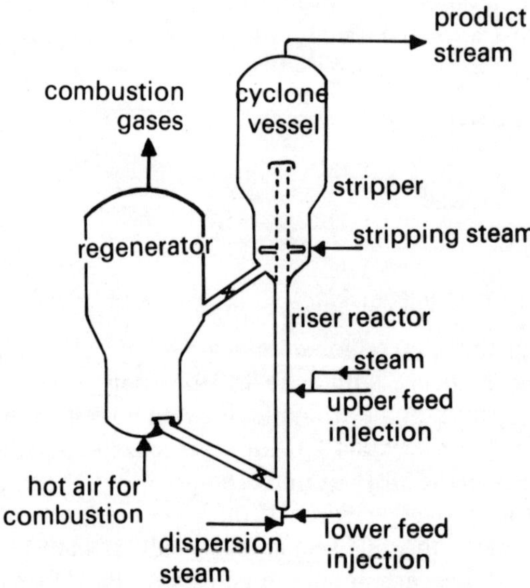

Figure 5.2 Schematic diagram of a fluid catalytic cracker with a riser reactor (adapted from Bryson *et al.*, 1972).

Such a process imposes very stringent catalyst requirements. It must combine cracking activity with high selectivity for the conversion of the feed to a high octane gasoline while not producing excessive amounts of carbon under cracking conditions. It must be capable of reasonably complete regeneration without loss of activity. It must be stable against continuous thermal shock, higher regeneration temperatures being followed by lower processing temperatures, as well as against deterioration by steam at high temperatures. In addition, the catalyst must not be susceptible to attrition, nor be excessively poisoned by nitrogen, sulphur, and increasingly, V and Ni compounds in the feedstock, and finally, must produce a gaseous product on regeneration containing a low percentage of CO.

Table 5.2 Typical data for the performance of different cracking catalysts

Development		*13% alumina catalyst*	*25% alumina catalyst*	*Low activity zeolite*	*High activity zeolite*	*High activity zeolite and CO combustion*
Conversion to product of b.p. <221°C	%	68	70	72	79	79
Yields						
C_2's	%wt	2.8	3.0	2.5	2.5	2.5
C_3's	%vol	8.3	8.7	8.0	9.2	9.2
C_4's	%vol	15.3	15.9	14.6	15.8	15.8
C_5/221° gasoline	%vol	52.0	52.7	56.7	62.8	64.2
Heating oil	%vol	27.0	25.0	23.0	16.0	16.0
Bottoms	%vol	5.0	5.0	5.0	5.0	5.0
Coke	%vol	5.9	5.9	5.9	5.9	4.8
Preheat temperature	°C	366	379	338	321	266

Taken from E. C. Luckenbach, *Chem. Eng. Prog.* **75** (Feb 1979) p. 56.

The improvement which has been made from the original 13% Al_2O_3 + silica catalyst to a modern high activity zeolite catalyst containing additives to aid CO combustion is illustrated in Table 5.2. The Y-zeolite employed is exchanged with 21% weight rare earth cations and constitutes up to 25% of the final catalyst. The average particle size is 63–65 microns, the surface area 150–250 $m^2 g^{-1}$ and the pore volume 0.2–0.5 $cm^3 g^{-1}$. To aid the conversion of CO to CO_2 trace amounts of platinum group metals are added. This development confers other benefits; more efficient energy utilisation because the preheat temperature of the feed can be reduced (see Table 5.2); lower residual carbon levels on the regenerated catalyst, 0.2% weight is now typical because the regeneration temperature is increased to 704°–760°C; and higher gasoline yields because of the lower carbon level and hence lower poisoning of the catalyst. The most recent development, necessitated by feeds containing vanadium and nickel, is the use of an antimony-based liquid injected into the feed to "passivate" these components as they are deposited on the catalyst. This reduces the hydrogen and methane in the product and the coke on the catalyst thereby maintaining, or even improving, gasoline yield with more refractory feedstocks. Overall, Table 5.2 clearly illustrates the benefits of the zeolite catalyst in an appropriately designed reactor. The conversion to product boiling below 221°C is improved from 68 to 79% while the selectivity to gasoline is increased from 52 to 64.2% mainly at the expense of heating oil. The lower coke yield is an added benefit.

The riser operating parameters are typically:

Temperature	500°–550°C
Pressure	atmospheric
WHSV	40
Catalyst : oil ratio	4

In regeneration the temperature reaches 700°–760°C.

Although a range of products can be produced in catalytic cracking, the process is generally operated to maximise gasoline production. It is capable of giving 65%w/w gasoline boiling up to 221°C with a RON of around 91. The level of olefins (ca. 30%) is, however, high. Olefins are less desirable as gasoline blending components and this is a limitation in the use of cat cracker product as a gasoline blending component.

5.4 Isomerisation

Catalytic reforming improves the octane number of material boiling in the 70°–200°C range. It is less effective in the range 30°–70°C (C_5–C_6). Although catalytic cracker gasoline in this boiling range has a very good octane number it is not always available. Isomerisation provides a means to improve the 30°–70°C cut so that the marketed gasoline has a good octane number throughout its boiling range.

The main hydrocarbons which boil in the 30°–70°C range are C_5 and C_6 paraffins. The RON values for normal straight chain paraffins are lower than for branched-chain isomers, and thus the aim of the isomerisation process is to convert *n*-pentane and *n*-hexane into branched-chain isomers. In a typical feed to an isomerisation unit, beside C_5 and C_6 paraffins, small quantities of butanes, naphthenes and benzene are present. The process should therefore maximise in the feedstock the production of isopentane and dimethyl butanes (DMB). The equilibrium concentration of the various isomers is very dependent on the reaction temperature, and the formation of both isopentane and 2,2-dimethyl butane is strongly favoured below 150°C. At higher temperatures the lower octane number singly-branched isomers predominate. Thermodynamic factors therefore provide a driving force to isolate catalysts which will provide near equilibrium conversions at the lowest possible temperature.

The historical development of catalysts for the isomerisation of light paraffin hydrocarbons is illustrated in figure 5.3. Friedel–Craft type catalysts meet the temperature requirement but their extreme activity results in poor selectivity

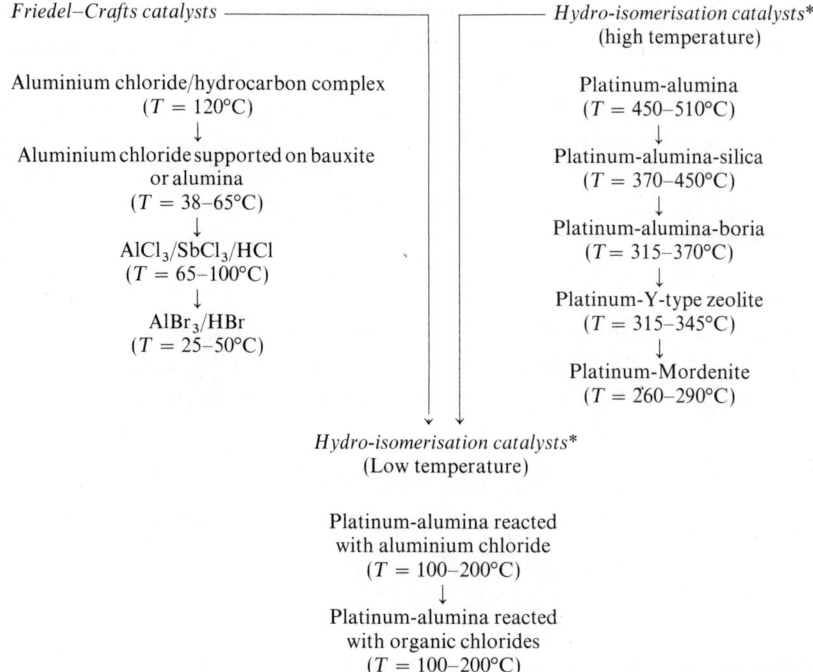

Figure 5.3 Historical development of catalysts for the isomerisation of light paraffin hydrocarbons (from Lawrence *et al.*, 1967).

and self-destruction of the catalyst by reaction with the feedstock and products unless special precautions are taken in feedstock pretreatment, control of operating conditions and use of suppressors to reduce the amount of side reaction occurring. The acidic nature of the Friedel–Craft halides can also lead to operational difficulties due to corrosion of reaction vessels and ancillary equipment. Furthermore, with Friedel–Craft catalysts side-reactions increase with molecular weight.

Dual-function hydro-isomerisation catalysts, usually precious metal loaded on a high surface area support, do not suffer from poor selectivity but their activity is much lower, and higher operating temperatures are required. Recycle operation is required to provide high yields of high octane isomers. The activity of these catalysts can be raised by increasing the acidity of the support thereby decreasing the reaction temperature. This is illustrated by the falling temperature requirement between alumina and mordenite.

An entirely new range of catalysts was developed in the 1960's which combined the selectivity and life of the hydro-isomerisation catalysts with much of the activity of the Friedel–Craft catalysts. These catalysts operate in the temperature range 100°—200°C to give near-equilibrium yields of pentane and hexane isomers. Catalyst selectivity is extremely high and only small amounts of C_1–C_4 paraffins are formed as byproducts. The first of these catalysts was formed by reacting anhydrous aluminium chloride with a platinum–alumina reforming catalyst at an elevated temperature; the second by controlled treatment of reforming catalysts with a specific range of non-metal containing organic chlorine compounds. Both these catalysts are sensitive to sulphur and the aluminium chloride activated catalyst is also sensitive to water. Hence the feedstock must be desulphurised (and dried).

Catalytic isomerisation of paraffins proceeds via a carbonium ion mechanism. The catalysts, which are active at low temperatures, are very strongly acidic and it is probable that the carbonium ion is generated from the paraffin by hydride ion abstraction:

$$CH_3{-}CH_2{-}CH_2{-}CH_2{-}CH_3 \rightarrow CH_3{-}\overset{}{\underset{+}{C}H}{-}CH_2{-}CH_2{-}CH_3 + H^-$$

Isomerisation takes place via an intramolecular rearrangement:

$$CH_3{-}\underset{+}{C}H{-}CH_2{-}CH_2{-}CH_3 \rightleftharpoons CH_3{-}\underset{\displaystyle CH_3}{\overset{+}{\underset{|}{C}}}{-}CH_2{-}CH_3$$

and proceeds by a chain propagation mechanism:

$$CH_3{-}\underset{\displaystyle CH_3}{\overset{+}{\underset{|}{C}}}{-}CH_2{-}CH_3 + CH_3(CH_2)_3CH_3 \rightarrow CH_3{-}\underset{\displaystyle CH_3}{\underset{|}{C}H}{-}CH_2{-}CH_3 + CH_3{-}\overset{+}{C}H{-}CH_2{-}CH_2{-}CH_3$$

An alternative mechanism can be postulated in high temperature isomerisation. The paraffin molecule is dehydrogenated at the platinum site to an olefin which passes to the acid site where it is converted to a carbonium ion. After intramolecular rearrangement the iso-olefin returns to a platinum site and is hydrogenated to an isoparaffin. A typical process uses a two-reactor system. Desulphurised C_5/C_6 is mixed with recycled hydrogen, heated and passed over the catalyst in the first reactor where any aromatics and olefins are saturated and isomerisation of normal paraffins to isoparaffins takes place. Reactor effluent is then cooled before entering a second larger reactor in which further isomerisation occurs at more favourable low temperature conditions.

Table 5.3 (opposite) illustrates the process achievements. The hydrogenated data indicate the conversion achieved in the first reactor. 67.8 % of the n-C_5 and 21.4 % of the n-C_6 is isomerised at this stage; benzene is hydrogenated and a small degree of cracking occurs. In the second reactor the temperature drops from 205°C to 135°C, the pressure remains at 34 atm (bar) and the space velocity decreases because of the greater catalyst charge in the reactor. The isomerisate data indicate the final conversions. Total conversion of the C_5 fraction is 75.8 % and of the C_6 fraction is 34.6 %. These figures represent 93.6 and 88.3 % of

Table 5.3 Typical recycle pilot plant data

Component		RON	Feedstock %wt	Hydrogenated %wt	Isomerisate %wt
C_3	Propane		—	0.2	0.2
C_4	*Iso*butane	100	—	1.6	1.9
	n-Butane		1.5	1.2	1.2
C_5	*Iso*pentane	92.3	22.7	34.1	37.9
	n-pentane	61.7	28.8	16.2	12.1
C_6	2,2-Dimethylbutane	91.8	1.1	8.4	14.0
	2,3-Dimethylbutane	103.5	2.3	3.9	4.0
	2-Methylpentane	73.4	14.2	13.2	11.9
	3-Methylpentane	74.5	8.7	8.0	6.6
	n-Hexane	24.0	10.0	5.8	4.0
Cyclic Hydro-carbons	Cyclopentane	101.3	1.4	1.4	1.2
	Methylcyclopentane	91.3	1.4	4.0	2.5
	Cyclohexane	83.0	5.7	1.9	2.6
	Benzene	111	2.2	—	—

Isomerisation conditions

	1st Reactor		2nd Reactor
LHSV (vol/vol h)	8		2
Reactor outlet temperature (°C)	205		135
Pressure (bar)		34	
H_2 : hydrocarbon		1.5 : 1	
Conversion at outlet (C_5/C_6)	67.8/21.4		75.8/34.6
RON (clear)			83.8

equilibrium conversion, respectively, for C_5 and C_6 fractions. The research octane number of the final product is 83.8.

While mainly used to isomerise C_5 and C_6 fractions for blending into gasoline, C_4 isomerisation is also commercially operated to provide a feedstock for alkylation. Alkylation, which is not discussed in detail, is the process in which an isoparaffin is combined with an olefin, or mixture of olefins, in the presence of a strong acid catalyst to form highly branched paraffins suitable for use as gasoline blending components. In practice, the isoparaffin is always isobutane, obtained from an isomerisation unit or hydrocracker, and the olefins mixtures of propylene, butenes and pentenes, byproducts of a catalytic cracker. Overall the chemistry is one involving carbonium ions e.g. for reaction of isobutene with 2-butene:

$$Me_3CH + H^+ \rightarrow Me_3C^+ \xrightarrow{C_4H_8} Me_3C\overset{Me}{\overset{|}{C}}H\overset{+}{C}HMe$$

$$\xrightarrow{Me_3CH} Me_3CCHMeCH_2Me + Me_3C^+$$

Only strong acids are able to catalyse the hydrogen transfer reactions which initiate the carbonium ion reactions—90 to 98% H_2SO_4 or HF are used commercially. In order to favour alkylation rather than polymerisation of the olefin, the concentration of isobutane must be kept as high as possible. Olefins are highly soluble in the acids whereas isobutane is not. This, together with the highly exothermic nature of the overall reaction, determines the method of operation. The reactor contains liquid acid and isobutane under violent agitation. The olefin stream is then slowly fed so that the isobutane is always in excess and can get into the acid phase and react with the carbonium ion intermediates and suppress polymerisation via further reaction with olefin. Reaction temperatures are kept low to ensure high levels of conversion (5 to 15°C with H_2SO_4).

The quality of the product from alkylation is judged by its RON. The totally debutanised alkylate from plants fed with a mixture of propylene, butenes and light pentenes will have values of 92–96. From butenes alone, the RON will be as high as 99. The boiling range of the product lies between 74 and 194°C.

5.5 Hydrotreatment

The major cost element in hydrotreatment is the amount of hydrogen consumed in the reaction. The hydrogen consumption, as a function of pressure, for various processes and feedstocks is illustrated in figure 5.4 below. The plot

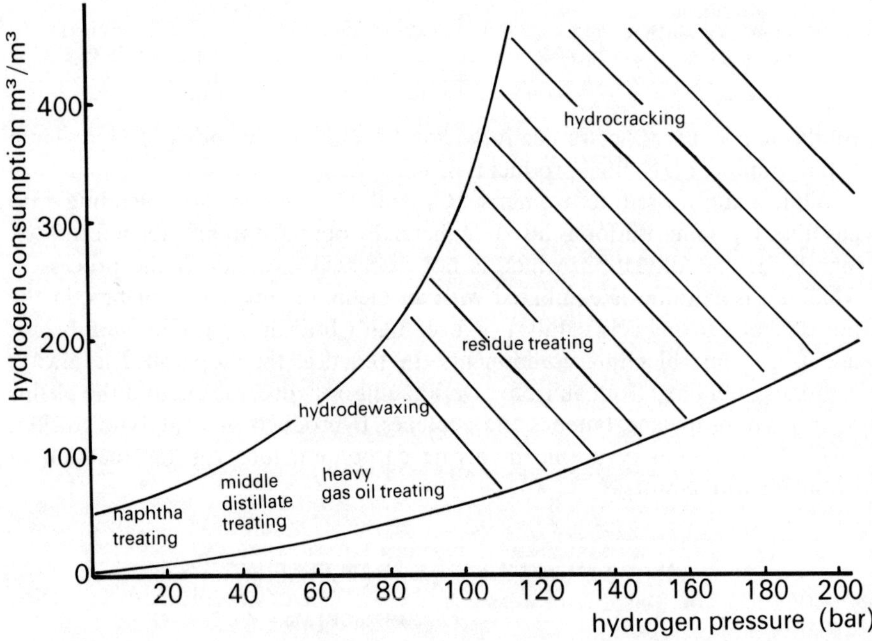

Figure 5.4 Hydroprocessing requirements (adapted from Weisz *et al.*, 10th World Petroleum Congress, **14**, 325, 1979).

shows that hydrogen consumption does vary enormously with the severity of treatment but is a particular limitation with residue hydrodesulphurisation (treating) and hydrocracking. Catalytic reforming, which is a net producer of hydrogen (see Table 5.4), can usually meet the hydrogen requirements of all other processes.

5.5.1 *Hydrodesulphurisation*

Hydrodesulphurisation (HDS) is the conventional means for the removal of sulphur compounds from crude oil fractions. The sulphur compounds are reacted with hydrogen in the presence of a hydrogenation catalyst with the resulting formation of one or more hydrocarbon molecules and hydrogen sulphide. There are many reasons why sulphur compounds should be eliminated. Chief among these are:

1. to remove a major poison in subsequent processing; important examples are the protection of the platinum catalyst in catalytic reforming and the palladium catalyst in hydrocracking;
2. to meet pollution legislation; sulphur concentrations in gas oils and fuel oils are limited in the industrialised world and cannot always be met by blending with low sulphur fractions;
3. to produce fractions having acceptable odour;
4. to reduce or eliminate corrosion during handling, refining, or use of petroleum products.
5. to decrease particulate formation with kerosine.

All crude oil fractions contain sulphur compounds. Crude oil itself contains 0.1 to 2.5 %wt depending on the source. When fractionated the sulphur content increases with the boiling point of the fraction—light gasoline (0–70°C) 0.001–0.02; naphtha (70–140°C) 0.002–0.02; kerosine (140–250°C) 0.01–0.2; diesel fuel (250–350°C) 0.1–1.4 and residue (>350°C) 0.3–4.1 %wt.

The sulphur compounds in petroleum can be categorised within the following compound types: thiols (alkane thiols), disulphides (dithia alkanes), sulphides (thia alkanes), thiophenes, benzothiophenes and dibenzothiophenes. Thiophenes, benzothiophenes and dibenzothiophenes will have associated alkyl groups of varying carbon number and complexity depending on the fraction. The compound type formed in a given fraction will depend on its boiling range but, in general, more thiols will be found in the lightest fractions and more dibenzothiophenes in the heavier fractions.

Chemical reactions which occur in hydrodesulphurisation are relatively simple. The product which is formed will be either a saturated hydrocarbon, plus H_2S, or an unsaturated hydrocarbon, plus H_2S. The major side reaction is cracking leading ultimately to the deposition of carbon. This occurs particularly with residue fractions which contain polynuclear aromatic molecules including nickel and vanadium atoms.

Hydrodesulphurisation reactions to saturated hydrocarbons and H_2S are exothermic. Thermodynamic equilibrium constants for this reaction are always positive between 200° and 400°C and fall in the following order, for the various compound types; disulphides > sulphides > thiols > thiophenes > benzothiophenes > dibenzothiophenes i.e. the removal of sulphur from disulphides is much easier than from dibenzothiophenes. Equilibrium constants do not favour the formation of unsaturated hydrocarbons in the same temperature range.

Actual reaction conditions reflect the thermodynamic constraints. The range of conditions are: pressure, 25–100 bar, temperature, 340–420°C, LHSV, 4–0.5. The most severe conditions are necessary with a residue feedstock.

The major requirements to be met by the hydrodesulphurisation catalyst are determined by the cost of hydrogen and its availability at a given location. If hydrogen has to be provided via steam reforming, its cost will be high and its consumption must be minimised. This will always be the case with residues. For all feedstocks the catalyst should combine high activity and selectivity for hydrodesulphurisation, with low selectivity for cracking reactions and for the hydrogenation of unsaturated bonds not formed via sulphur cleavage, which are the main additional sources of hydrogen consumption. The catalyst should also exhibit long term stability (i.e. no loss of surface area with time, or during regeneration) and minimum sensitivity to poisons such as carbon, vanadium and nickel.

The main catalyst which meets these requirements is molybdenum disulphide supported on alumina and promoted with either cobalt or nickel—the so-called CoMo or NiMo catalysts, which were discovered in the 1930's. The components are normally deposited as oxides and a typical CoMo catalyst will be composed of ~3.0 %wt cobaltic oxide and ~12 %wt molybdenum trioxide supported on γ-Al_2O_3 with a surface area of ~300 $m^2\,g^{-1}$. The pore diameter for the alumina will be specified for residue desulphurisation to enhance stability by minimising access of the polynuclear aromatic molecules to the active sites within the pores. Modern catalysts are presulphided. Hydrogen is dissociated only on MoS_2 and the reaction rate for both hydrogenolysis and hydrogenation has been shown to depend on the concentration of MoS_2 (at an equivalent concentration of promoter). The role of the promoter itself has never been unambiguously ascertained.

The rate of hydrodesulphurisation of different isomers depends strongly on the exact location of alkyl side chains. Thus, under the same reaction conditions, the fractional conversion of the following compounds based on dibenzothiophene is very dependent on the steric shielding of the sulphur atom.

compound	fractional conversion
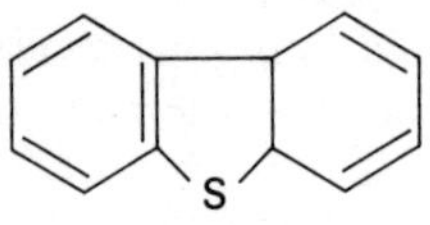 dibenzothiophene (structure)	0·166
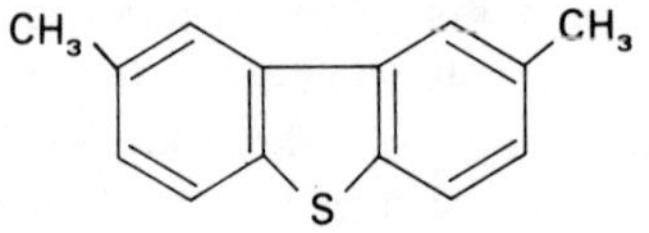 dimethyldibenzothiophene, CH_3 / CH_3 (structure)	0·241

S CH$_3$ 0·047

S CH$_3$ CH$_3$ 0·023

The model developed to account for these kinetics has thiophene and H_2S competing for one type of site, and hydrogen absorbed on a second type. Two mechanisms may be postulated. In the first the primary step after the adsorption of the reactants is S—C bond scission in thiophene followed by hydrogen transfer to the unsaturated carbon atom; in the second the primary step is hydrogenation to tetrahydro thiophene prior to C—S bond scission. Thus:

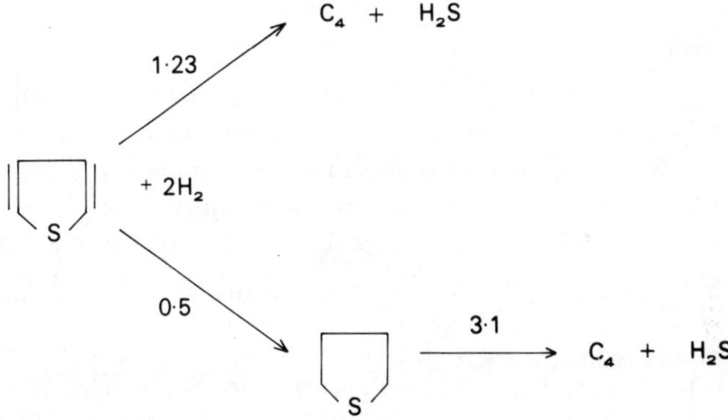

The relative rates of reaction show that the initial step is C—S bond scission and this is supported by the detection of butadiene in the product at higher space velocities.

The achievement of the hydrodesulphurisation process with feedstocks in all boiling ranges is shown in Table 5.4. As the boiling point of the feed increases so the degree of desulphurisation economically feasible falls while the temperature, pressure, volume of catalyst and quantity of hydrogen required all increase to give the final sulphur levels given in the last column.

Hydrodesulphurisation is found in all refineries. It is particularly necessary to prevent poisoning of the new high stability bimetallic reforming catalysts and within the last ten years has seen extensive study for treatment of residues. This is particularly important in countries where the major source of crude oil is from the Middle East oil field. Where the major source is North Africa and the North Sea, residue desulphurisation is not required. However, prior to the advent of North Sea crude oil, some improvement in catalyst performance was achieved

Table 5.4 Hydrodesulphurisation processing achievements

Cut	*S content* (%wt)	% *desulphurisation*	*Temperature* (°C)	*Pressure* (bar)	*Catalyst volume* (m^3)	*Hydrogen requirement* (m^3/m^3)	*Final sulphur* (%wt)
Naphtha	0.05–0.2	99	320	15	10–15	−9	down to 1 ppm.
Gas oil	0.2–2.0	90	350	30	25–30	−27	0.02–0.2
Vacuum gas oil	1.5–3.0	90	380	50	30–40	−71	0.15–0.30
Atm. residue	2.0–5.0	75–80	400	70–100	200–250	−178	0.5 –1.0
Cf. catalytic reforming						+210	

by tailoring the pore size distribution to minimise the access of asphaltenes to the active site within the pore, by controlling the impregnation conditions to ensure a uniform distribution of cobalt and molybdenum oxides, and by changing the shape of the catalyst extrudates to reduce diffusion limitations. These improvements are used in plants operating in Japan.

5.5.2 *Hydrocracking*

Hydrocracking converts light and heavy gas oil into more valuable lower boiling range products by reaction with high pressure hydrogen over a cracking catalyst. It is a balancing process to meet market demands in a given location. It gives a refinery greater flexibility, thereby improving its overall efficiency. The flexibility of hydrocracking is illustrated by the possible product spectra. By suitably optimising both the catalysts and the process conditions, products range from LPG, gasoline, chemical feedstocks, including steam cracker feed and benzene, toluene and xylene (BTX) precursors, to jet fuel, middle distillate heating oils, lubricating oils, and catalytic cracking feedstocks. In theory any desired petroleum fraction can be produced by hydrocracking. Furthermore, hydrodesulphurisation is a special form of hydrocracking in which the aims are solely to crack out the heteroatoms without major changes in the hydrocarbon skeleton and hence the overall boiling range.

The very flexibility of the process makes it difficult to describe uniquely. There are two fundamentally different approaches to process design in hydrocracking. In one, feedstocks are hydrotreated to remove components which will poison the catalyst in the hydrocracking stage which follows. In the other approach, hydrocracking catalysts are employed which accomplish substantial cracking in the presence of sulphur and nitrogen compounds or of the hydrogen sulphide and ammonia resulting from hydrotreatment.

The feed will normally be a gas oil defined by its boiling range, gravity (an indication of its C : H ratio) plus sulphur and nitrogen contents. Typical ranges are:

Boiling point	°C	165–575
Sulphur	%wt	0.5–4.5
Nitrogen	ppm wt	200–2700

Ideally, the feedstock should be low in nitrogen, free of organically combined metals, such as vanadium, and as close as possible to the carbon number of the desired product. If these requirements are not met, the hydrotreatment step must be under more severe operating conditions. This causes the hydrogen consumption to increase and the catalyst life to decrease because of enhanced poisoning. The increased severity is necessary because hydrodenitrogenation is more difficult than hydrodesulphurisation. The enhanced poisoning arises from vanadium and nickel containing metalloporphyrins and asphaltenes. Not only vanadium but also carbon, due to the low C:H ratio in asphaltenes, is deposited on the pretreatment catalyst, blocking reaction sites and decreasing its active life. For these reasons feedstock specifications may also include the concentration of metals and some criterion of asphaltene content such as Conradson carbon or heptane insolubles.

The catalyst employed in hydrocracking is dual functional with a hydrogenation component supported on strongly acidic sites. The chemistry of hydrocracking is that which is expected from carbonium ion principles, described under catalytic cracking, plus the chemistry of hydrogenation. It may be summarised as follows:

1. Two types of reaction are envisaged for the production of the carbonium ions to initiate the hydrocracking process. In the first either a paraffin or a cycloparaffin is dehydrogenated to an olefin which moves to a Brönsted acid site picking up a proton. Alternatively the saturated hydrocarbon reacts directly with the Brönsted site to form the carbonium ion and molecular hydrogen.

2. Cracking reactions of paraffins and alkyl side chains occur as in catalytic cracking.

3. Because polycyclic aromatic hydrocarbons are hydrogenated to naphthenes, refractory aromatics are converted into readily cracked naphthenes. Subsequent reaction leads either to the rupture of the end ring, resulting in a side chain attached to the ring structure, or to the rupture of the centre rings resulting in two ring structures with attached side chains. The relative importance of these reactions controls the proportion of paraffins and ring structures in the products. By this route refractory aromatics can be recycled to extinction.

4. Olefins appearing in the gasoline from catalytic cracking appear as paraffins in hydrocracking. Since tertiary carbonium ions are favoured, this means that the desirable high octane branched chain paraffins are produced. The high octane number of the light hydrocracker gasoline boiling below almost 85°C is due primarily to the isoparaffins produced in this way.

5. Alkyl benzenes will be hydrogenated but single rings resist further cracking. This has a deleterious effect on gasoline in the 100°–200°C boiling range. The naphthenes produced have a lower octane number than the corresponding aromatics so the hydrocrackate has a lower octane number than that from catalytic cracking. This naphtha is customarily catalytically reformed to regenerate the aromatics.

6. The secondary polymerisation reactions of catalytic cracking which may

finally lead to coke are halted almost completely with feeds which do not contain asphaltenes.

7. Where there is no pretreatment, reactor sulphur and nitrogen compounds will be converted to H_2S and ammonia.

8. Hydrogen is required in all these reactions. The quantity will depend on the degree of conversion of the feed which the final product specification demands. This cannot normally be met by hydrogen from other processes such as catalytic reforming, so hydrocrackers usually have their own dedicated hydrogen plant.

The reaction of hydrogen with hydrocarbon is exothermic, and the degree of exothermicity depends on the amount of hydrogen consumed. Estimated heats for different types of hydrogenation reaction observed in hydrocracking are listed below:

Type of reaction	*Heat released* kJ/m^3 H_2 consumed
Hydrocracking	−2.2
Desulphurisation	−2.5
Saturation of olefins	−5.5
Saturation of monoaromatics	−3.0

Taken from Galbreath *et al.*, 8th World Petroleum Congress, **4**, 129 (1971).

Thus for any type of hydrocracking the amount of heat released will be greater than 2.2 kJ/m^3 consumed. In addition the heat released when hydrocracking a lower boiling range feedstock will be greater than that released from a higher boiling range feedstock for three reasons: greater hydrogen consumption, complete saturation of the products, and use of unsaturated feedstocks. Because of the exothermicity of the reactions a cold recycle gas quench is used in the reactor to absorb the heat of reaction.

The operating conditions employed in hydrocracking are a compromise which depend on the nature of the feedstock, product requirements, catalytic fouling and hydrogen consumption within the overall requirement to achieve the most favourable yield structure without incurring unreasonable processing costs. Operating temperatures and pressures lie within the ranges 316–482°C, and 68–204 bar respectively. Lower temperatures and pressures (316°–400°C; 68–136 bar) are used for gas oil hydrocracking; residues containing asphaltenes and higher concentrations of sulphur and nitrogen require more severe conditions.

To meet the requirements of the chemistry and feedstock composition the following factors are important in catalyst design: activity for cracking, activity for hydrogenation, selectivity for desired products, porosity, ability to operate in the presence of high boiling aromatics, sensitivity to sulphur compounds, nitrogen compounds and water vapour as well as on-stream stability and regenerability. Nitrogen is a key component both in the combined form as in the feedstock, and as ammonia. Both forms act as hydrocracking catalyst poisons because the alkalinity neutralises the acid cracking sites, the combined form more severely. The problem can be eliminated by a pretreatment step followed by NH_3 removal before true hydrocracking occurs. The economic balance then lies between increased hydrocracking catalyst activity and higher process capital costs.

All hydrocracking catalysts consist of a hydrogenating metal on a cracking base. The nature of the components depends on the ultimate product.

LPG catalysts. The typical catalyst used to produce LPG from gasoline is believed to be NiS on $SiO_2—Al_2O_3$.

Gasoline catalysts. Gasoline will be produced from gas oil. Hydrodenitrogenation, relative to hydrodesulphurisation, discussed previously, requires a more acid support and a higher loading of metals. NiMo, or CoMo, as sulphides, are supported on Si–Al (with ~2% F), or η-Al_2O_3 or silica stabilised Al_2O_3 with 0.1–10% F. The hydrocracking step itself is performed over a Pd/Y-type zeolite. The loading of the Pd is ~0.5% weight and it is introduced to the zeolite via ion exchange rather than impregnation to obtain a more uniform dispersion. This will minimise the extent and rate of coke formation during hydrocracking. The hydrogen form of the Y-zeolite was chosen because of its high acidity, which is not irreversibly poisoned by ammonia, its pore size (~10 Å), which admits hydrocarbon molecules from the feedstock to the active sites, and its crystallinity and high Si : Al ratio which impart heat and steam stability during regeneration. Even with the enhanced tolerance of the Y-zeolite to ammonia, without pretreatment higher operating temperatures are required, which will cause greater coking and light hydrocarbon production, thereby decreasing the gasoline yield.

Middle distillate, jet fuel and lube oil catalysts. The catalysts used to produce these products consist of molybdenum or tungsten, promoted by nickel or cobalt, as sulphides, supported on alumina or silica-alumina of suitable acidity to meet the product specification, i.e. the catalyst needs to promote the correct combination of hydrogenation, isomerisation and controlled hydrocracking.

Product yields achieved in hydrocracking vacuum gas oil to either gasoline, middle distillate or lube oils are shown in Table 5.5. To produce gasoline the feed is recycled to extinction; hence the two-stage process. This product has the highest hydrogen requirement. Although the light gasoline has a research octane number of 87, the heavy gasoline with the paraffins, naphthenes and aromatics distribution shown will have an octane number of only about 60. Hence the requirement for further upgrading via catalytic reforming. The hydrogen requirement for middle distillate and lube oil is also significant. In fact these product specifications clearly illustrate the dilemma of the hydrocracking process. It does have a high degree of flexibility but it is a multistage process requiring its own hydrogen plant which produces products which frequently need further upgrading.

Table 5.5 Yields and properties of hydrocracker products

Product maximised	*Gasoline*	*Middle distillate*	*Lube oil*
Feedstock	VGO*	VGO	Vacuum distillate
Crude source	Middle East	Arabian Light	Kuwait
Boiling range (°C)	350–550	350–550	—
Number of stages	2	1	1
H_2 consumption (m^3/m^3)	325	235	200
Yields (%wt on feed)			
Light gasoline, C_5–80°C	25.5	7.0	1.5
Heavy gasoline, 80°–175°C	61.25	22.0	5.0
Middle distillate, 175°–480°C	—	51.0	36.6
Lube oil	—	—	55.0
RON, light gasoline	87	80	76
Heavy gasoline (%wt)			
Paraffins	44	43	40
Naphthenes	46	50	50
Aromatics	10	7	10

(Taken from *Hydrocarbon Processing* **57**, p. 117, Sept 1978.)

* vacuum gas oil

5.6 Steam cracking

The primary source of ethylene and propylene, and some higher olefins and dienes, is the thermal cracking of light hydrocarbons in the presence of steam, hence the term steam cracking. This process is non-catalytic (indeed, the involvement of a catalyst is carefully avoided) but it must be remembered that steam cracking provides the major source of all our non-aromatic building blocks.

The primary objective is to maximise the yield of ethylene, the largest tonnage chemical building block. To this end ethane is the feedstock of choice. In the U.S.A., ethane and propane, co-products of crude oil production, have been freely available and have accounted for over 90% of ethylene production. Elsewhere in the world, particularly in Western Europe and Japan, these materials are less widely available and so naphtha has been the dominant feedstock.

Steam cracking and catalytic cracking have only formal similarities. Steam cracking is operated under more severe conditions and proceeds via a free radical rather than a carbonium ion pathway. The severity of the reaction is related to kinetic and thermodynamic constraints. Thermodynamically it must be operated in the region that favours dehydrogenation reactions of alkanes to alkenes, but not alkenes to alkynes. This can best be seen for the reactions of ethane and ethylene:

$$C_2H_6 \rightleftharpoons C_2H_4 + H_2 \qquad \Delta G = 0 \text{ at } \sim 720°C$$
$$C_2H_4 \rightleftharpoons C_2H_2 + H_2 \qquad \Delta G = 0 \text{ at } \sim 1150°C$$

Thus, steam crackers for the production of ethylene must operate in the region 750 to 1100°C (other considerations restrict commercial operation to 750 to 900°C) while processes aimed at acetylene production must operate at temperatures above 1200°C.

Mechanistically, the reaction proceeds via a conventional free radical chain pathway which consists of three main parts:

(a) *Initiation.* The primary process is homolysis of C—C bonds, e.g.

$$C_{10}H_{22} \rightarrow C_2H_5\cdot + C_8H_{17}\cdot$$

These reactions have high activation energies, ca. 220 kJ/mol. C—H bond scission is not significant.

(b) *Propagation.* Small radicals, e.g. ethyl, can propagate the chain via either formation of ethylene and a hydrogen atom or abstraction of a hydrogen atom from a higher hydrocarbon to give a new hydrocarbon radical and ethane. Larger radicals are very short-lived and decompose to smaller radicals via a series of β-elimination reactions producing ethylene or an α-olefin, e.g.:

$$CH_3CH_2CH_2CH_2CH_2\cdot \rightarrow CH_3CH_2CH_2\cdot + C_2H_4$$
$$CH_3\dot{C}HCH_2CH_2CH_3 \rightarrow CH_3CH{=}CH_2 + C_2H_5\cdot$$

Overall, the initiation and propagation reactions serve to reduce the molecular weight of the product.

(c) *Termination.* The radical chain is terminated by radical combination or disproportionation reactions. These reactions have zero activation energy and being bimolecular are favoured at high radical concentrations.

In addition to these "primary reactions", some secondary reactions take place. These become more important as the reaction proceeds to high conversion. They include olefin polymerisation, alkylation, and condensation of aromatics (present in the feed) leading to the production of coke.

In practice, therefore, the yield of ethylene is maximised by operating at:

(i) high temperature, 750–900°C.
(ii) short residence times, typically 0.5 to 1 s, depending on temperature.
(iii) a low partial pressure of hydrocarbon. This minimises chain termination reactions and secondary reactions such as ethylene polymerisation. Steam is used as the "inert" diluent at up to 1:1 wt/wt level depending on feedstock. Steam also has some beneficial effects: it removes coke via the water gas reaction ($C + H_2O \rightarrow CO + H_2$), improves heat transfer, and renders heat transfer surfaces passive.

In steam cracking of naphtha, feedstock diluted in preheated steam is passed through tubes approx. 50–200 mm in length by 80–120 mm in width in a heated furnace. The products which emerge are rapidly cooled, from ca. 850 to 300°C, to avoid secondary reactions and at the same time raising high pressure steam. Water and high boiling hydrocarbons are then condensed and the cracked gases purified and separated in a complex distillation train.

Naphtha is itself a complex mixture of hydrocarbons. The cracked product is even more complex and poses certain problems to the chemical manufacturer. No one of the various products is there in major proportion, e.g. ethylene, the largest single chemical, forms only around 30% of the total product. Thus, the economic disposal of the various fractions creates problems and must be tackled in a carefully integrated manner. These problems are discussed further in chapter 7.

As the light hydrocarbon fractions (naphtha) are attaining premium values, so attention is turning increasingly to the cracking of other feedstocks, LPG, heavier oil fractions (e.g. gas oil), and even crude oil itself. Liquefied petroleum gas (LPG) which was previously largely flared at the well-head in regions (such as the Middle East) which are relatively remote from the major petrochemical complexes, is becoming particularly attractive.

5.7 Concluding remarks

Any large petroleum refinery will contain only some of the processes described in this chapter. The combination chosen will depend on the normal source of crude oil fed to that refinery and the local market requirements for petroleum products. In addition, the refinery will contain a range of separation units either for initial fractionation (e.g. atmospheric and vacuum distillation units) or for final product adjustment to meet specifications (e.g. furfural extraction of lubricating oils). This chapter has discussed the major catalytic petroleum processes. None is of recent origin but all have a listing of continuous development both in the catalysts employed and the process engineering. This development will continue as refiners seek to process an ever-widening range of feedstocks incorporating more and more high-boiling components.

REFERENCES

Reforming

1. M. J. Sterba and V. Haensel, *Ind. Eng. Chem. Prod. Res. Dev.*, **15**, 2, 1976.
2. F. G. Ciapetta and D. N. Wallace, *Catalysis Rev.*, **5** (1), 67, 1971.
3. J. H. Surfelt, *Adv. Chem. Eng.*, **5**, 37, 1964.

Cat cracking

1. A. W. Chester, A. B. Schwartz, W. A. Storer and J. P. McWilliams, *Chemtech*, 50, 1981.
2. V. Haensel, *Advances in Catalysis*, **3**, 179, 1951.
3. L. C. Bryson, G. P. Huling and W. E. Glausser, *Hydrocarbon Processing*, **52** (5), 85, 1972.
4. S. M. Jacob, B. Gross, S. E. Voltz, V. W. Weckman, *AICHE Journal*, **22**, 701, 1976.
5. E. C. Luckenbach, *Chem. Eng. Prog.*, **75** (2), 57, 1979.
6. P. B. Weisz, A. B. Schwarts and V. W. Weekman, 10th World Petroleum Congress, **14**, 325, 1979.

Isomerisation

1. P. A. Lawrance and A. A. Rawlings, 7th World Petroleum Congress, **4**, 135, 1967.
2. P. Buiter, P. Van 't Spijker and D. Van Zooneu, 7th World Petroleum Congress, **4**, 125, 1967.
3. R. E. Davis, 1978 NPRA Annual Meeting, March 19–21, 1978, San Antonio, Texas, AM-78-53.
4. P. C. H. Mitchell, Chemical Society Specialist Periodical Reports, *Catalysis*, **1**, 204, 1977.
5. S. C. Schuman and H. Shalit, *Catalysis Rev.*, **4**, 245, 1970.
6. J. B. McKinley, *Catalysis*, **5**, 405, 1957.
7. J. Winsor, *The Chemical Engineer*, No **261**, 193 May, 1972.
8. J. C. Vlugter and P. Van 't Spijker, 8th World Petroleum Congress, **4**, 159, 1971.
9. R. B. Galbreath and R. P. Van Driesen, 8th World Petroleum Congress, **4**, 129, 1971.

Hydrodesulphurisation

1. P. Grange, *Catalysis Rev.*, **21**, 135, 1980.

CHAPTER SIX

COAL- AND NATURAL GAS-BASED CHEMISTRY

R. PEARCE AND M. V. TWIGG

6.1 Raw materials and resources and routes to fuels and chemical building blocks

6.1.1 *Raw materials and resources*

Oil is not the only source of organic chemicals. Hydrocarbons occur in other forms, e.g. coal, natural gas, and living plants (biomass). Each has had and will have a role to play. The oldest parts of the organic chemicals industry had their origins in the nineteenth century and were based on the fermentation of natural products. As the industry grew the raw material base shifted to coal. The first half of the twentieth century saw a rapid growth of processes based largely on coal tar reaching the peak of their development during World War II.

The emphasis in raw materials then shifted to oil and, in the U.S.A. in particular, to natural gas. After early beginnings in the U.S.A. in the 1920's, the stimulus provided by World War II for such strategic materials as gasoline and synthetic rubber marked the start of the modern petrochemical industry. In less than forty years the industry grew in a meteoric fashion, growth far outstripping rises in general economic activity, to reach the dominance we see today. Oil is now the raw material of 90 % of organic chemicals and accounts for some 60 % of overall fossil fuel consumption. These changes in the raw materials base are nicely illustrated by reference to the U.K. chemical industry. In the early 1950's, around 80 % of all chemicals were obtained via fermentation or from coal. By the late 1970's this figure had fallen to less than 10 % reflecting both a decline in absolute values for these materials and rapid increase (5 to 7 fold) in the total output of the industry. This situation cannot continue in the long term. Oil, currently meeting such a major portion of world energy demand, represents only a relatively small proportion of recoverable hydrocarbon reserves. We cannot reliably predict the price or availability of the various sources of hydrocarbons over the next few years or beyond: nor can we predict rates of consumption with any degree of certainty. To attempt to do so would be folly. The events of the 1970's have shown that almost all predictions do not stand the test of time even in the short term. We can, however, take a closer look at world reserves and relate these to current patterns of consumption. Some longer term trends become clear.

Figure 6.1 shows world reserves of coal, oil and gas broken down geographically. It can be seen that coal accounts for some 70% of total reserves and is distributed regionally in roughly the same manner as world population. Oil, on the other hand, is more limited (with some 20% of the reserves) and is distributed very unevenly. The major reserves and the main regions of consumption do not coincide. Clearly oil cannot continue to meet such a large fraction of world energy demand in the longer term; rates of production must eventually decline as supplies dwindle. The age of oil, the age of cheap energy, will be short-lived in historical terms, spanning perhaps a mere half century.

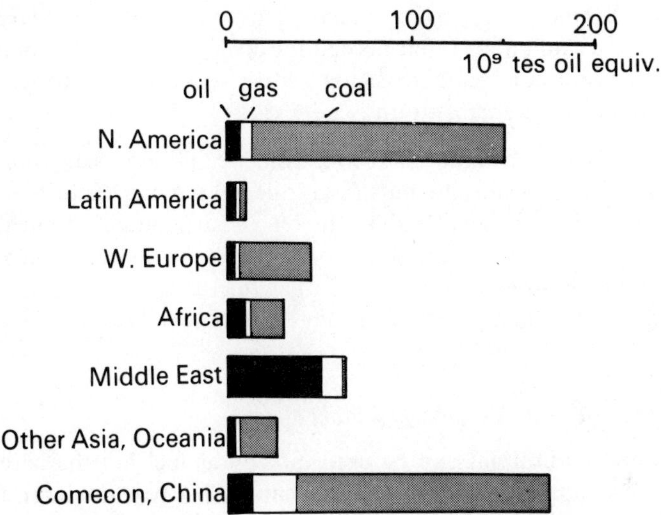

Figure 6.1 Ultimately economically recoverable fossil fuel reserves (adapted from *Oil Industry Petroleum*, B.P. Co., 1977). Oil and gas resources in Latin America and all resources in the Soviet Bloc may be underestimated.

At current rates of consumption, coal is the only serious alternative and is likely to regain the dominance it once held. The change to coal as a source of organic chemicals will not be a rapid one, however. Estimates vary widely, but it has been predicted that even by the turn of the century the chemical industry will be taking feedstocks only 10 to 30% of which will be coal-derived. Among the reasons for this apparently slow rate of change are:

(a) *Commercial readiness.* Improved coal conversion technologies are even now only in early stages of commercial development and will not be ready before the late 80's or early 90's.

(b) *Replacement timescale.* The oil-based industry could not be replaced overnight: it has been estimated that in the U.S.A. around 5% of the GNP must be invested over a 25-year period to effect the total changeover, clearly a daunting prospect.

(c) *Incentives for change.* Oil is so admirably suited to the production of organic chemicals that it may be reserved for these "more mobile uses". In any case, much of the present petrochemicals plant will not be worn out or require replacement before the turn of the century.

(d) *Logistical constraints.* Even where extensive coal deposits are available, the mining and transportation industries will need to undergo massive expansion to cope with the increased coal demand. This will necessitate large capital expenditure over and above that needed for coal processing. In addition, strip mining on the enormous scale envisaged is likely to meet with strong opposition from environmental pressure groups.

Natural gas has, with oil, supplanted the older coal-based technologies. It will continue to play a significant but lesser role as a premium fuel or chemical feedstock. Reserves are smaller than for oil and it is often obtained as a co-product in oil extraction. In the short to medium term a strong incentive exists for making greater use of the co-product gas (methane with associated ethane, propane, and butane) that is wastefully flared at the well head.

6.1.2 *Routes to fuels and chemical building blocks*

Both coal and natural gas can be used directly as fuel but they are not in a suitable molecular form for chemical synthesis. They require transformation into a more reactive or chemically purer state. Natural gas (CH_4) is available in reasonable purity but is chemically relatively unreactive. By the processes of steam reforming (reaction with water) and/or partial oxidation it can be converted to synthesis gas (mixtures of H_2 and carbon oxides, CO with lesser amounts of CO_2).

Coal is a heterogeneous black solid which has been produced over the millennia by anaerobic degradation of vegetable matter under high temperatures and pressures. The predominant organic fraction is a complex mixture of partially hydrogenated, condensed aromatic molecules, rich in heteroatoms, N, O, and S. Drastic alteration is required to convert it into a usable state. This change can be effected in two ways:

(a) partial cleavage of carbon–carbon and carbon–heteroatom bonds with hydrogen to give lower molecular weight, more highly hydrogenated species (liquefaction);
(b) total fragmentation by conversion to a gas rich in CO and H_2 (gasification).

Thus, synthesis gas is an intermediate common to the processing of both coal and natural gas and provides one of the unifying themes of this chapter. The other, the main theme, is the discussion of routes to fuels and chemical building blocks from a non-oil base. Our main concern is with the popularly-titled "coal-based chemical economy".

Coal is so readily and generally available that there must be good reasons for preferring the alternatives. Its disadvantages can be summarised as:

(a) As a solid it is considerably more difficult and more costly to mine and transport, and more difficult to handle at processing plants.
(b) As a chemical feedstock it has a very low H:C ratio and very high levels of heteroatoms, particularly sulphur. Coals are classified according to "rank", ranging from the least altered brown coals and lignites (typical H:C = 1:0.85) through the bituminous coals (1:0.7) to the most highly altered, carbon-rich anthracites (1:0.3). Considerable adjustments are needed to give ratios of around 2:1 more typical of chemical intermediates and end products (cf. values of 4:1 for CH_4 and 1.8:1 for oil). Sulphur is present at levels of 1–3 %, and being a potent catalyst poison, must be removed.
(c) It is very heterogeneous. Its composition shows wide variation not only from source to source but from seam to seam and even within an individual particle. Non-combustible solids are invariably present. This can present problems both in abrasion of processing equipment and in the need for its recovery and disposal.
(d) On combustion, oxides of sulphur and nitrogen are released to the environment, contributing to atmospheric pollution. This has become a political issue in many areas, particularly the U.S.A., and has been a strong stimulus for the various "clean fuels from coal" projects. Removal of sulphur is an important target in coal conversion, but is by no means a trivial operation. Treatment of oil fractions, which are liquids, is reasonably straightforward (cf. 5.5.1). Coal, a solid, presents more of a problem. Desulphurisation is only possible after conversion to a liquid or gaseous form.
(e) Coal-derived liquids, solids and tarry residues, and even coal itself, contain potent carcinogens. The hazards of the large scale adoption of coal liquefaction processes are only now becoming fully apparent and may have a powerful negative influence on the oil to coal conversion timescale.
(f) Coal has a relatively low calorific value. As a fuel, around 1.5 te of coal is equivalent to 1.0 te of oil. Thus, larger quantities must be processed for the same output.
(g) Coal-based processes are thermally inefficient. This is associated with factors such as (i) the often greater number of processing steps to the intermediates and end products, and (ii) the need for improvement in the H:C ratio which can only be achieved by sacrificing part of the coal to provide additional hydrogen.
(h) Coal-based processes are more capital intensive. As a consequence of a

combination of the factors above, plants for coal-based processes cost at least 2–3 times their oil-based equivalents.

The strengths of coal are its general availability and its potentially lower cost. Cost is however related to its desirability as a feedstock or fuel. Given sufficient quantities of natural gas and oil as alternatives, the cost of coal will be largely tied to their cost. It is only when global scarcity of oil and gas occurs, real or politically-managed, that the differential will increase significantly. More locally, incentives for the adoption of coal-based processes can exist where there is abundant cheap coal and limited oil reserves, and where there exists a strong desire to achieve independence from external energy supplies. South Africa, a coal-rich, oil-poor, and politically isolated country has already moved strongly in this direction: and the U.S.A., energy-hungry, coal-rich, and becoming depleted of oil and gas is likely to respond similarly in the coming decades.

Scheme 1 summarises routes from coal and natural gas to chemical building blocks and fuels. Primary processes are the conversion to synthesis gas and the generation of primary coal liquids. Secondary processes involve the transformation of synthesis gas into hydrocarbons (Fischer–Tropsch processes) and methanol and the upgrading of coal liquids. Synthesis gas to methane (methanation) from coal is included here when shortages of natural gas require the provision of a manufactured, high-energy, pipeable fuel gas (SNG = synthetic natural gas). Tertiary processes involve the conversion of methanol to hydrocarbons (Mobil process). (At this stage it may be unclear as to why a need exists to convert methanol to hydrocarbon fuels—this will be examined in more detail in 8.2.3.) Thermal efficiencies of some processes are shown in Table 6.1.

Table 6.1 Thermal efficiencies of coal-based conversions

Process	*Thermal efficiency‡* (%)
Gasification	
to SNG (CH_4)	60–70
to CO/H_2	70–90
Liquefaction	60–70
Fischer–Tropsch*	50–60
Mobil†	60–65

* To broad spread of products from SNG to waxes.
† To gasoline.
‡ This takes no account of product value. Recycle of products, e.g. CH_4 in the Fischer–Tropsch synthesis, to improve the yield of more commercially desirable materials can drastically lower overall thermal efficiency.

One thing is certain, a future coal-based industry will not look like that based on oil. It will, however, have to meet virtually the same demands of product type and quality which will place severe constraints on the various process options.

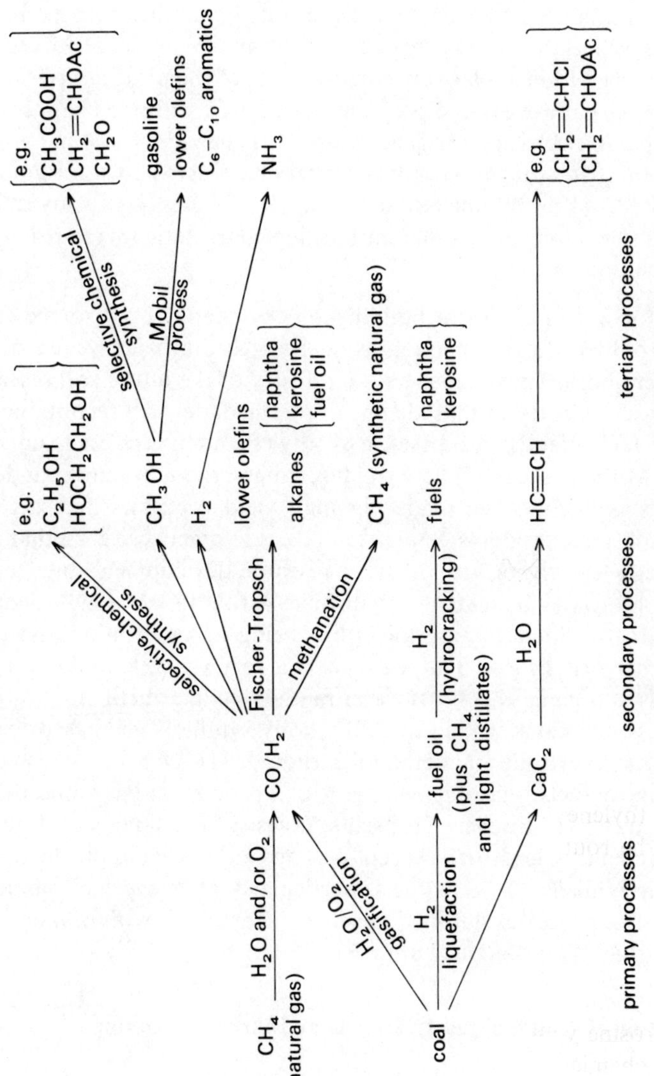

Scheme 1 From coal/natural gas to fuels and chemical building blocks

(i) The chemical industry will still be called upon to provide the same end products and will therefore demand either the same chemical building blocks or the development of wholly new synthetic pathways.
(ii) The industries that supply transportation fuels will still be called upon to provide a similar pattern of products to those used today unless totally new automotive concepts become widely adopted. Specifications of the various fuels, as well as the broad range of lubricating oils, are largely fixed and can only be met within a relatively narrow range of chemical compositions.
(iii) Allied to (ii) above are the problems associated with the changeover from oil to coal in electric power generation. This change is by no means a trivial operation. Modern furnaces have been constructed to use liquid hydrocarbons. Coal needs different handling techniques, has a lower calorific value, and exhibits quite different burning characteristics requiring a larger combustion zone.

The production of chemical building blocks cannot be divorced from the production of fuels. In this chapter, as in the preceding one, we are obliged to consider them both. Thus a coal-based industry of the future will resemble the present day oil industry in that chemical intermediates will be obtained as byproducts of fuel oriented processes, e.g. ethylene as byproduct from Fischer–Tropsch or Mobil processes. There will be changes, though—an expanded range of chemicals is likely to be made via high yield processes dedicated to the production of single products. Some are in current practice, e.g. methanol, while others, e.g. ethylene glycol, are still at the research/development stage (cf. 8.3.2).

Each of the routes in scheme 1 will have a role to play in the coal-based economy. Most of the attention looks like being on gasification and products derived from synthesis gas. Synthesis gas is a very versatile building block for the synthesis of organic chemicals either in one step (e.g. methanol synthesis) or via multistage reactions which are still wholly synthesis gas based (e.g. acetic acid and acetic anhydride via methanol carbonylation, cf. 8.5.1). It also provides a broad range of fuels—gasoline via the Mobil process and gasoline, diesel, and aviation kerosine via Fischer–Tropsch synthesis. Coal liquids look much less desirable as chemical feedstocks except perhaps for aromatics production. Their primary role is likely to be as fuels, particularly as heavy fuel oils in power generation. The processes outlined in scheme 1 are dealt with in more detail in the sections that follow and in chapter 8.

6.2 Production of synthesis gas, hydrogen, and carbon monoxide

6.2.1 *Introduction*

Synthesis gas and its constituents, carbon monoxide and hydrogen, are each separately required as chemical feedstocks: for example, synthesis gas of appropriate $CO:H_2$ ratio is used in methanol synthesis and in hydroformylation, carbon monoxide in acetic acid synthesis via methanol carbonylation, and hydrogen in ammonia synthesis. It can be generated via steam reforming or partial oxidation of gaseous or liquid hydrocarbons and via coal "gasification", a term that embraces a more complex set of reactions. Synthesis gas is the preferred raw material for the production of pure hydrogen and carbon monoxide. In terms of volume produced, hydrogen is by far the more important.

The hydrogen content of synthesis gas can be increased by chemical transformation of carbon monoxide via the water gas shift reaction. At low temperatures the position of the equilibrium in (1) lies well to the right-hand side:

$$CO + H_2O \rightleftharpoons CO_2 + H_2 \tag{1}$$

This is the preferred method for large-scale production. For the majority of chemical applications, hydrogen must be of high purity, and carbon oxides, in particular carbon monoxide, must be reduced to very low levels (ppm). Hydrogen is also obtained as a valuable by-product in a number of chemical processes, e.g. in chloralkali production via electrolysis of brine and in catalytic reforming of hydrocarbons. Both shipping and storage are problematical, so except on the small scale it is produced where consumed. By-product hydrogen that finds no immediate chemical outlet is generally used as fuel. Strong incentives exist, therefore, for the common siting of hydrogen-producing and hydrogen-consuming processes.

Hydrogen can be produced directly via electrolysis of water. This gives a very pure product but, at present, is appropriate only on the small scale or in areas where cheap electricity is available (e.g. hydroelectric generation). In the longer term electrolytic production may become more significant as the costs of hydrocarbon feedstocks increase and as relatively cheap nuclear or solar energy becomes more widely available.

Both carbon monoxide and hydrogen can be obtained via separation of the components of synthesis gas. Methods of separation include:

(a) *cryogenic fractionation* (Linde process). This involves partial condensation of the carbon monoxide and residual methane at ca. $-180°C/40$ bar followed by fractional distillation. Carbon dioxide and water must first be removed to low levels since both will solidify and cause blockages at the low temperatures used.

(b) *diffusion* (Monsanto Prism process). This involves preferential diffusion of hydrogen through a specially prepared, supported membrane.

(c) *selective adsorption* on zeolites or in solutions containing copper salts. Carbon monoxide is reversibly bound to copper (I) as a simple carbonyl complex, and separation and recovery utilise pressure/temperature swing techniques. Adsorption is conducted at low temperature and high pressure (e.g. 25°C, 20 bar) and recovery at high temperature and low pressure (e.g. 100°C, 1 bar). Early processes used aqueous ammoniacal copper (I) solutions. More recently organic solvents have found favour, e.g. the Tenneco "Cosorb" process using copper (I) chloride/aluminium chloride in toluene.

The generation of synthesis gas, the production of hydrogen via the water gas shift reaction, and its purification by removal of carbon oxides are considered in more detail in the sections that follow. It will then be shown how these separate catalytic processes are linked together in a complex and highly interdependent manner in the synthesis of ammonia (6.3) and methanol (6.4).

6.2.2 *Steam reforming of hydrocarbons*

The term "steam reforming" describes the highly endothermic reaction of superheated steam with hydrocarbons in the presence of a catalyst to give a mixture of hydrogen and carbon oxides, e.g. as in (2). Equilibrium conversion

$$CH_4 + H_2O \rightleftharpoons CO + 3H_2 \qquad \Delta H = +205\ \text{kJ/mol} \tag{2}$$

is increased by increasing temperature, increasing the steam : hydrocarbon ratio, and decreasing the pressure. The steam reforming of natural gas and low boiling hydrocarbons has been established industrially since the 1930's and has become a major process for the production of synthesis gas and hydrogen. In Britain, steam reforming of liquid hydrocarbons preceded that of natural gas. In the late 1950's and early 1960's steam reforming of naphtha, readily available at low cost from new petroleum refineries, rapidly displaced synthesis gas generation from coal. In its turn it was to be replaced in the 1970's by North Sea natural gas which is now the preferred feedstock. In those parts of the world where natural gas is not available, steam reforming of higher hydrocarbons is still widely practised.

The catalysts used in steam reforming are sensitive to a number of poisons and it is important that these are removed from the feed. Those encountered include sulphur, arsenic, halogens, phosphorus, lead and copper, of which the most important is sulphur. The principles of the hydrodesulphurisation process have been described in section 5.5.1. The purified feed, mixed with steam, then passes to the steam reformer where it encounters a supported nickel catalyst—typical reaction conditions are 700–830°C/15–40 bar. The catalyst is contained in a number of reaction tubes heated externally by combustion of natural gas or liquid hydrocarbons—it is customary to use the same feed for reforming and heating. Large modern plants may contain up to 400 tubes, each about 10 cm internal diameter and 13 m long. Because of the extreme conditions (high pressures and wall temperatures of up to 950°C) the tubes must be constructed of high performance alloy, and this accounts for much of the total cost of a primary reformer.

Steam reforming catalysts are required to operate under demanding conditions (high temperatures in the presence of steam) and must therefore be robust. In natural gas reforming, nickel catalysts impregnated on a refractory support such as α-alumina or calcium aluminate are used, most commonly in the form of Raschig rings.

Higher hydrocarbons are more difficult to reform than natural gas because of their tendency to deposit carbon on the catalyst, and special catalyst formulations are required. The most successful commercial catalyst contains an alkali metal oxide (typically K_2O) promoter which plays a complex role in preventing carbon formation. The alkali neutralises acid sites on the catalyst on which hydrocarbon can be catalytically cracked, and, at the temperatures employed, catalyses the reaction of steam with carbon to give carbon monoxide and hydrogen (cf. the beneficial effects of alkali metal catalysts in coal gasification, 6.5.2).

Modern steam reformers are operated at high pressures (up to 40 bar). This reduces plant capital costs and improves overall efficiency, but has the undesirable effect of limiting overall conversion to synthesis gas. With rising pressure the methane content of the product gas at thermodynamic equilibrium increases for a constant reaction temperature. Materials of construction limit reaction temperature in externally heated reforming tubes to ca. 830°C. The product gas from the reformer may therefore contain up to 10% methane, depending on the steam : methane ratio used. Where the ultimate objective is the production of ammonia, it is highly desirable that the methane level in the gas should be as low as possible (see below). Reforming must therefore be carried out at higher temperatures. The gas from the first reforming reactor, the *primary reformer*, is led into a separate lined reactor filled with heat-resistant nickel catalyst. This is called the *secondary reformer*. A portion of the gas is combusted with air, raising the temperature to over 1000°C and driving the reforming reaction to completion. The methane content is reduced to ca 0.2% v/v. The correct nitrogen : hydrogen ratio can be achieved in this way without the need for separate production of oxygen and nitrogen.

6.2.3 *Partial oxidation of hydrocarbons*

Partial oxidation is a non-catalytic process. It differs from catalytic steam reforming in the manner in which heat is supplied to drive the steam reforming reaction. In partial oxidation this heat is generated by combustion of a portion of the feedstock *in situ* and in this respect it shows strong similarities to the secondary reforming step discussed above. Typical reaction conditions are 1200–1500°C, 30–80 bar with a preheated feed consisting of the hydrocarbon, water and substoichiometric quantities of oxygen. Successful processes have been developed by BASF, Lurgi, Texaco, Hydrocarbon Research, and Shell and are in operation throughout the world.

Advantages are that a range of hydrocarbon feedstocks can be used, ranging from methane to heavy oil fractions. In contrast to catalytic steam reforming, it is not necessary to free the feed from catalyst poisons, although these must still be removed in a later process stage. In the future, as less desirable feedstocks (e.g. heavy fuel oil residues) have to be used, partial oxidation may well gain in importance. In practice, the disadvantages are that overall these processes are thermodynamically less efficient (not an important consideration with heavy feeds), they require an associated air separation plant to supply oxygen, and soot formation can be troublesome.

6.2.4 *The water gas shift reaction*

This reaction was first used to realise more fully the hydrogen potential of water gas (the gas obtained via the reaction of steam with coke)

$$CO + H_2O \rightleftharpoons CO_2 + H_2 \qquad \Delta H = -42\ \text{kJ/mol} \tag{3}$$

and hence its name. Conversion of carbon monoxide is favoured at low temperatures and at high partial pressure of water. The reaction is kinetically limited and requires the provision of a catalyst to establish equilibrium. Catalysts based on iron and chromium oxides (Fe_3O_4 supported on Cr_2O_3) have been in use for almost seventy years. They are relatively insensitive to poisons. Indeed, iron sulphides, which can be formed from hydrogen sulphide impurities present in coal-derived synthesis gas, are effective catalysts. They are, however, not very active and must be operated at relatively high temperatures (400–500°C). In consequence conversions are thermodynamically limited: typically 2–4% of carbon monoxide remains in the reaction gas. Catalysts of this kind are known as *high temperature shift catalysts*.

An increased conversion of carbon monoxide is highly desirable in the manufacture of pure hydrogen. In this case more active catalysts capable of operation at lower temperatures are required. For the development of catalysts of this type, known as *low temperature shift catalysts*, the problems of selectivity and deactivation by poisoning had to be overcome. Selectivity is crucial: the thermodynamically preferred conversion to methane and higher hydrocarbons at these temperatures must be precluded. Active catalysts are based on the oxides of copper and zinc or copper, zinc, and aluminium. Copper-based catalysts are very sensitive to poisoning by sulphur and thus require a very pure synthesis gas feed. Such a pure feed was provided in the early 1960's when steam reforming of hydrocarbons became widely practised and it is only since this time that low temperature shift catalysts have been in general use. These are active in the temperature range 190° to 260°C and give carbon monoxide levels of only ca. 0.1% v/v.

It is not practical to carry out the shift reaction in a single-stage adiabatic reactor since the reaction is moderately exothermic. The inlet temperature to the reactor has a lower limit of around 200°C, set by the dew-point of the feed gas, and, with a typical adiabatic temperature rise through the reactor of up to 100°C, exit levels of carbon monoxide will be at best around 1%. A better arrangement is to carry out the reaction in two stages with inter-stage cooling: one operating at high temperature (400° to 500°C) using the conventional iron oxide chromia catalyst and the other operating at low temperature using the copper-based catalyst. Partial conversion in the high temperature shift reactor reduces the level of carbon monoxide in the feed to the low temperature shift reactor and limits temperature rise.

6.2.5 *Removal of carbon oxides*

Production of pure hydrogen requires removal of carbon dioxide and residual carbon monoxide—CO_2 levels are relatively high (15–35%) unlike CO levels (as low as 0.1%). Purification occurs in two stages. CO_2, an acid gas, is readily removed by solvent extraction. The many different processes available are of two types: firstly, using aqueous alkaline solutions, making use of the equilibrium between CO_3^{2-}, CO_2 and HCO_3^- (CO_2 is converted to HCO_3^- at low temperature and elevated pressure and is released at high temperature and low pressure); secondly, using pressurised washing with organic solvents. These include methanol at -40°C (Lurgi Rectisol process), dimethylethers of polyglycols (Allied Selexol process), sulpholane and di-isopropylamine (Shell Sulfinol process) or *N*-methyl pyrrolidone (Lurgi Purisol process). These are also effective in removing hydrogen sulphide and carbonyl sulphide in the gas and are widely used for the clean-up of synthesis gas obtained via partial oxidation and coal gasification. Removal via adsorption on molecular sieves has also been developed.

Residual trace CO and CO_2 is usually removed by conversion to methane over a supported Ni catalyst at 250–350°C (cf. 8.2.2). The methanation reactions,

$$CO + 3H_2 \rightleftharpoons CH_4 + H_2O \qquad \Delta H = -205\,\text{kJ/mol}$$

and

$$CO_2 + 4H_2 \rightleftharpoons CH_4 + 2H_2O \qquad \Delta H = -165\,\text{kJ/mol}$$

are the converse of steam reforming and are strongly favoured at low temperatures. Levels of carbon oxides in hydrogen streams can routinely be reduced to a few ppm; this is essential when the hydrogen is used for ammonia synthesis, and in organic hydrogenations, since CO is a powerful catalyst poison. In ammonia synthesis, it is also undesirable that the purified gas should contain appreciable quantities of methane—while inert to the catalyst, methane will build up in the recycle gas. Levels are controlled by a continuous purge of part of the recycle gas—a higher level of methane in the feed gas will require a higher level of purge. Methanation is well suited to removing final traces of carbon oxides in plants using a low temperature shift stage. Plants without this stage (i.e. using partial oxidation) have higher CO levels, and here, alternative methods of CO removal include selective oxidation to CO_2 (more easily separated when present at low levels) using a precious metal catalyst, and selective extraction by washing with liquid N_2. Liquid N_2 washing is useful in ammonia synthesis: at this stage the correct hydrogen–nitrogen balance can be obtained.

The move towards the use of feedstocks containing high levels of catalyst poison (in particular sulphur) seems inevitable. As a consequence, processes generating synthesis gas via partial oxidation, followed by a shift catalyst resistant to sulphur poisoning, could be attractive in terms of plant operability. Some argue, however, that economic advantages over alternative processes using present catalysts are not high, because at some stage in either approach most of the impurities have to be removed.

During recent years there has been considerable academic interest in the chemistry of soluble transition metal carbonyl complexes which are active catalysts for the carbon monoxide shift reaction. It remains to be seen if homogeneous systems using noble metal complexes will be developed.

6.3 Ammonia manufacture—an example of the integration of catalytic processes

A modern ammonia plant producing synthesis gas by steam reforming natural gas, illustrates how a number of individual catalytic reactions can be brought together to form an integrated single stream plant. In such a complex, the performance of the individual catalytic processes is influenced by those upstream and, in turn, influences the performance of those downstream. Careful integration of process design is required, and the influence of one unit on the operation of the other must be fully evaluated.

Notable features of an integrated single stream plant are that no intermediates are stored; they are produced sequentially at their respective rates of consumption. Similarly, heat produced in various stages is coupled with heat demands so as to make the overall plant efficient in energy usage. For instance, in steam reforming large amounts of high temperature combustion gases leave the furnace. These are used to raise steam for driving machines such as compressors used in the plant, while lower pressure steam (some of which is obtained from downstream exothermic heat of reaction) is used to preheat reactants to reaction temperature, recover carbon dioxide from saturated solution, etc. The sequence of stages in a typical ammonia plant is shown in figure 6.2. Seven of these stages are catalytic reactions, six of which operate at 20–50 bar, and produce the necessary hydrogen/nitrogen mixture. These have already been described and will not be discussed further.

Ammonia synthesis is exothermic, and favoured by high pressure. Highest equilibrium concentrations of ammonia are therefore obtained at high pressure and low temperature.

$$N_2 + 3H_2 \rightleftharpoons 2NH_3 \qquad \Delta H = -46\ \text{kJ/mol} \tag{4}$$

Maximum operating pressure is ultimately limited by plant capital cost, and minimum operating temperature by catalyst activity. Large plants (~1000 te/day) use a centrifugal compressor to increase the pressure to 200–350 bar for the synthesis reaction, which is run at 400°–500°C. The potassium-promoted iron catalyst (a 1000 tonne/day plant has some 100 tonne of catalyst which has a life up to 10 years) is contained in a special high-pressure reactor, which forms part of the "synthesis loop". Gas leaving the reactor is cooled to condense ammonia, in which form it is removed. Unconverted hydrogen and nitrogen is recirculated, together with additional fresh gas, around the synthesis loop.

The small amount of methane and argon present in the feed gas becomes

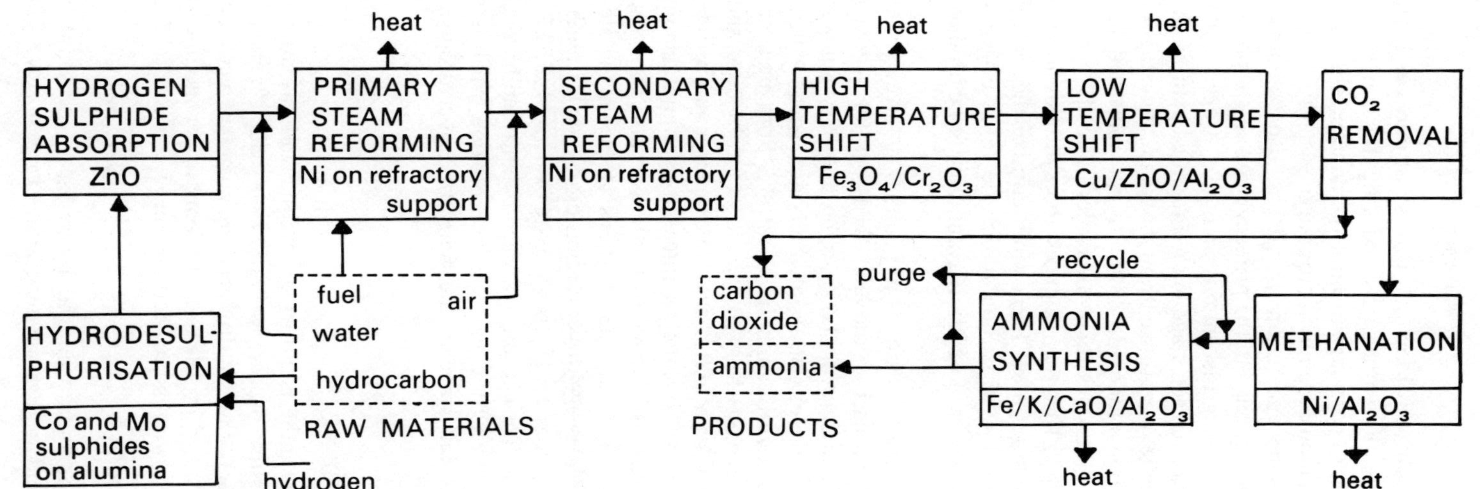

Figure 6.2 Simplified block diagram of modern single stream ammonia plant showing individual catalytic stages and heat sources.

concentrated in the synthesis loop, and this is not only costly to circulate, but also reduces the rate of synthesis by reducing the partial pressure of the reactants. It is therefore necessary to limit the concentration of these "inerts". This is done by continuously removing (or purging) gas from the synthesis loop. In addition to methane and argon, purge gas also contains hydrogen and nitrogen. This is either recycled to the primary reformer, or physically separated (cryogenic or pressure swing adsorption), and the hydrogen used in the process.

The interdependence of each stage in an integrated plant is illustrated by the importance of the low-temperature carbon monoxide shift stage on overall plant efficiency. Each carbon monoxide molecule *not* converted to hydrogen via reaction with water consumes three molecules of costly hydrogen when it is converted to methane. This then reduces the synthesis loop efficiency, and demands increased purge rates. The overall consequence in a large plant of improving the conversion of carbon monoxide to hydrogen by only 0.1% enables more than 350 te/yr of additional ammonia to be produced at marginal cost.

The concept of integration extends to plants themselves. An ammonia plant produces ammonia and carbon dioxide—feedstock for a urea plant; so it is logical for ammonia and urea plants to be adjacent. Similarly, nitric acid is obtained *via* catalytic oxidation of ammonia, which when reacted with additional ammonia affords ammonium nitrate. Both urea and ammonium nitrate are manufactured on a very large scale for fertiliser use, and it is common practice for ammonia, urea, nitric acid, and ammonium nitrate plants to be located close together, so forming an integrated fertiliser manufacturing site. Integration of ammonia and methanol synthesis is also highly beneficial when methane is used as feedstock. The synthesis gas from the steam reforming contains an excess of hydrogen over carbon monoxide for methanol synthesis (theoretical carbon oxides to H_2 ratio ex-reformer = 1:3; MeOH synthesis requires a ratio of 1:2). The carbon/hydrogen balance can be nicely restored by utilising the otherwise very low-value carbon dioxide which is removed after the low-temperature shift stage in the generation of hydrogen for ammonia synthesis.

6.4 Methanol synthesis

The first commercial synthesis of methanol from synthesis gas was by BASF in 1923. This high-pressure process, using a zinc oxide/chromia catalyst, soon gained acceptance and rapidly displaced manufacture via wood distillation (wood spirit). It remained in sole use with only minor modifications until the 1960's when the ICI Low Pressure Methanol process was developed.

$$CO + 2H_2 \rightleftharpoons CH_3OH \qquad \Delta H = -92\ \text{kJ/mol} \tag{5}$$

Conversion of carbon monoxide and hydrogen to methanol increases with increasing pressure and *decreasing* temperature. Table 6.2 gives illustrative data.

Table 6.2 Mol % CH_3OH at equilibrium

	Pressure (bar)			
T (°*C*)	50	100	200	300
250	33	52	73	82
300	8	24	48	62
350	3	7	25	40
450	<1	2	4	7

In this respect it resembles the synthesis of ammonia. Indeed, methanol plants use a similar method of gas recycle in a synthesis loop, and experience gained in large-scale, high-pressure ammonia synthesis was used in the development of the early methanol plants. In commercial practice a balance must be struck between temperature and pressure. The temperature must be high enough so that a suitable reaction rate is obtained, while the pressure must be great enough to ensure that an adequate conversion to methanol is not prevented by a low equilibrium concentration. Thus it can be seen that the advantage of high activity catalysts is their ability to operate at low temperatures where high conversions are possible at lower pressures.

The original zinc oxide/chromia catalyst, although relatively inactive, requiring operation at about 350°C and pressures up to 350 bar, had the advantage of being able to operate with relatively impure (by modern standards) synthesis gas, obtained via reaction of coal-derived coke with steam. Other catalysts were known to have high initial activities, but suffered rapid deactivation by poisons present in the feed gas. With the introduction of the steam reforming of hydrocarbons, synthesis gas of higher purity became available. It was then possible to consider the development of high-activity catalysts in the hope that when suitably formulated they would have adequate lifetimes in the purer feed gas. These ideas were fully realised in the development of the now highly successful ICI low pressure methanol process. This uses a copper-based catalyst (such as $CuO \cdot ZnO \cdot Al_2O_3$ or $CuO \cdot ZnO$) which is active in the temperature range 230–270°C. Operation at these lower temperatures permits a significant reduction in working pressure, to 50–100 bar, for similar equilibrium conversions. This has significant economic advantages, primarily associated with lower compression and equipment costs (section 4.4.3). Since the introduction of this process, no new high-pressure plants have been built.

Copper-based catalysts do suffer from a number of drawbacks, and although their superior activity had been realised a decade or more beforehand, it was these problems that had prevented their early adoption. These include: (a) their sensitivity to poisons in the feed gas, in particular chlorine and sulphur which must be reduced to very low (ppm) levels; (b) their sensitivity to thermal degradation which demands control of reactor temperature; and (c) their greater cost and potentially shorter lifetime compared to the original zinc oxide/chromia catalysts.

Selectivity in methanol synthesis catalysts is very important. Higher alcohols, ethers, and particularly alkanes, are thermodynamically preferred. The selectivity of copper-based catalysts is accordingly remarkable; overall yields in excess of 99 % are realisable. Side products, which include dimethyl ether, higher alcohols, alkanes, aldehydes, ketones and esters, are separated, together with the water formed during the synthesis, by distillation.

Where hydrogen-rich synthesis gas is used, as produced by steam reforming of natural gas, the hydrogen, carbon balance is restored by the addition of carbon dioxide. Mechanistically the situation is more complicated than that shown in (6), the catalyst also catalysing the water gas shift reaction.

$$CO_2 + 3H_2 \rightleftharpoons CH_3OH + H_2O \qquad \Delta H = -50\,\text{kJ/mol} \tag{6}$$

An alternative means of obtaining the correct balance is to include carbon dioxide, along with steam, in the primary reforming stage. This approach (7) can have process design advantages, and does *not* require the use of a different reforming catalyst.

$$CH_4 + CO_2 \rightleftharpoons 2CO + 2H_2 \tag{7}$$

Methanol synthesis is moderately exothermic—cf. (5) and (6), and because of the sensitivity of copper-based catalysts to deactivation via sintering, it is important that the temperature of the catalyst bed is carefully controlled. This can be achieved in one of several ways:

(a) by maintaining levels of methanol in the recycled gas;
(b) by limiting conversion per pass to low levels, i.e. less than is achievable at thermodynamic equilibrium;
(c) by injecting cold reactant gas between catalyst beds in the reaction, there being several beds in series;
(d) by incorporation of heat exchangers in the reactor;
(e) by use of a fluidised catalyst bed.

The most successful approach has been (c) and a number of plants each producing 1000 te methanol per day are in reliable operation.

It is somewhat surprising that such a major industrial process appears to have been based on wholly empirical research. Details of the mechanism of methanol synthesis are somewhat sparse and fragmented. For $CuO \cdot ZnO$-based catalysts it has been proposed that activity is associated with a Cu^{I}-rich zinc oxide phase. Both Cu^{I} and zinc oxide are necessary for the formation of active centres. The Cu^{I} sites bind carbon monoxide in a non-dissociative manner while the zinc oxide activates molecular hydrogen. Intermediates in the synthesis are poorly understood. One possibility has features common to the proposed mechanism for the synthesis of ethylene glycol using homogeneous rhodium catalysts (8.3.2), and involves formyl (Cu—CHO), hydroxycarbene (Cu=C(OH)H), and hydroxymethyl (Cu—CH_2OH) species. Other proposals consider O-bonded surface species, e.g. carbonate, bicarbonate, formate and methoxide.

Since the introduction of the ICI low pressure process, scales of production have increased dramatically. Plant sizes of 10^6 te/yr are possible and sizes up to 10^7 te/yr have been proposed. Present consumption is directed largely to formaldehyde (ca. 50%—see section 11.3.1), an essential ingredient in many thermosetting resins. Other uses are as solvents and as a reagent in selective synthesis, e.g. for methylamines, acetic acid. This pattern of consumption is likely to expand considerably in the coming years. Indeed, growth rates for methanol are predicted to far outstrip those of other major chemical intermediates over the next ten years. Its attraction is that it can be produced with high purity and very high selectivity from any carbon/hydrogen source via synthesis gas; and being a liquid it is easily stored and transported. Some growth areas for methanol use, many of which are discussed in detail elsewhere in the book, are:

(a) direct use as a fuel for power generation or transportation;
(b) as a chemical building block, displacing routes based on petroleum fractions, e.g. in synthesis of acetic acid, acetic anhydride and vinyl acetate (cf. 8.5.1 and 8.5.2);
(c) indirectly as a fuel via conversion to MTBE (ButOMe), a gasoline octane booster, or via zeolite-catalysed conversion to gasoline (cf. 8.2.3);
(d) indirectly as a chemical building block by selective conversion over novel zeolites to lower olefins or aromatics;
(e) in protein production as a feedstock for the growth of micro-organisms, e.g. the ICI Pruteen® process.

In many ways methanol can be regarded as liquid synthesis gas, a view supported by the adoption of methanol cracking at high temperature and low pressure for the small scale production of very pure synthesis gas and hydrogen.

6.5 Coal-based processes

6.5.1 *Liquefaction*

Coal has a hydrogen to carbon ratio of less than unity. To make a desirable range of liquids with ratios of 1.5 to 2 or more, it is necessary either to remove carbon or to add hydrogen. Pyrolysis achieves the first end and hydrogenation the second. One or both of these methods is used in modern coal liquefaction processes but neither is particularly new. Coal pyrolysis is a very old technology which probably reached the commercial scale of operation during the industrial revolution. The main aims are the production of coke, for steelmaking, smokeless fuels and town gas. Liquefaction via hydrogenation had its origins in the latter part of the nineteenth century, was transformed into a commercial process in Germany during the period 1910–1920, and reached its climax of development in Germany during World War II, spurred on by the need for self-sufficiency in liquid fuels.

The rapid expansion in use of oil in the decade that followed eliminated commercial coal liquefaction. Attention was focused on oil-based processes which were cheaper and simpler to operate. This almost total preoccupation with oil was relatively short-lived and even by the 1960's the appreciation of the more limited extent of oil reserves created a renewed incentive for research into coal conversion methods. What is new in the more recent methods of coal liquefaction is the application of more sophisticated chemical engineering and

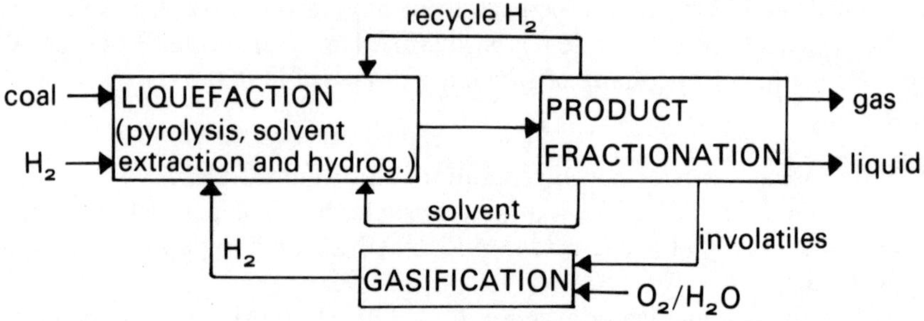

Figure 6.3 Catalytic coal liquefaction.

the use of improved catalysts. The key problems to be tackled are still the same, however: poor selectivity to the more desirable fractions and relatively high hydrogen consumption.

There are many different approaches to coal liquefaction. They differ in the way they combine the fundamental reaction processes of pyrolysis, solvent extraction, and hydrogenation. Two typical configurations are shown in figure 6.3 (e.g. as in the H-Coal or Solvent Refined Coal processes) and figure 6.4 (as in the Exxon Donor Solvent process). In more detail the main steps are:

(i) *Pyrolysis*. The first step in all the processes is pyrolytic breakdown of the coal particles which takes place at reasonable rates above 400°C. Free radicals are formed during this process. These can either combine one with another, leading eventually to coke, or abstract hydrogen from neighbouring molecules to produce liquid or gaseous products. The major product from pyrolysis is a highly carbonaceous char with lesser amounts of volatile gases and liquids. The liquids are complex mixtures of pyridines, phenols, and aromatic hydrocarbons, e.g. BTX and polycyclic species such as naphthalene and anthracene. Individual compounds can be isolated by complex separation methods. For many years coal tar liquids provided a valuable source of aromatic chemicals, the majority of which are now almost wholly derived from oil. The liquid yield, typically 5 %, is difficult to increase significantly. With the declining demand for coke for steelmaking, output of these products is likely to fall. Pyrolysis alone cannot therefore satisfy future demands for these compounds.

Pyrolysis can be catalysed. Some mineral residues within the structure of certain coals appear to be active as natural catalysts, but catalysis by design presents more of a problem: to be effective the catalyst must be brought into intimate contact with the matrix of the whole coal particle. Added solids cannot achieve this but liquids or gaseous species may be effective. Catalysts of any kind can, however, intercept the mobile or volatile initial products—see (iii) below.

(ii) *Solvent extraction.* At elevated temperatures, the lower molecular weight components can be extracted from the coal into a solvent. The use of a solvent enables a larger proportion of the coal organic material to be separated from the char. Most conveniently, recycled coal liquids are used for this purpose; they are highly aromatic and have good solvent properties.

An alternative approach being pioneered by the British National Coal Board is the use of supercritical fluids for extraction. Toluene, with a critical temperature of around 400°C, is most suitable. Advantages are claimed for the supercritical method in the more easy separation of liquids from the solid material and recycle of the working solvent.

(iii) *Hydrogenation.* Addition of hydrogen to increase the H:C ratio in the coal-derived liquids may be effected in a number of ways. Two commonly adopted options are (a) to effect hydrogenation in the primary liquefaction reactor, or (b) carry out partial hydrogenation in the primary reactor and complete the reaction in a separate stage. There are advantages and disadvantages to both. During the pyrolysis stage, free radicals and labile intermediates are generated. Such transient species are best intercepted *in situ* by hydrogen, preferably in the presence of a catalyst. In the absence of a trap, they may undergo further reactions among themselves giving higher proportions of less desirable products, e.g. coke. On the other hand, better control of the extent of hydrogenation may be possible in a separate hydrogenation reactor where conditions can be optimised. Molecular hydrogen is most commonly used. An alternative and successful approach is to provide H_2 via a hydrogen-donating solvent, e.g. partially hydrogenated aromatics like tetralin. This forms the basis of the Exxon Donor Solvent process (figure 6.4).

Hydrogenation requires the presence of a catalyst to achieve relatively mild reaction conditions. Catalysts effect both hydrogen addition reactions to sites of unsaturation (hydrogenation) and carbon–carbon or carbon–hetero atom bond cleavage (hydrogenolysis). Catalysts may be present in the coal itself in the form of mineral impurities or may be added. The demands placed upon catalysts in the primary liquefaction reactor are severe. They must be capable of tolerating high levels of sulphur and must not be irreversibly fouled by char and mineral matter that rapidly form on their surfaces. Conditions are less severe in

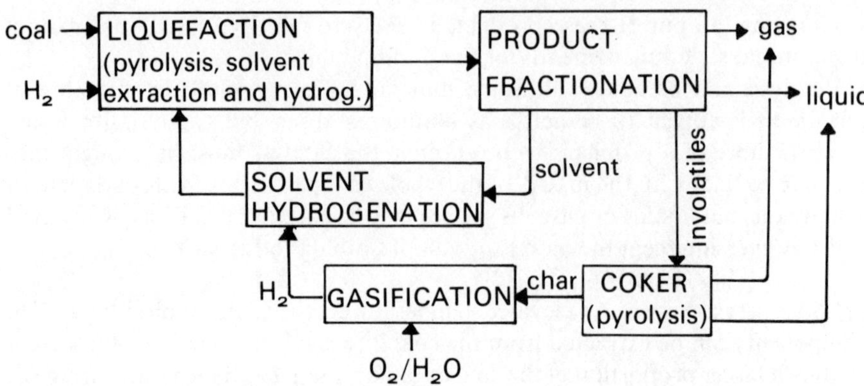

Figure 6.4 Catalytic coal liquefaction with donor solvent.

a separate hydrogenation reactor where they are milder and more controlled; better catalyst lifetimes are achievable here. The high S-levels rule out the majority of noble metal catalysts—most successful have been the conventional cobalt molybdate hydrodesulphurisation catalysts.

The process of liquefaction gives products with lower levels of heteroatoms N, O and S than were present in the original coal (via conversion to e.g. H_2S). Hydrogenation in a separate reactor is more effective in reducing these levels in that conditions can be specifically tailored. With coal liquids, nitrogen removal is presently a problem. The nitrogen is bound up in polycyclic aromatic molecules and is only labilised after extensive hydrogenation of the aromatic rings. Sulphur bound in this way is readily removed by hydrogenolysis without competing ring hydrogenation. As yet there are no known catalysts that can effect selective cleavage of such C—N bonds as they do C—S bonds, and the development of one will have a major impact on coal liquefaction technologies. To be commercially successful, any scheme must also be able to deal effectively with the char product. This is usually carried out separately where it can be burnt to provide process heat, or subjected to reaction with water and oxygen to generate hydrogen for the hydrogenation steps.

The major organic product is a heavy fuel oil fraction, which may be upgraded via hydrocracking with lesser amounts of a naphtha cut and light gases. The liquids have a higher aromatics content than their equivalent oil-derived fractions, and are especially rich in two- and three-ring polycyclic species. Thus as chemical feedstocks they are suitable primarily as sources of aromatic hydrocarbons.

Some ten modern liquefaction processes, such as the H-Coal®, Exxon Donor Solvent (EDS) and Solvent Refined Coal (SRC) process, are under intensive development. Many have reached the pilot-plant or "demonstration" plant stage. This latter is an intermediate stage of development (6000 te coal/day) en route to large-scale (ca. 30 000 te/day) commercialisation. Some coal liquefaction processes are compared in Table 6.3.

Table 6.3 Typical coal liquefaction processes

Process	Typical catalyst	Conditions		Hydrogenation		Comments
		T (°C)	P (atm)	Prime source	Method	
Bergius (original process developed early 1900's)	Iron oxide	465	200	H_2	in liquefaction reactor	Most severe conditions; catalyst discarded
Solvent Refined Coal (Gulf Oil)	Minerals in coal (none added)	450	140	H_2	in liquefaction reactor	Recycle of portion of product liquid to reactor; lack of hydrogenation specificity
H-Coal® (Hydrocarbon Research Inc.)	$CoO—MoO_3/Al_2O_3$	450	200	H_2	in liquefaction reactor	Catalyst ages rapidly
Exxon Donor Solvent	Minerals in coal in liquefaction reactor $CoO—MoO_3/Al_2O_3$ in separate hydrogenation reactor	450	140	Tetralin in liquefaction reactor. Recycled after hydrogenation in separate reactor		Further hydrogenation of product liquids in separate reactor—catalyst deactivation slow. Typical product yields 0.3 to 0.4 te liquids/te coal feed
National Coal Board supercritical extraction	In separate hydrogenation reactor	350–450	100–200	H_2 in separate reactor		Supercritical gas extraction of portion of coal with PhMe as solvent

6.5.2 *Coal gasification*

In contrast to liquefaction, interest in and commercial practice of coal gasification has never fallen into total decline. Today there are three distinct systems in operation: Lurgi, Koppers–Totzek, and Winkler Gasifiers (see Table 6.5), and there are many variants at the research and development stage.

The main products of gasification are CO, H_2, CH_4 and CO_2; H_2S and COS are the main co-products which, being potent catalyst poisons, must be removed before further processing. In modern gasifiers, there are two main modes of operation:

(a) maximisation of the yield of methane when the final objective of the gasification is the production of high-energy fuel gas for pipeline transmission. Conversion to methane is only partially complete in the gasifier. An additional methanation stage (6.2.5, 8.2.2) is required to give methane-rich gas;
(b) maximisation of the yield of CO and H_2 when the gas is to be used as a chemical feedstock. For conversion into organic chemicals methane is undesirable: it is chemically unreactive and will build up in any recycle streams.

Coal in a suitably prepared form, steam and oxygen are fed into the gasification reactor where they enter into an extremely complex set of reactions. These include pyrolytic breakdown of the coal into coke and lighter hydrocarbons, steam reforming and partial oxidation of these products to give CO and H_2, water gas shift, and methanation. The steam reforming reaction is

Table 6.4 Optimisation of key steps in coal gasification/methanation

Reaction step	*Heat* (ΔH_{298} *kJ/mole*)	*Optimum reaction temp. (K)*
$2C+2H_2O \rightarrow 2CO+2H_2$	+262.6	1250
$CO+H_2O \rightarrow CO_2+H_2$	−41.1	600
$3H_2+CO \rightarrow CH_4+H_2O$	−206.1	650
Overall:		
$2C+2H_2O \rightarrow CH_4+CO_2$	+15.4	

particularly important. It is the main reaction generating synthesis gas but is highly endothermic ($\Delta H \sim +130$ kJ/mol). The heat to drive it is generally supplied by *in situ* combustion of a portion of the feed (partial oxidation), but it may also come from external sources. This can be by burning of the coal itself, but in the longer term, as nuclear power becomes cheaper, heat may be taken from nuclear reactors. The costs of driving the steam reforming reaction can be significant, but, in theory, where methane is the desired end product they can be drastically reduced or even eliminated. The major problem is an inefficient thermal balance. The overall reaction sequence generating methane is essentially thermoneutral. Unfortunately, the exothermic heat of the water gas shift and methanation reactions is generated at lower temperatures and cannot be used efficiently to drive the reforming reaction (Table 6.4). A more efficient thermal balance is theoretically achievable by conducting the reaction at temperatures down to 650 K. This would have a major impact on gasifier economics but, being kinetically limited at these temperatures, an efficient catalyst is required.

Catalysis has so far made little impact in gasification and yet, as shown above, the incentives are considerable. The demands placed upon a catalyst are severe both in terms of reaction conditions and the multiplicity of reactions it is desired to catalyse. The problem of contacting the catalyst efficiently with the coal is common to both liquefaction and gasification. It is most severe in gasification where a gas–solid reaction is to be catalysed. To be effective it must be highly mobile. The catalyst must be highly tolerant of sulphur, and to a lesser extent nitrogen. This eliminates the noble metals most effective in catalysing hydrocarbon reactions, e.g. Ni, Ru, Pt, which are all deactivated at the sulphur levels prevailing in most coals. The catalyst will be difficult to recover and regenerate and may have to be dispensed with after one or relatively few cycles. Catalysts which are liquid under reaction conditions may combine with mineral impurities present, largely SiO_2 and Al_2O_3, and solid catalysts are likely to be heavily contaminated with both mineral matter and char or heavy tars. Cost will therefore be a key constraint. In gasifiers directed towards methane production, the catalyst must be multi-functional and effect simultaneous catalysis of the steam reforming, water–gas shift, and methanation steps. It is also desirable that catalysts effective in hydrocracking of high molecular weight species be found. This reaction, exothermic and producing the desired methane, could contribute usefully to the overall heat balance but it is severely kinetically limited.

Catalysts that meet some of the above requirements include alkali metal salts and some metal halides, e.g. K_2CO_3, $ZnCl_2$, $FeCl_3$ all of which are molten under reaction conditions. Unfortunately high levels are often needed for useful degrees of catalysis and this has been a major limitation on their more general adoption. Of these, potassium carbonate is perhaps the most promising, and Exxon have recently announced development of a gasification process operating at temperatures as low as 700°C which gives a product gas consisting predominantly of methane. The mode of action of this catalyst is as yet not fully understood. Where synthesis gas is the desired end product the situation is simpler. At high temperatures CO and H_2 are favoured over other products. Improvement could be made by operating at higher pressures (smaller equipment and less downstream compression costs) and by catalysis of the Boudouard reaction

$$C + CO_2 \rightleftharpoons 2CO \qquad \Delta H = +172\ \text{kJ/mol}$$

which would increase levels of CO at the expense of CO_2 and reduce the excess levels of steam generally used which must be condensed at a later stage.

Table 6.5 Gasification processes (commercially proven)*

Process	*Conditions*	*Typical products (vol. %)*				*Comments*
		CH_4	H_2	CO	CO_2	
Lurgi	Fixed bed reactor ~1000°C; 30 bar	12	37	18	32	Production of by-product heavy tar (~1%) restricts coal to "non-caking" types. "Slagging Gasifier" under development by British Gas Corporation to enable the difficult "caking" coals to be handled
Koppers–Totzek	Entrained bed reactor ~1800°C; 1 bar	—	34	51	12	Can handle all coals; high temperatures destroy heavy organic tars. "Shell–Koppers" pressurised version (15–30 bar) under development
Winkler	Fluidised bed reactor ~900°C; 1 bar	3	42	36	18	Higher pressure process (15 bar) under development
Texaco	Entrained bed reactor ~1200°C; 20–80 bar					Commercially successful process for partial oxidation of fuel oil to synthesis gas being developed to handle coal as coal/water or coal/oil slurries

* The field of coal gasification is in a very active state of development. Nearly 20 other processes at various stages of development have been described; see A. Verma, *Chemtech*, 1978, 372 and 626.

REFERENCES

Fuels and chemical building blocks from coal

(a) *Some recent general reference works*

L. L. Anderson and D. A. Tillman, *Synthetic Fuels from Coal. Overview and Assessment*, Wiley-Interscience, New York, 1980.

Chemistry of Coal Utilisation, H. H. Lowry, ed., Vols. I and II, 1945; First Suppl. Vol. 1963; Second Suppl. Vol. 1980 (M. A. Elliott, ed.), Wiley-Interscience, New York.

N. Berkowitz, *Introduction to Coal Technology*, Academic Press, New York, 1979.

J. A. Cusumano, R. A. Dalla-Betta, and R. B. Levy, *Catalysis and Coal Conversion*, Academic Press, New York, 1978.

(b) *More selective reviews*

M. L. Gorbaty *et al*., *Science*, 1979, **206**, 1029 (review of research opportunities in coal liquefaction and gasification).

T. Wett, *Oil Gas J*., 1980 (June 16th), **55** (and see also other papers in this issue).

Authors various, *Chem. Eng. Progr*., 1980 (March) (collection of papers on recent developments in "synfuels").

J. Gibson, *Chem. Brit*., 1980, **26** (chemicals from coal).

K. Weissermel, *Angew. Chem. Int. Ed. Engl*., 1980, **19**, 79 (chemicals from coal).

B. W. Wojciechowski, *Hydrocarbon Proc*. 1980 (May), 237 (liquefaction).

J. L. Johnson, *Kinetics of Coal Gasification*, Wiley-Interscience, New York, 1980.

J. Miron and I. Skeist, *Process Economics Internat*., 1979/80, **1**, 12 (chemicals from coal).

A. Ziegler and R. Holighaus, *Endeavour*, 1979, **3**, 150 (general survey).

B. C. Gates, *Chemtech*, 1979, **9**, 97 (liquefaction).

S. P. S. Andrew, *Chem. Engineer*, 1979 (June), 414 (economic aspects).

R. Shinnar, *Chemtech*, 1978, **8**, 686; A. Stratton, *Chem. Ind*., 1978, 551 (economic aspects).

P. G. Caudle, *Futures*, 1978, 361 (chemicals from coal).

H. Perry, in *Advances in Energy Systems and Technology*, P. Auer, ed., Academic Press, New York, 1978, Vol. 1, p. 243 (general survey).

A. Verma, *Chemtech*, 1978, **8**, 372 and 626 (gasification).

P. H. Spitz, *Chemtech*, 1977, **7**, 295; G. A. Mills, *Chemtech*, 1977, **7**, 418 (economic aspects).

B. W. Wojciechowski, *Hydrocarbon Processing*, 1980, July, 153 (coal processing using nuclear heat).

Future Sources of Organic Raw Materials, CHEMRAWN I, L. E. St-Pierre and G. R. Brown, eds., Pergamon Press, 1980.

Steam reforming

J. R. Rostrup-Nielsen, *Steam Reforming Catalysts*, Danish Technical Press Inc., Copenhagen, 1975.

J. P. van Hook, *Catalysis Rev*., 1980, **21**, 1.

Partial oxidation

G. E. Weismantel and L. Ricci, *Chem. Engineering*, 1979, Oct. 8th, 57.

Hydrogen production and purification

R. I. Berry, *Chem. Engineering*, 1980, July 14th, 80 (economic aspects).

L. C. Bassett, *Chem. Eng. Progress*, 1980, March, 93 (small scale production processes).

Chem. Engineering, 1979, Dec. 3rd, 88 and 90; H. Knieriem, *Hydrocarbon Processing*, 1980, July, 65; R. Zeller, *Chem. Economy and Eng. Rev*., 1979, **11** (12), 22; J. S. Davis and J. R. Martin, *Chem. Eng. Progress*, 1980, Feb., 72 (articles on CO, H_2, CO_2 separation technologies).

L. Lloyd and M. V. Twigg, *Nitrogen*, 1979 (118), 30 (low temperature shift catalysts).

D. S. Newsome, *Catalysis Rev*., 1980, **21**, 275 (water gas shift catalysts).

Methanol synthesis

P. J. Denney and D. A. Whan, Chem. Soc. Spec. Period. Rep., *Catalysis*, 1978, **2**, 46.

K. Klier *et al*., *J. Catalysis*, 1979, **56**, 407; 1979, **57**, 339.

Ammonia synthesis

M. Appl, *Nitrogen*, 1976 (100), 47 (historical survey).

G. W. Bridger, R. E. Gadsby and D. E. Ridler, in *A Treatise on Dinitrogen Fixation*, R. W. F. Hardy, F. Bottomley and R. C. Burns, eds., Wiley, 1979, Chapter 6 (Process Reviews).

G. Ertl, *Catalysis Rev*., 1980, **21**, 201.

CHAPTER SEVEN

MANAGING ROUTES FROM BUILDING BLOCKS TO FINAL PRODUCTS

C. R. HARRISON

7.1 From building blocks to final products

In chapters 5 and 6 the formation of the basic chemical building blocks from parent carbon source was described. In this chapter we shall consider some factors which affect the next phase of chemicals manufacture, the conversion of the building blocks into downstream chemical products.

Only a limited number of building blocks are required: given ethylene, propylene, butadiene, benzene, toluene, xylenes and methanol, the vast majority of organic products can be manufactured. The relative demand for these building blocks and their major uses (figure 7.1) show the dominant position of ethylene among the other large-scale building blocks. Ethylene is the principal feedstock for the manufacture of some 30% of all petrochemicals. In the first part of this chapter we shall consider the consequences of the method of manufacture of ethylene and the flow of ethylene into its downstream derivatives from which the highly integrated and interdependent nature of the process steps will become apparent. The aromatics (benzene, toluene and xylenes—BTX) are important building blocks in their own right and we shall consider the formation of nylon 6,6 from benzene as illustrating the degree of chemical complexity in the synthesis of both parts of the polyamide molecule. There is often a choice of reaction sequence or even building block for the manufacture of a particular chemical so in the final part of this chapter factors governing this will be explored.

7.2 The ethylene derivative tree

7.2.1 *Managing ethylene manufacture*

Ethylene is manufactured by steam cracking of a widely varying range of hydrocarbon feedstocks (see 5.6). In Japan and Western Europe, naphtha feedstock dominates whereas in the U.S.A. ethane and propane are currently the main cracker feedstocks.

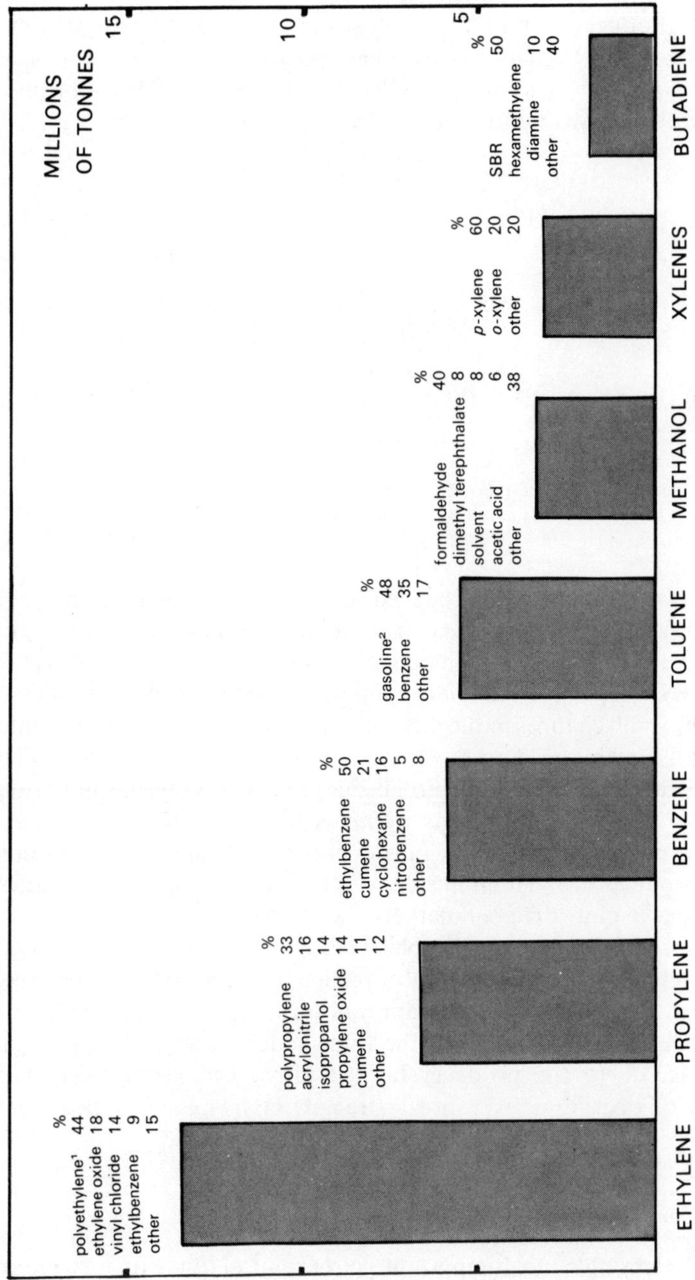

Figure 7.1 Production of basic chemical building blocks, U.S.A. 1979 (from *Chem. Eng. News*, 5 May 1980 production figures). End use distributions are approximate. [1]High- and low-density polyethylene. [2]Isolated toluene used for petrol blending.

Cracking of ethane/propane is a relatively clean process, particularly ethane which yields 85% ethylene under typical conditions. Naphtha cracking on the other hand produces a multiplicity of products in addition to ethylene. To ensure profitability of a naphtha cracking operation, outlets must be available for all the co-products as well as ethylene. A full-range naphtha gives the following product distribution (Table 7.1).

Table 7.1 Cracking yield from full range naphtha*

	%
Ethylene	32
Propylene	13
Butadiene	4.5
Aromatics (BTX)	13.5
Other†	37

*A crude oil fraction, boiling range 60–200°C.

†Other products include tail gas (methane + hydrogen), gasoline and fuel oil.

For each tonne of ethylene produced more than two tonnes of by-products must be disposed of on the most favourable economic terms for optimum profitability. By changing cracking conditions the cracker operator has only a limited ability to optimise the relative amounts of each cracker product and in practice only small changes in the distribution are attainable. Normally ethylene production is maximised. A wide variety of products therefore needs to be sold ranging in value from premium petrochemicals such as ethylene and propylene to fuel value components such as pyrolysis fuel oil. The need to maximise profitability by co-product sales introduces the dilemma that demands for particular products are seldom in step with each other since such a wide range of applications in quite different markets are involved.

It is not surprising, given the complex nature of the product distribution, that ethylene producers will usually wish to reduce their uncertainties by involving themselves in some of the downstream transformations of ethylene (and other cracker products). This also allows the value added by upgrading the building blocks to accrue to the producer himself rather than some second party. Integration of operations is a great strength in managing a petrochemical complex.

7.2.2 *Ethylene—the ubiquitous petrochemical*

Ethylene, as the single most important petrochemical, finds its way into many high tonnage intermediates (Table 7.2). The predominant outlet for ethylene is polyethylene, a bulk polymer whose two variants (high and low density

Table 7.2 Ethylene utilisation pattern

Direct products	*W. Europe % (1979)*	*U.S.A. % (1979)*	*End uses*
Polyethylene	55	44	
Vinyl chloride	18	12.5	PVC, solvents
Ethylene oxide	11	16	(see Table 7.3)
Ethylbenzene/styrene	7	7.5	Polystyrene copolymers
Miscellaneous	9	20	Vinylacetate, acetaldehyde, ethanol etc.

polyethylene) are manufactured by quite different processes and possess complementary properties. The properties of these polymers together with a detailed discussion of the catalysis involved in their manufacture will be covered in chapter 10.

Oxychlorination of ethylene to vinyl chloride (11.8) provides the monomer for a second commodity plastic, polyvinyl chloride. This is an exceptionally versatile material which can be used in rigid and flexible applications finding uses as diverse as gramophone records, footwear, plastic pipe and cable coating.

The initial product of ethylene oxidation is ethylene oxide, the third important outlet for ethylene. The oxidation reaction provides an interesting study in catalysis and is covered in 11.2. Ethylene oxide itself finds few direct uses; small quantities are used as fumigants and antibacterials, but most ethylene oxide is further converted to other products (Table 7.3).

Table 7.3 Ethylene oxide percentage utilisation
(1979, U.S.A. total production 2.4×10^6 te; 1978, W. Europe 1.4×10^6 te)

	U.S.A. (1979)	*W. Europe (1978)*
Ethylene glycol	66	48
Surfactants	13	20
Ethanolamines	6	8
Glycol ethers	8	11
Others	7	13

Hydrolysis of ethylene oxide to ethylene glycol is the predominant conversion pathway. Ethylene glycol is used in two main areas: as an important (30–45%) component of antifreeze formulations for motor vehicles and as the alcohol component for polyester manufacture (50–55%). Polyester is one of the dominant bulk artificial fibres accounting for almost half of synthetic fibre production, with trade names such as "Terylene" (ICI), "Dacron" (du Pont) and "Trevira" (Hoechst). The polymer is poly(ethyleneterephthalate) consisting of the repeat structure unit below.

$$\left(CH_2-CH_2-O-CO-C_6H_4-CO.O \right)_n$$

Components are thus formally terephthalic acid and ethylene glycol. In addition to the fibre outlet, polyethylene terephthalate is becoming increasingly employed in film and plastic bottle manufacture.

Ethylene oxide is an important constituent of nonionic surfactants, providing the hydrophilic component balancing the hydrophobic influence of long chain alcohols, alkyl phenols or fatty acids and amines. Surfactants of very different properties can be obtained by treating the hydrophobe with varying amounts of ethylene oxide. In practice, the molar quantity of ethylene oxide utilised varies from three to thirty moles per mole of hydrophobe. Its end uses range from low-foam detergent raw materials through to dispersion agents and emulsifiers.

Ring opening of ethylene oxide by ammonia yields ethanolamines as a mixture of mono-, di- and trisubstituted products

$$\underset{\diagdown\,O\,\diagup}{CH_2{-}CH_2} \xrightarrow{NH_3} \begin{array}{l} HOCH_2CH_2NH_2 \\ (HOCH_2CH_2)_2NH \\ (HOCH_2CH_2)_3N \end{array}$$

Ethanolamines are important as acid gas extraction media in industrial gas purification; H_2S, HCl and CO_2 are removed in this way. They also find uses as the hydrophilic components in detergents and in fact act as mild cleansing agents in their own right.

Ring opening of ethylene oxide by alcohols (usually methanol, ethanol, propanol and *n*-butanol) yields glycol ethers

$$\underset{\diagdown\,O\,\diagup}{CH_2{-}CH_2} + ROH \rightarrow HOCH_2CH_2OR$$

Glycol ethers are used extensively as solvents in the paint industry, in ink manufacture and as components of hydraulic fluids (such as brake fluids).

There are strong links between the aromatics and the ethylene derivative trees; the linking of ethylene glycol and terephthalic acid (or its esters) in polyester manufacture has already been described. A second example is provided by ethylbenzene manufacture where ethylene is used to alkylate benzene (in either liquid or vapour phases) to give ethylbenzene. Catalysts for the reaction usually have some acid functionality; Friedel–Craft catalysts for the liquid phase route, and zeolites or supported Lewis acids for the vapour phase process.

Apart from its small solvent usage, ethylbenzene is exclusively used as a feedstock for styrene manufacture. The dehydrogenation of ethylbenzene direct to styrene provides the major route. In addition, some styrene is available, via phenylethanol, as a byproduct of propylene oxide manufacture by the Oxirane process (7.6.2). Styrene is the third bulk monomer for thermoplastic manufacture following ethylene and vinyl chloride. Polystyrene is a cheap, easily processed plastic which is widely used in household goods, toys and packaging. Expanded polystyrene foams are excellent insulators and find applications in refrigeration as well as packaging. Styrene is also a component of many important copolymers such as styrene/acrylonitrile (SAN) and acrylonitrile/butadiene/styrene (ABS) as well as the well known styrene/butadiene rubbers (SBR).

The above constitute the major outlets for ethylene. Other significant consumers such as ethanol, acetaldehyde and vinyl acetate are important commodity chemicals in their own right and they show the same diverse usage patterns which we have already analysed for the major outlets. Ethylene is just one of the cracker products and, as we have seen, its usage pattern is complex. Managing the flow of all the cracker products to their downstream end uses is thus by no means simple when each product, and end use within that product, is subject to differing market pressures on demand.

7.3 The manufacture of nylon

7.3.1 *Introduction*

Nylon is one of the three dominant synthetic fibres (the other two are polyester and acrylics). Production of nylon in the U.S.A. (1979) was 1.2×10^6 te representing 30% of synthetic fibre manufacture. Nylon is increasingly employed as an engineering plastic, accounting for 10% of nylon consumption. The routes used to manufacture nylon polyamides are complex and involve a larger number of process steps than for most commodity chemicals. The efficient processing of nylon fibres also sets a very much lower limit on impurities in the monomer components than is usual for large tonnage materials.

There are many possible nylon polyamides depending on choice of amine and acid; they are named by considering the number of carbon atoms in the combining monomer or comonomer units. Exemplifying this, the nylon from caprolactam is named nylon 6, whereas the nylon formed from hexamethylene diamine and adipic acid is nylon 6,6, and that from hexamethylene diamine and dodecanedioic acid, nylon 6,12.

(caprolactam ring: C=O, NH) $\quad\quad$ $-\!\!\!\![CONH(CH_2)_5CONH(CH_2)_5]\!\!\!\!-_n$

Nylon 6

$H_2N(CH_2)_6NH_2 + HOCO(CH_2)_4COOH$ $\quad\quad$ $-\!\!\!\![CONH(CH_2)_6NHCO(CH_2)_4]\!\!\!\!-_n$

Nylon 6,6

$H_2N(CH_2)_6NH_2 + HOCO(CH_2)_{10}COOH$ $\quad\quad$ $-\!\!\!\![CONH(CH_2)_6NHCO(CH_2)_{10}]\!\!\!\!-_n$

Nylon 6,12

Nylon was the first wholly synthetic fibre to be produced and the first two variants commercialised (nylon 6,6 and nylon 6) remain dominant. Many of the other nylon polyamides have been synthesised, and a few have been commercialised in speciality outlets such as nylon 6,12 and nylon 12 (from dodecanolactam).

Some of the more esoteric nylon polyamides are thought to have particularly good properties for fibre applications. For example nylon 4 is a cotton substitute, but two factors have precluded its widespread adoption: difficulty in devising a cheap enough synthetic route, and the expense involved in retooling a large portion of the fibre industry to process a material having different properties from those currently employed. The high entry fee for new technology (4.4) has thus prevented commercialisation.

7.3.2 *The manufacture of nylon 6,6 from benzene*

In order to illustrate the complexity of the reaction sequence a particular variant of nylon 6,6 manufacture will be examined, where both diacid and diamine components are derived from benzene. Adipic acid is manufactured first, a portion retained for incorporation into the final polyamide and the remainder converted to hexamethylene diamine, the amine comonomer. The reaction steps employed are as follows:

$COONH_4$–$(CH_2)_4$–$COONH_4$ → $CONH_2$–$(CH_2)_4$–$CONH_2$

O OH KA

$COOH$–$(CH_2)_4$–$COOH$ adipic acid

CN–$(CH_2)_4$–CN adiponitrile

Nylon 6,6

CH_2NH_2–$(CH_2)_4$–CH_2NH_2 hexamethylene diamine

The best route to adipic acid is currently the oxidation of cyclohexane. To obtain adequate selectivities to the acid a two-step process is employed; first, cyclohexane is oxidised to a mixture of cyclohexanone and cyclohexanol (called KA, *K*etone *A*lcohol) and this mixture oxidised further to adipic acid. The oxidation of cyclohexane to KA is carried out in the liquid phase with cobalt or manganese naphthenate catalysts. To maximise selectivity to KA, conversions are limited to $<10\%$. One process utilises boric acid as additive which further enhances selectivity to KA to around 90%.

The unseparated KA mixture is then further oxidised with nitric acid over a vanadium/copper catalyst to give adipic acid (96%). This process utilising a corrosive acid oxidant is currently preferred, although a variant catalysed by Cu/Mn acetate has also been commercialised.

Part of the purified adipic acid is set aside for nylon manufacture as acid comonomer, and the remainder converted to adiponitrile and hexamethylene diamine. The amination reaction, yielding initially diammonium adipate then (via dehydration) adipic acid diamide, and finally adiponitrile, is traditionally catalysed by boron phosphate in the vapour phase at 350°C. Overall selectivities of up to 90% adiponitrile are feasible. Vaporisation of adipic acid leads unavoidably to some decomposition and a melt process with a homogeneous catalyst (usually phosphoric acid) has been developed as an alternative. Since pure monomers are of particular importance for ease of fibre processing, adiponitrile is extensively purified before hydrogenation to hexamethylene diamine (95%) over a heterogeneous iron catalyst at 150°C/300 bar. Other hydrogenation catalysts such as heterogeneous Co/Cu at high pressure or Ni catalysts at low (30 bar) pressure in the liquid phase may also be employed. Whereas cyclohexane is the preferred feedstock for adipic acid manufacture the position is less clear for adiponitrile/hexamethylene diamine where propylene or butadiene may alternatively be employed. The choice of a route to adiponitrile is discussed later in this chapter (7.8).

To make nylon 6,6, adipic acid and hexamethylene diamine are mixed in water or methanol solution from which the pure 1:1 adduct (nylon salt) may be crystallised. The condensation is then carried out by heating the salt to around 250–270°C and removing the water formed. Careful control of the processing conditions in this final step is required to allow the optimum molecular weight distribution in the polyamide. Sometimes acetic acid is used to modify chain length by termination. The resulting polymer can then be melt-spun to fibre.

The manufacture of nylon 6,6 shows that the comonomers can both be made on a large scale in high purity by a quite complex reaction sequence from the same chemical feedstock. Catalysis has a key role in the "enabling technologies" which allow the economic operation of the linked process pathways. Without suitable catalysts, nylon 6,6 (or for that matter nylon 6) would not exist as useful and cheap bulk fibres.

7.4 Two different types of complexity

We have examined two quite different areas in the first half of this chapter. Olefin production and processing result in the need to balance production and disposal of a wide and disparate range of products with relatively simple technology. Nylon production requires fewer products but these must be synthesised in high purity by more complex reaction sequences. A further complication arises when a choice of feedstocks or chemical pathways to a particular product is possible and we shall now discuss this.

7.5 The choice of a chemical route from the basic building blocks to downstream products

There is often a choice of chemical pathways for the manufacture of a particular product. The route selected will sometimes not be the most obvious or even the simplest route available. The choice will be based on an analysis of a complicated series of interrelated influences, the most important of which are:

(i) cost and availability of the process raw materials at the plant location
(ii) commercial value of byproducts of the process, and the possibility of an integrated producer using byproducts internally
(iii) total capital cost of the plant
(iv) energy usage
(v) pollution or other environmental problems.

In this section, routes to four established bulk chemicals will be described, their relative strengths and weaknesses identified and the reasons for some routes becoming obsolete pointed out.

7.6 Routes to propylene oxide

Propylene oxide is an important chemical intermediate accounting for approximately 15% of propylene consumed for chemical use. Total U.S. production was 1.01×10^6 te in 1979. The two major outlets for propylene oxide are in the manufacture of propoxylated polyhydric alcohols for urethane foams (60%) and for propylene glycol formation (30%). Propylene glycol is a component in many unsaturated polyester resins.

7.6.1 *Chlorhydrin route*

Until 1969 essentially all propylene oxide was made by the chlorhydrin route which is based on classical organic chemistry dating back almost a century. Propylene is reacted with an aqueous chlorine solution to form a mixture of chlorhydrins, which is then dehydrochlorinated without separation with a slurry of excess calcium hydroxide to yield propylene oxide.

$$Cl_2 + H_2O \rightarrow HOCl + HCl$$

$$CH_3CH{=}CH_2 + HOCl \rightarrow CH_3CH.(OH)CH_2Cl + \underset{\sim 10\%}{(CH_3CH.ClCH_2OH)}$$

$$2CH_3CHOHCH_2Cl + Ca(OH)_2 \rightarrow 2CH_3\underset{\diagdown O \diagup}{CH{-}CH_2} + CaCl_2 + 2H_2O$$

Raw materials are propylene, chlorine and lime, and overall yields of propylene oxide are believed to be around 90% based on propylene. The major organic by-products are dichloropropane (9%) and bis(chloropropyl)ether (1%). Dichloropropane finds some use as a speciality solvent or as a feedstock for perchlorinated products.

The major disadvantage suffered by this route is the downgrading of chlorine to calcium chloride during the reaction. Two tonnes of calcium chloride are produced for every tonne of propylene oxide, worse still, the chloride is discharged as a 6% solution in water which provides severe disposal problems. A more attractive variant of the process involves integration of propylene oxide production with chlorine manufacture by brine electrolysis. Here sodium hydroxide from the electrolysis replaces lime for dehydrochlorination and the resulting sodium chloride solution is recycled for chlorine production, which largely eliminates halide disposal problems.

7.6.2 *Peroxidation route*

In the 1960's Halcon International and Atlantic Richfield independently discovered and developed through a joint company, Oxirane, a cooxidation route to propylene oxide. A hydrocarbon, generally isobutane or ethylbenzene, is converted to its hydroperoxide and this hydroperoxide used to add an oxygen atom to propylene. For isobutane the route involves:

$$CH_3-\underset{CH_3}{\overset{CH_3}{\underset{|}{\overset{|}{C}}}}-H + O_2 \rightarrow CH_3-\underset{CH_3}{\overset{CH_3}{\underset{|}{\overset{|}{C}}}}-OOH \left(+ CH_3-\underset{CH_3}{\overset{CH_3}{\underset{|}{\overset{|}{C}}}}-OH \right)$$

$$CH_3-\underset{CH_3}{\overset{CH_3}{\underset{|}{\overset{|}{C}}}}-OOH + CH_3CH{=}CH_2 \rightarrow CH_3\underbrace{CH-CH_2}_{O} + CH_3-\underset{CH_3}{\overset{CH_3}{\underset{|}{\overset{|}{C}}}}-OH$$

The first step is simple uncatalysed oxidation of isobutane in the liquid phase with air or oxygen. The conditions are chosen to maximise the formation of hydroperoxide and minimise formation of alcohol arising from its decomposition. The reaction is carried out at 120–150°C/30 bar and 60% selectivity to tertbutyl hydroperoxide is attained at 25% conversion. Tertiary butanol is the main by-product. The crude product is then reacted with propylene in the liquid phase with a molybdenum catalyst at 100–120°C and yields propylene oxide in 90% selectivity at 10% conversion. Starting materials are isobutane, propylene and oxygen and the co-product is tertiary butanol. This is produced in large quantities (theory 2.5 te/te propylene oxide, in practice 3 te/te) and for a viable process this must be disposed of in the most economic way. Tertiary butanol itself has value as a gasoline component, as does its methyl ether which also is finding increasing use as a substitute for lead as an octane improver. Tertiary butanol may be used as a feedstock for methylmethacrylate manufacture (11.5.2) and may also be dehydrated to isobutene which has value as an alkylation feedstock. Finally, isobutene may itself be hydrogenated to isobutane for recycle to the peroxidation stage.

As well as using isobutane, Oxirane also operate a process based on ethylbenzene as oxygen carrier.

$$C_6H_5CH_2CH_3 + O_2 \rightarrow C_6H_5CHOOH.CH_3$$

$$C_6H_5CHOOH.CH_3 + CH_3CH{=}CH_2 \rightarrow CH_3\underset{\diagdown O \diagup}{CH{-}CH_2} + C_6H_5CHOH.CH_3$$

In this case 1-phenylethanol is the co-product, which is normally dehydrated to styrene over a titanium dioxide catalyst. Styrene is a major product (theory 1.8 te/te propylene oxide, in practice 2.5 te/te) of the reaction.

Both variants of the Oxirane process could more correctly be described as tertiary butanol or styrene manufacturing processes with a propylene oxide co-product. In these circumstances, when coproduct formation dominates, sale of tertiary butanol and its derivatives or styrene is vital for profitable operation. Both are currently made by other well-developed routes so further pressure is put on the economics of the process if demand for propylene oxide is not in step with demand for tertiary butanol or styrene.

7.6.3 *Other routes*

The chlorhydrin and peroxidation processes dominate world propylene oxide manufacture, but in Japan Daicel operate a unit based on peracetic acid oxidation of propylene. Peracetic acid is obtained by acetaldehyde oxidation and the coproduct of propylene oxide production is acetic acid. This route is also vulnerable to cheap acetic acid available by methanol carbonylation (8.5.1). Other processes at the development stage include a peracetic acid route based on hydrogen peroxide and acetic acid as peracid source (by Propylox in Belgium) and a hydrogen peroxide/isobutane route (by Bayer–Degussa). Chlorhydrin routes carried out wholly in electrochemical cells are being investigated by several groups, notably Kellogg and Bayer.

7.6.4 *Comparison of routes*

The absence of a direct oxidation process for propylene analogous to that for ethylene (11.2.1) shows itself in the problems faced by the indirect routes. Indeed many chlorhydrin plants are converted from obsolete chlorhydrin ethylene oxidation units. The chlorhydrin route is dogged by the problems of disposal of the calcium chloride byproduct (see 7.6.1) which can require a capital investment of the same order as that for the synthesis section merely to ensure a pollution-free disposal of halide. Integration with a chloralkali unit, as practised by Dow in Germany, effectively solves this problem, so for a propylene oxide manufacturer who also makes chlorine this is the most attractive option.

The peroxidation routes rely heavily on economic disposal of tertiary butanol and its derivatives or of styrene. Since the same plant cannot use both starting materials, the choice between the two variants will depend on the availability of isobutane or ethylbenzene at the plant location and on the co-product demand in the particular area. In Japan and Spain, ethylbenzene has been selected, whereas in Holland isobutane was chosen, so the choice depends very much on local circumstances. The rapid adoption of methyl tertiary butyl ether (MTBE) as a gasoline additive could increase the amount of isobutane-based technology in operation.

At present peroxidation technology is thought to have a slight economic edge over the integrated chlorhydrin route but both are very vulnerable to the invention of a direct process for oxidation of propylene to propylene oxide. Current work in this area is still limited to around 50% selectivity to propylene oxide, which is not yet economic.

7.7 Routes to phenol

Phenol manufacture is a particularly interesting area in which to study the choice of chemical routes since until recently there were five distinct and operated routes to phenol, excluding the now largely obsolete extraction from coal tar production. Phenol is a moderate tonnage commodity chemical (U.S. production 1979 1.3×10^6 te) which has a wide end-use pattern dominated by phenolic resins. In Western Europe the distribution is: phenolic resins (domestic appliances etc.) 38%; caprolactam (to nylon) 20%; bisphenol A (epoxy resins, polycarbonates) 13%; alkyl phenols (surfactants, antioxidants) 7%; adipic acid (nylon) 6%; and other uses (including plasticisers) 16%.

Table 7.4 Routes to phenol

Feedstock	*Intermediate*	*Designation*
Benzene	Isopropyl benzene	Cumene process
Benzene	Benzene sulphonic acid	Sulphonation route
Benzene	Chlorobenzene (basic hydrolysis)	Dow process
Benzene	Chlorobenzene (oxychlorination)	Hooker–Raschig process
Toluene	Benzoic acid	Dow/DSM process
Benzene	Cyclohexane	Scientific Design process

7.7.1 *Cumene route*

Currently the dominant route for synthetic phenol production is based on cumene (isopropylbenzene); this process now accounts for 90% of phenol manufactured in Western Europe. Benzene is alkylated with propylene over an acid catalyst to give cumene. The basis of the BP–Hercules process is the oxidation of cumene to its hydroperoxide and subsequent acid catalysed cleavage to phenol and acetone. Conversion is limited to 40% at the peroxidation stage to control formation of by-products, predominantly acetophenone and dimethylphenyl methanol. In the BP–Hercules process the cleavage reaction is carried out with sulphuric acid in acetone. Overall selectivity to phenol (and acetone) is 90%. The major by-products of the hydroperoxide cleavage are α-methylstyrene and mesityl oxide.

$$C_6H_6 + CH_3CH{=}CH_2 \longrightarrow C_6H_5{-}CH(CH_3)_2$$

$$C_6H_5{-}CH(CH_3)_2 + O_2 \longrightarrow C_6H_5{-}C(CH_3)_2{-}OOH$$

$$C_6H_5{-}C(CH_3)_2{-}OOH \xrightarrow{H^+} C_6H_5{-}OH + (CH_3)_2C{=}O$$

7.7.2 *Sulphonation process*

This was the first route to synthetic phenol and was developed in Germany late in the last century. The process is multistep and reaction takes place through the scheme below. The water formed in the initial sulphonation must be removed, usually as its azeotrope with benzene, to drive the reaction to completion. The neutralisation of the sulphonic acid, step (ii), produces large amounts of by-product sodium sulphate through the presence of excess sulphuric acid from the sulphonation. The fusion of sodium benzene sulphonate with sodium hydroxide is carried out batchwise at 310–320°C. The overall yield of phenol is 85–90%, with organic by-products as tarry materials. The process is an overall producer of sodium sulphite, which may be sold to the paper industry, and sodium sulphate. Although the chemical operations involved are simple it is difficult to scale manufacture much above 4000 te/yr owing to the batch nature of several of the stages.

$$\text{(i)}\quad C_6H_6 + H_2SO_4 \longrightarrow C_6H_5SO_3H + H_2O$$

$$\text{(ii)}\quad 2\,C_6H_5SO_3H + Na_2SO_3 \longrightarrow 2\,C_6H_5SO_3Na + SO_2 + H_2O$$

(iii) $C_6H_5SO_3Na + 2NaOH \longrightarrow C_6H_5ONa + Na_2SO_3 + H_2O$

(iv) $2\,C_6H_5ONa + SO_2 + H_2O \longrightarrow 2\,C_6H_5OH + Na_2SO_3$

Overall $C_6H_6 + H_2SO_4 + 2NaOH \longrightarrow C_6H_5OH + Na_2SO_3 + 2H_2O$

7.7.3 *Chlorobenzene routes*

In the Dow chlorobenzene process, benzene is treated with chlorine in the presence of an iron salt to minimise polychlorination, and the resulting chlorobenzene hydrolysed with sodium hydroxide at high temperatures and pressures (350–400°C/300 bar) to give sodium phenate which after acidification yields phenol in overall 90–95% yield. The main by-products are dichlorobenzene (from over-chlorination) and diphenyl ether. Both products have some commercial value. Operation of this route is particularly relevant when integration with a conventional chloroalkali unit is possible since the sodium chloride product may be recycled to produce fresh chlorine.

In the Hooker–Raschig process, chlorobenzene is formed by oxychlorination of benzene with air/HCl at 240°C over a $CuCl_2$–$FeCl_3$/alumina catalyst. Over-chlorination is controlled by limiting conversion of benzene to 10–15%. Under these conditions 90% selectivity to chlorobenzene is achieved with 6–10% dichlorobenzenes. In the Raschig process, hydrolysis of chlorobenzene to phenol is achieved with a calcium phosphate catalyst at 450°C. The Hooker variant uses a modified hydrolysis catalyst which converts any dichlorobenzene to monochlorobenzene and hence phenol, with the result that chlorinated by-products are greatly reduced. HCl is recycled after reconcentration. Overall selectivities to phenol are 70–85% at 10–13% benzene conversion.

7.7.4 *Toluene route*

Toluene is initially oxidised to benzoic acid (90%) in the liquid phase with air and a cobalt catalyst at 120°C. In the second step, the purified benzoic acid is oxidatively decarboxylated to phenol at 220–250°C in the presence of steam and air with a copper/magnesium catalyst system. Overall selectivity to phenol is 70–75% and the major by-products are tars formed by over-oxidation which need to be removed from the reaction mixture since they act as inhibitors. This process is operated on a limited scale in the U.S.A. and Holland. Reaction is thought to take place through intermediate formation of cupric benzoate and oxybenzoylation to benzoyl salicylic acid which is cleaved to phenol by hydrolysis and decarboxylation. Recently CE Lummus have brought a vapour phase route from benzoic acid to phenol to the development stage. This is claimed to offer improvements in capital cost and in reduced tar formation over the conventional liquid phase route.

$2Cu(II)(OCOPh)_2$

C_6H_5—CO.O—C_6H_5 ($-CO_2$); H_2O (−C_6H_4COOH)

COOH

C_6H_4—O—CO—C_6H_5 + $2Cu(I)(OCOPh)$ → OH / C_6H_5

H_2O; $-CO_2$

COOH / C_6H_5 + COOH / C_6H_4 / OH

7.7.5 *Cyclohexane routes*

The Scientific Design Company has offered for licence a benzene-based route to phenol involving cyclohexane as intermediate. The key step is dehydrogenation of the cyclohexanol/cyclohexanone mixture (KA) to phenol; the remaining steps are well-developed for the synthesis of nylon intermediates (7.3). Dehydrogenation is carried out over a Pt or Co/Ni catalyst at 400°C. Separation of phenol from cyclohexanone is difficult, however, owing to azeotrope formation, so an expensive solvent extraction of phenol is necessary. Overall yields of phenol based on benzene of 80–85% are possible.

$$C_6H_6 + 3H_2 \longrightarrow C_6H_{12}$$

$$C_6H_{12} + O_2 \longrightarrow C_6H_{10}O + H_2O$$

$$+\tfrac{1}{2}O_2 \longrightarrow \text{OH}$$

$$\text{O} \longrightarrow \text{OH} \quad +2H_2$$

$$\text{OH} \longrightarrow \text{OH} \quad +3H_2$$

7.7.6 *Which route to phenol?*

Although the sulphonation, Dow and Hooker–Raschig processes are all declining in importance and the Scientific Design process via cyclohexane is not thought to be currently in operation, there are five fully-developed routes to phenol. All are indirect, involving at least one other intermediate stage, indicating that a direct route from benzene to phenol, if it could be achieved economically, would be a most important development.

The sulphonation route, which is still used in Italy and the G.D.R. has the advantage that it can be operated on a small scale (below 5000 te/yr) which is sometimes useful where markets are limited. If by-product sodium sulphite can be sold to a nearby paper mill then the process may be attractive. Unfortunately, several of the steps are batch operations and the disposal of sodium sulphate produced is a major problem. Where large-scale phenol production, 40–250 000 te/yr, is required, sulphonation is not a viable option owing to scale-up difficulties. This factor, together with the unavoidable formation of inorganic by-products, has led to the decline of the sulphonation process.

The Dow chlorobenzene route has the disadvantage of using expensive chlorine as reagent. This disadvantage can be overcome, as with propylene oxide production, by integration of phenol manufacture with chloralkali technology for brine electrolysis. The chlorine is converted to sodium chloride in the phenol process and this may be recycled to further electrolysis. This requires input of electrical energy, so will only be viable where electricity is cheap and/or the phenol manufacturer is also in the chlorine business. Disposal of dichlorobenzenes is generally not a major problem since they find a market as solvents. In the Hooker oxychlorination route, by-product make is very much reduced so there is no need to dispose of significant quantities of dichlorobenzenes. Use of hydrogen chloride rather than the more expensive chlorine has advantages but the system is very corrosive so that extra capital investment is required for corrosion-resistant plant for the oxychlorination and hydrolysis sections. The need to reconcentrate the HCl for recycle also means that the process is a large consumer of energy. It is significant that Hooker have shut down most of their oxychlorination capacity and now operate the cumene route.

The cyclohexane route of Scientific Design could be economic if it were integrated with caprolactam manufacture via cyclohexanol/cyclohexanone. The fact that the only plant known to operate this process (in Australia) has apparently been converted to the cumene route indicates that the route is currently uncompetitive.

In most countries of the world toluene has historically been cheaper on a molar basis than benzene although the floor value for toluene has been set by the cost of its dealkylation to benzene. The Dow/DSM toluene route to phenol has thus a distinct raw material cost advantage over the benzene routes. The low overall selectivity to phenol (70–75%) and the problem of poisoning of the catalyst by tarry by-products materially reduce this advantage, and as a result this technology has not gained wide acceptance. In Holland, however, DSM started up a new plant based on toluene as recently as 1976 indicating that under some circumstances the route is competitive. The vapour phase process developed by Lummus could make a toluene-based route more competitive.

Thus at the moment the cumene route dominates synthetic phenol production and will continue to do so for as long as the acetone co-product finds a ready market. If outlets for acetone decline, for instance if methyl methacrylate manufacture changes from the acetone cyanohydrin route to an isobutene feedstock, then the cumene route could lose some of its edge.

7.8 Routes to adiponitrile

Adiponitrile, $NC(CH_2)_4CN$, is used almost exclusively as an intermediate in the manufacture of hexamethylene diamine. Adiponitrile can be reduced, with hydrogen, at 300 atmospheres/150–180°C with an iron catalyst to yield hexamethylene diamine. Hexamethylene diamine is of great importance as the diamine component of nylon 6,6 (and also nylon 6,10 and nylon 6,12). Total U.S. capacity for hexamethylene diamine was 0.46×10^6 te in 1979.

Four distinct routes to adiponitrile have been developed based on three different raw materials—

Raw material	*Intermediates*	*Process*
Cyclohexane	Cyclohexanol/cyclohexanone; adipic acid	Classical adipic acid route
Propylene	Acrylonitrile	Monsanto—acrylonitrile EHD
Butadiene	Dichlorobutenes	Du Pont—indirect butadiene
Butadiene	3-Pentenenitrile	Du Pont—direct butadiene

7.8.1 *Amination/dehydration of adipic acid*

This route, based ultimately on benzene, involves adipic acid synthesis as described in 7.3. The amination/dehydration reaction is carried out either in the vapour phase at 300–350°C over boron phosphate in the presence of a large excess of ammonia or in a melt at 250°C using a soluble phosphoric acid catalyst. In the former case difficulties in volatilising the adipic acid cause decomposition and limit yield, but nonetheless selectivities to adiponitrile of up to 90 % are possible. Reaction proceeds by successive formation of diammonium adipate and adipic acid diamide, finally to yield adiponitrile.

COOH, COOH $\xrightarrow{2NH_3}$ $COONH_4$, $COONH_4$

$\downarrow -2H_2O$

CN, CN $\xleftarrow{-2H_2O}$ $CONH_2$, $CONH_2$

7.8.2 *Electrohydromerisation of acrylonitrile* (*Monsanto EHD process*)

This route is based on propylene, from which acrylonitrile may be obtained by ammoxidation (11.5.2). The overall cathodic reaction is formally

$$2H_2C{=}CHCN + 2e^- + 2H^+ \rightarrow NC(CH_2)_4CN$$

Cell design is critical for success of this process and the more recent patents claim electrolysis in a two phase medium, an aqueous phase consisting of support electrolyte and a solubility-limited quantity of acrylonitrile, and organic phases consisting of acrylonitrile and adiponitrile. Selectivities to adiponitrile are claimed to be 90 % with propionitrile and bis(cyanoethyl)ether comprising the major by-products. Monsanto started the first plant based on this technology in 1974 with the intention of converting all their capacity previously based on cyclohexane.

7.8.3 *Indirect hydrocyanation*

In 1959 du Pont commercialised a route to adiponitrile based on butadiene. Chlorination of butadiene in either liquid or vapour phase yields 3,4-dichloro-butene or 1,4-dichlorobutenes. When the mixed dichlorobutenes are treated with HCN in the liquid phase with a copper catalyst 1,4-dicyanobutenes are obtained in 95 % selectivity. The terminal dinitrile is formed by direct substitution in the case of the 1,4-dichloro-2-butene and by substitution and allylic rearrangement of 3,4 dichloro-1-butene. Hydrogenation of the internal double bond is effected at 300°C in the presence of a palladium catalyst and yields adiponitrile in 95% selectivity. Overall selectivity in this three stage route is 85–90%. Du Pont currently manufacture part of their ADN requirements by this route.

$$\text{butadiene} + Cl_2 \longrightarrow \begin{cases} ClH_2C\text{–}CH\text{=}CH\text{–}CH_2Cl \ (cis) \longrightarrow NCCH_2\text{–}CH\text{=}CH\text{–}CH_2CN \\ CH_2\text{=}CH\text{–}CHCl\text{–}CH_2Cl \\ ClH_2C\text{–}CH\text{=}CH\text{–}CH_2Cl \ (trans) \longrightarrow NCCH_2\text{–}CH\text{=}CH\text{–}CH_2CN \end{cases}$$

$$NCCH_2CH{=}CHCH_2CN + H_2 \longrightarrow NCCH_2CH_2CH_2CH_2CN$$

7.8.4 *Direct hydrocyanation of butadiene*

Du Pont also operate a direct hydrocyanation route to adiponitrile. The process is carried out in the liquid phase under mild conditions and selectivities of up to 95% to adiponitrile are possible. The reactions involved are discussed in more detail in 9.5. Du Pont are reportedly phasing out their chlorination route in favour of the direct hydrocyanation.

7.8.5 *Choice of a route to adiponitrile*

The routes to adiponitrile described above involve the use of three distinct base feedstocks, benzene, propylene and butadiene. All three feedstocks are currently in use in Europe so clearly the choice is not overwhelmingly in favour of any route.

Butadiene, a by-product of steam cracking of hydrocarbon liquids, is predicted to increase in availability over the next ten years as the cracking of liquid feedstocks increases. The direct hydrocyanation of butadiene achieved by du Pont is clearly superior to the earlier chlorination route, avoiding the need for integration with a brine electrolysis unit to produce chlorine and utilise sodium chloride by-product. The cost of hydrogen cyanide is of obvious importance in such a process and the availability of cheap HCN as a by-product from, for example propylene ammoxidation, could be advantageous in some circumstances.

The propylene based route has the advantage that the ammoxidation to acrylonitrile occurs readily and cheaply and since electricity charges represent only 4–7% of the cost of manufacture this is not limiting. Availability of propylene should not be a problem for the future.

The benzene based route seems to have most problems since it is multistage and energy intensive with low conversion in some steps, particularly cyclohexane oxidation. Furthermore the availability of aromatics is predicted to be tight for the medium term as pressures to reduce lead in gasoline result in increased demands for aromatics in petrol. The gasoline situation may change, however, leading to increased benzene availability as a chemical feedstock, so that a benzene-based route to adiponitrile would become more attractive. In the U.S.A. the benzene based route to adiponitrile is not practised, the market is dominated by du Pont and Monsanto processes based on butadiene and propylene. This could represent the shape of things to come elsewhere in the world.

7.9 Acetylene as a chemical feedstock

The rise and fall of acetylene as the building block of choice for synthesis of many commodity petrochemicals provides an interesting case study in how feedstock relativities can change with time.

7.9.1 *Acetylene 1890–1960*

Industrial production of acetylene began in the 1890's primarily for use as an illuminant and in metal welding and cutting. The first industrially important chemicals prepared directly from acetylene were the tri- and per-chloroethylenes and acetaldehyde which appeared during World War I. The real expansion in the use of acetylene for the production of a broad range of organic chemicals took place in the 1930's and 40's. By the end of World War II all of the processes shown in figure 7.2 were well established. Throughout the period up to the early 1960's the use of acetylene continued to grow. Thereafter it went into a rapid decline.

The advent of large-scale production of ethylene, propylene and butadiene by steam cracking of oil fractions or natural gas resulted in the rapid displacement of acetylene from its dominant position as a chemical building block. The reasons for this changeover were:

(i) Acetylene was originally based on coal as carbon source which made its manufacture tedious and expensive. In the late 50's routes from oil and natural gas were developed but this improved technology came too late and acetylene was unable to compete on cost with cheap olefins from large scale cracking of oil fractions or natural gas.
(ii) With the growth of the worldwide trade in petrochemicals, acetylene's position became untenable, its explosive properties rendering storage and long

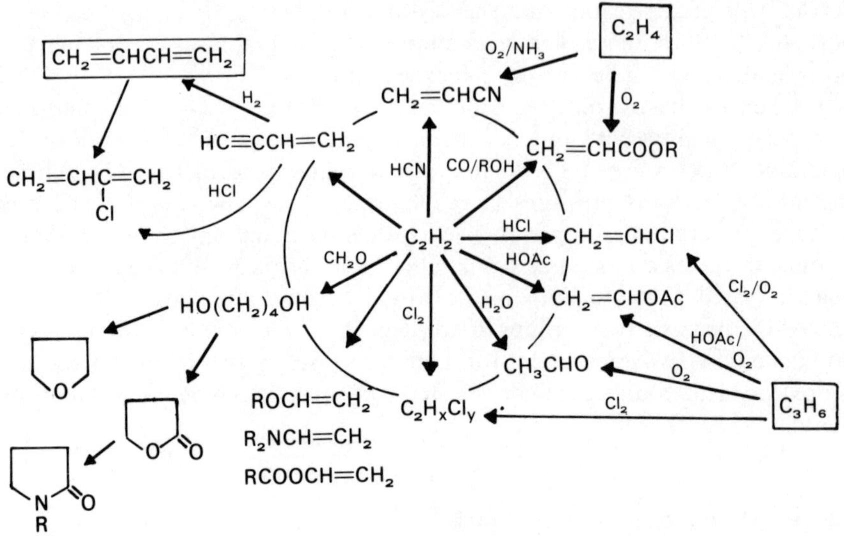

Figure 7.2 Chemical intermediates from acetylene or from ethylene, propylene and butadiene.

distance transportation difficult. The light olefins on the other hand can be stored and transported by pipeline, ship or tanker over large distances without major problems.

Acetylene was always difficult to handle and now became a relatively expensive building block; before it could be replaced, a new set of technologies had to be developed for production of the required downstream chemicals from light olefins. Acetylene is a very reactive chemical in comparison with ethylene and propylene so the catalyst science required is quite different. The new technologies for olefins were based largely on oxidation rather than the addition chemistry used in acetylene processing. The key role of catalysis in allowing the change from acetylene to light olefins will be apparent from later chapters and from Table 7.5 which compares acetylene chemistry with the now dominant olefin-based routes. So dominant are these new technologies that they have supplanted acetylene in all but a few, relatively small-scale, areas.

7.9.2 *Acetylene—the present and future*

From the previous discussion, it could be concluded that acetylene as a chemical feedstock is soon to be relegated to the history books. This is not so. Certain strengths keep it alive and will make it increasingly attractive in certain areas over the coming decades. We can summarise those circumstances under which acetylene-based processes can be operated successfully in the four categories below. There is again no best route for all situations; local factors can have a powerful influence.

Table 7.5 Acetylene-based processes

Feed	*Product*	*Process conditions* C (conversion) S (selectivity)	*Catalyst*	*Comments*	*Sections covering olefin-based routes*
CH_3COOH	$CH_2{=}CHOCOCH_3$	200°C/~1 bar; gas phase; 97% S	Zn $(OAc)_2/C$	Hg(OAc) more active but losses to environment make its use unacceptable	11.3.8
HCl	$CH_2{=}CHCl$	100–180°C/~1 bar; gas phase; 97% C; 99% S	$HgCl_2/C$	Catalyst losses acceptable	11.8.
H_2O	CH_3CHO	70–100°C/<1.5 bar; aq H_2SO_4 95% S at 25% C	Hg^{2+} stabilised by Fe^{3+}	Hg lost in product recovery; process environmentally unacceptable	11.3.6
HCN	$CH_2{=}CHCN$	80°C/~1 bar; aq. phase; 80% S	Cu^+		11.7.2
CH_2O aq.	$HOCH_2{\equiv}CCH_2OH$	100°C/5 bar; liquid phase; 80% S	$CuC_2{-}B_{12}O_3/SiO_2$	$HC{=}CCH_2OH$ byproduct recycled; butanediol obtained via liquid phase hydrogenation (Ni/200°C/200 bar—95% S)	—
CO/H_2O	$CH_2{=}CHCOOH$	200–225°C/100 bar; liquid phase in e.g. THF (BASF process)	$Ni(CO)_4$	Use of solvent minimises hazards of acetylene under pressure	11.5.2

(i) *Where there are no suitable alternatives.* The preparation of butane-1,4-diol, a component of certain polyurethanes and polyesters, provides a good example; although, even here, attempts are being made to define propylene based routes:

$$CH_3CH{=}CH_2 \rightarrow CH_2{=}CH{-}CHO \rightarrow CH_2{=}CHCH(OR)_2 \xrightarrow[Rh]{CO/H_2} OHCCH_2CH_2CH(OR)_2 \xrightarrow{H_2} HO(CH_2)_4OH$$

$$CH_3CH{=}CH_2 \rightarrow CH_2{=}CHCH_2OH \xrightarrow[Rh]{CO/H_2} OHCCH_2CH_2CH_2OH \xrightarrow{H_2} HO(CH_2)_4OH$$

$$CH_3CH{=}CH_2 \rightarrow CH_2{=}CHCH_2OAc \xrightarrow[Rh]{CO/H_2} OHCCH_2CH_2CH_2OAc \xrightarrow{H_2} HO(CH_2)_4OH$$

That via allyl alcohol is being tried on the commercial scale by Daicel in Japan at the time of publication of this book.

(ii) *Where the acetylene part of the final product is small* and therefore relatively insignificant with respect to raw materials costs; and/or the product is a speciality chemical. Examples are in the synthesis of vinyl esters of higher carboxylic acids, vinyl ethers of high alcohols, and *N*-vinyl compounds.

(iii) *Where local or political factors dominate.* Acetylene can be made from any of the common raw material bases, coal, oil or gas. Acetylene thus represents a way for oil-deficient countries with adequate reserves of coal to achieve a partial independence of imported, and potentially insecure, oil by exploitation of indigenous resources. Countries such as South Africa and India operate acetylene-based processes very successfully and are continuing to construct new plants.

(iv) *Where large and fully paid-off plants exist* that can adequately cope with the product demand. A change to a non-acetylene-based route will require investment in a wholly new plant. There may well be an advantageous trade-off between higher raw material costs for acetylene and higher fixed charges associated with the new plant (see also 4.5).

REFERENCES

1. *Olefin production* A good review is given by L. F. Hatch and S. Matar, *Hydrocarbon Processing*, January 1978, 135–139 and March 1978, 129–139.
2. *Synthetic fibre technology* J. E. McIntyre, *The Chemistry of Fibres*, E. J. Arnold, London 1972.

3. *Routes to propylene oxide* More information in K. H. Simmrock, *Hydrocarbon Processing*, November 1978, 105–113, and R. Landau, G. A. Sullivan and D. Brown, *Chem. Tech.*, October 1979, 602–607.
4. *Routes to phenol* A. P. Gelbein and A. S. Nislick review a vapour phase route from benzoic acid to phenol in *Hydrocarbon Processing*, November 1978, 125–128, and cite key references to the other technologies for phenol manufacture.
5. *Routes to adiponitrile* The electrohydrodimerisation of acrylonitrile is described in detail by D. E. Danly in *Chemistry and Industry* (1979) 439–447.
6. *Chemicals from acetylene* The definitive work is *Acetylene: its properties, manufacture and uses*, S. A. Miller, Benn, London 1965.

CHAPTER EIGHT

CARBON–CARBON BOND FORMATION. I: CARBONYLATION

D. T. THOMPSON

8.1 Introduction

The stimulus for this chapter is the necessity in the future to develop non-oil-based routes to chemicals from synthesis gas. Syn gas could be obtained from coal gasification; at the present time it can also be produced by steam reforming natural gas (methane) or from liquid hydrocarbons (6.2.2). Much of the work described here is thus in areas where rapid changes are occurring. We consider C—C bond formation using carbon monoxide and/or hydrogen to effect the synthesis of useful fuels and organic chemical intermediates and end products, including paraffins, alcohols and organic materials with CO-containing structures; some of the more important raw material/reagent/product combinations are indicated in Table 8.1.

The early part of the chapter (8.2) is concerned principally with the use of syn gas to produce liquid and gaseous hydrocarbon fuels and involves C—O bond scission. The related technology of the direct liquefaction of coal using syn gas in

Table 8.1 Examples of important chemical transformations involving use of CO or CO/H_2

Starting materials	*Catalyst*	*Products*	*Comments*
CO/H_2	Co, Fe, Ru	Hydrocarbons up to C_{20+}	Petrol, diesel, aviation fuels
CO/H_2	Ni, Ru	CH_4	Synthetic natural gas (SNG)
CO/H_2	Metal (e.g. Fe, Co, Ni) or oxide catalysts	Alcohols etc.	Direct synthesis of higher primary alcohols
CH_3OH	ZSM-5 zeolite	Hydrocarbons up to C_{10}	Synthetic gasoline
CH_3OH CO/H_2	Co	Ethanol CH_3CH_2OH	Non-oil based ethanol source
CO/H_2	Rh	Ethylene glycol $(CH_2OH)_2$	Antifreeze Polymer intermediate
$RCH{=}CH_2$ CO/H_2	Co, Rh	Aldehydes Alcohols	Plasticisers Detergent intermediates
CH_3OH/CO	Rh/I_2	Acetic acid (CH_3CO_2H)	Best route to acetic acid

the presence of steam and cobalt/molybdenum catalysts has already been discussed (cf. 6.5.1). The remainder of this chapter involves reactions in which the C—O bond remains intact. Some of the examples are useful for assessing the comparative merits of homogeneous and heterogeneous catalysis and also indicate the implications of careful choice of Group VIII metal catalysts and promoters.

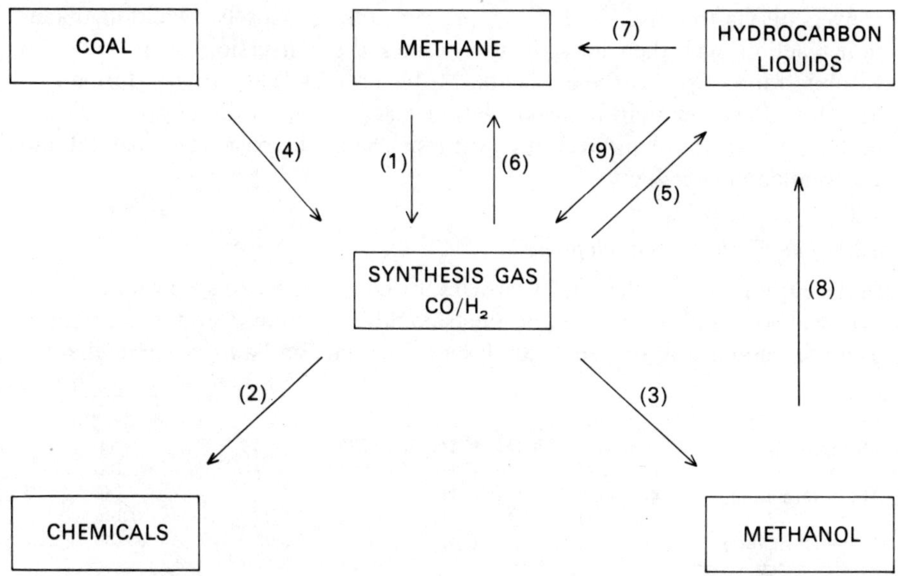

Scheme 1 Current and prospective interconversions of fuel and chemicals.

The relationships between the various fuels and chemicals discussed in this chapter are indicated in scheme 1. This diagram helps us to consider possible changes in routes for the future. At present the routes which are utilised are methane to methanol and other chemicals via syn gas, (1)+(2); (1)+(3), and hydrocarbon liquids to syn gas and chemicals and methanol, (9)+(2); (9)+(3). Only in special circumstances are routes (4)+(2), (4)+(3), (4)+(5), (4)+(6), and (7) operated, i.e. in the conversion of coal to syn gas which is then used to make chemicals, methanol, hydrocarbon liquids (Fischer–Tropsch) or methane, or the cracking of liquid hydrocarbons to methane. Use of route (1)+(3)+(8), i.e. methane via syn gas and methanol to hydrocarbons, is to be made shortly in New Zealand (8.2.3). The reader will find these routes more fully described in this chapter; the preferred routes will doubtless change in the future. Special note should be taken of the fact that the largest use of methane and liquid hydrocarbons is still as a source of heat, in competition with coal. Until this situation changes the price of hydrocarbons will remain tied directly to the price of coal in a free economy. Routes (4)+(6), (4)+(5), (4)+(3)+(8), (4)+(3)+(8)+(7), and (4)+(5)+(7) all represent interconversion of fuels and exploitation is dependent on overall thermal efficiency.

8.2 Hydrocarbons from synthesis gas

This section is concerned with the production of liquid and gaseous hydrocarbons from carbon monoxide and hydrogen. Firstly, it deals with direct conversion via the Fischer–Tropsch process, and the closely related methanation reaction, and then goes on to consider the conversion of methanol to gasoline range hydrocarbons via the Mobil process. The latter is formally a dehydration rather than a carbonylation reaction but is discussed here since both routes, direct or indirect, use synthesis gas as the basic raw material and give similar end products.

8.2.1 *The Fischer–Tropsch process*

In the early 1920's Fischer and Tropsch showed that a broad spectrum of liquid hydrocarbons and oxygenated products could be produced from synthesis gas using promoted iron or cobalt catalysts.[1] The reaction had great potential for

		ΔH approx. (kJ/mol)
Alkane formation	$(n+1)H_2 + 2nCO \rightleftharpoons C_nH_{2n+2} + nCO_2$	−200
	$(2n+1)H_2 + nCO \rightleftharpoons C_nH_{2n+2} + nH_2O$	−160
Alkene formation	$nH_2 + 2nCO \rightleftharpoons C_nH_{2n} + nCO_2$	−180
	$2nH_2 + nCO \rightleftharpoons C_nH_{2n} + nH_2O$	−140
Alcohol formation	$2nH_2 + nCO \rightleftharpoons C_nH_{2n+1}OH + (n-1)H_2O$	−140

the production of synthetic liquid fuels from coal and from the outset generated considerable interest, especially in Germany. During the 1930's and 1940's development proceeded rapidly stimulated by an oil-to-coal price ratio of up to 10:1 then ruling in Western Europe and a need for self-sufficiency in transportation fuels. By the end of World War II some nine plants were in operation in Germany with capacities of up to 10^5 te/yr, relatively small in scale by present-day standards but significant at the time. The success was to be short-lived. By the early 1950's the greater availability of crude oil had eroded the cost advantage of coal, and the multistep nature and relative thermal inefficiency of the Fischer–Tropsch process had become an embarrassment. With one exception all plants ceased operation. In South Africa the coal-to-oil price differential was maintained, and by the mid-1950's SASOL, the South African Coal Oil and Gas Corporation, was operating its first plant with a capacity of 2.10^5 te/yr. The experience gained in the operation of this plant and the growing uncertainties in oil supplies in the 1970's led to the decision to construct two further and much larger plants. SASOL II, with ten times the capacity of the first plant is operational now (scheme 2), and the other, similar, plant (SASOL III) will be ready in the mid-1980's.

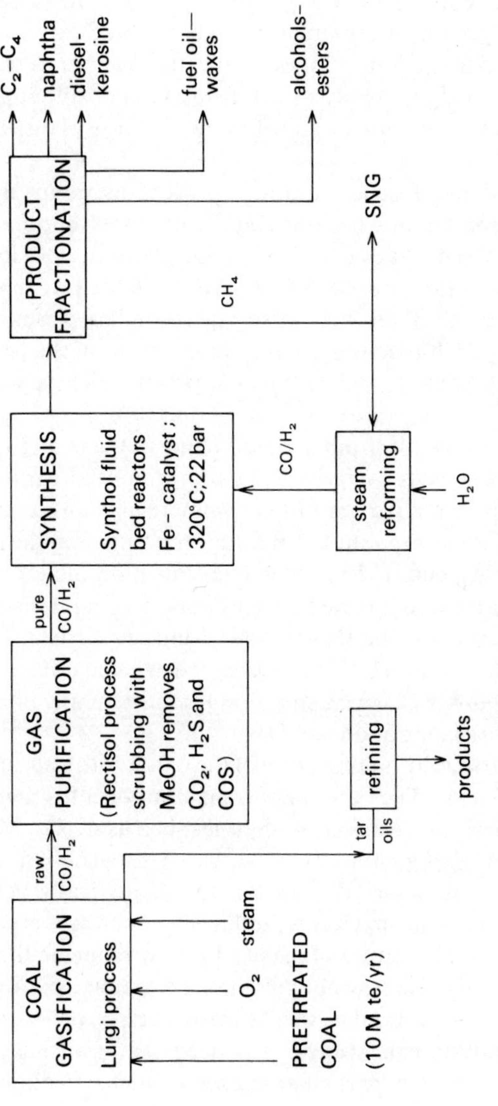

Scheme 2 Fischer–Tropsch process—Sasol II plant.

The interruptions in the supply of crude oil occasioned by the troubles in the Middle East during the 1970's, with their associated rapid price escalations, have also turned thoughts back to coal conversion technologies in many other parts of the world. All the various coal-based processes, including direct liquefaction (6.5.1) and indirect conversion via synthesis gas and methanol (8.2.3), could compete favourably with oil in situations where there exists abundant cheap coal, as in South Africa; but, of these, only the Fischer–Tropsch process is commercially well-tried and ready now. The others are still at a developmental stage and it may take a decade or so before really large plants are in successful operation.

The products of the Fischer–Tropsch process as commercially operated consist of a broad spectrum of hydrocarbons with lesser amounts of oxygenated species such as alcohols and esters. The hydrocarbons are predominantly linear alkanes along with methyl branched alkanes and alkenes. Aromatic compounds occur in only small amounts and arise via secondary reactions. The major influences on the overall reaction are the composition of the catalyst, temperature, pressure, $H_2 : CO$ ratio, and method of operation. These will be considered briefly.

A large number of metals, from a broad band in the centre of the transition series, form active catalysts. Activity is most pronounced with Fe, Co, Ni, and Ru. Of these, Fe and Co have been used commercially, although more recently considerable interest has been shown in catalysts containing the more active, but considerably more expensive, Ru. Some late transition metals, e.g. Pd and Cu, are essentially inactive in hydrocarbon synthesis. The main distinction between the two groups appears to be their mode of binding of CO. Metals active in hydrocarbon synthesis adsorb CO *dissociatively* whereas those such as Pd and Cu, which give products in which the C—O bond remains intact, e.g. MeOH, show only *associative* adsorption of CO.

Metal, usually produced via reduction of the oxide, forms the major part of the various catalyst systems. The other important component is the promoter(s) for which two main roles can be discerned. Species such as Al_2O_3, MgO, and Cr_2O_3 increase activity by giving catalysts of increased surface area. Others, such as alkali metal oxides, act more directly on the catalytically active sites. They increase activity, without markedly influencing surface area, and exert a profound influence on the nature of the products by reducing the hydrogenating power of the catalyst. Increasing amounts of alkali in iron catalysts give products of increased molecular weight with greater proportions of olefins. Alkalised iron catalysts can also be produced that give high selectivities to oxygenated compounds, in particular higher primary alcohols which can be obtained as over 50% of the total product. In fact, the initial discoveries of Fischer and Tropsch gave a product of this type, and it was only several years later that they were able to direct the synthesis to the then more desirable hydrocarbons.

Another transition metal, manganese, while itself not particularly active, exerts a major influence on catalyst selectivity. Fe/Mn catalyst formulations can give a product containing up to 80% olefins in the range C_2–C_4 and are being actively pursued *inter alia* by workers at Ruhrchemie as a direct route to lower olefins from synthesis gas.[2]

The influences of temperature, pressure, and $H_2:CO$ ratio are still not fully resolved. As a very general rule, the molecular weight of the product is reduced by increases in temperature, increases in $H_2:CO$ ratio, and decreases in total pressure. The first two relate to the same effect—the more effective removal via hydrogenation of the surface intermediates. These features are most pronounced with Ru-based catalysts. At high temperature and low pressure, methane is preferred, whereas at low temperature and high pressure (up to 1000 bar) polymethylene waxes of high molecular weight (up to 10^5) can be produced. These waxes have very similar properties to low-molecular-weight, high-density polyethylene prepared using Ziegler–Natta catalysts (10.7).

Two types of reactor systems have been well-tried and can give quite different product distributions. The first SASOL plant has both fixed-bed and recirculating fluid-bed reactors which are operated under slightly different conditions, 220–250°C/25 bar/1.8 : 1, $H_2:CO$/ca. 70% CO conversion per pass versus 320°C/20 bar/6 : 1, $H_2:CO$/ca. 85% CO conversion per pass. In the fluid-bed reactor some catalyst passes overhead with the gas stream, is separated in a cyclone, and then recirculated to the reactor with fresh feed.

The fixed bed produces predominantly straight-chain, high-boiling hydrocarbons, the main percentage fractions being typically C_5–C_{11} gasoline range 33, diesel oil 17, heavy paraffin oil and higher waxes 40% w/w. The fluid bed, on the other hand, produces lower-molecular-weight material (72% in the gasoline range) richer in olefins and containing greater amounts of oxygenated compounds, 14% versus 4% in the fixed bed. The fluid bed is generally preferred in that it gives a greater proportion in the gasoline range and avoids formation of a large amount of the less desirable waxes. Its sole use does, however, lead to a shortfall in the diesel fraction and so the two systems have been operated in parallel. It should be noted that the gasoline range product is unsuitable for use as produced. The preponderance of linear alkanes gives a very low octane number and it requires uprating in a separate reaction stage.

Slightly different catalyst formulations are used for the two systems. That for the fixed-bed reactor is obtained by precipitation of hydrated iron oxide in the presence of Cu and K promoters, while that for the fluid bed is produced by melting iron oxide and promoters (oxides of K, Al, Si, and Mg) in an electric arc furnace at 1500°C and then milling the cooled melt to a fine powder. It is interesting to note that catalysts of the latter type were originally formulated for ammonia synthesis. The lower molecular weight and the higher degree of unsaturation of the product from the fluid bed appears to relate to the effects of higher temperature and higher $H_2:CO$ ratio already alluded to and to the lower porosity of the fused catalyst. This minimises contact time and suppresses secondary reactions of the α-olefins once desorbed from the catalyst surface.

One feature of these reactions that remains *invariant* is the molecular weight distribution of the products. A considerable accumulation of data both from commercial operations and laboratory experiments has shown that, with very few exceptions, the molecular weight distribution is in full accord with that predicted by Schulz–Flory kinetics. The reaction can thus be described by a simple "growth", or polymerisation, mechanism. All mechanistic proposals must therefore incorporate the classic features of such a process (see section 8.6).

The essential features of the three most generally cited mechanisms are summarised on pp. 171–3. The key steps of initiation, propagation, and termination have been separately identified. These mechanisms all differ in the nature of the initiator and the "monomeric" (chain building) unit. For example, the surface carbide mechanism uses a metal-methyl as initiator with surface bound methylene as "monomer", while the hydroxycarbene mechanism uses this species in both roles. The carbonyl insertion mechanism takes $(CO+2H_2)$ as "monomer" which is incorporated in the chain by a series of discrete reactions.

Another difference is the mode of adsorption of CO. In the surface carbide mechanism CO is *dissociatively* adsorbed, whereas, in the others it is *associative*, i.e. C—O bond breaking takes place as part of the chain propagation/termination process and does not precede it. In the scheme the detailed nature of the reaction steps and the bonding of the organic fragments to the metal surface have not been identified. For the latter, bonding to two or possibly three metal sites seems likely and may be an essential feature of the reaction. This may account for the present lack of authenticated examples of synthesis gas to hydrocarbon conversion with soluble metal species. Perhaps it is only among metal cluster complexes that the chemistry can take place, and it is with these complexes that success with homogeneous catalysts is likely to be realised.

The precise nature of the reaction mechanism is still the subject of much controversy and debate. No one mechanism can account for all the facts. All have their strengths and weaknesses. The surface carbide mechanism, a variation of which was originally proposed by Fischer and Tropsch, has received strong support from some recent studies on SiO_2-supported Ni, Co, and Ru catalysts involving isotopic labelling. Thus, by preforming a ^{13}C-rich surface carbide layer and treating this with synthesis gas, multiple incorporation of the label was found in alkane products above methane.[3] It is incapable, however, of accounting for the production of oxygenates without modification. The additional element needed involves CO insertion into the metal-alkyl bond of the growing chain to give a metal-acyl which is no longer able to propagate the growth. It is removed from the surface via hydrogenation to give e.g. primary alcohols or carbonyl-containing species.

The other mechanisms suffer from problems of credibility as to the nature of the surface-bound species. Moieties such as formyl, hydroxycarbene, and hydroxymethyl were for many years considered most unlikely since their chemistry among organometallic complexes was virtually non-existent. How-

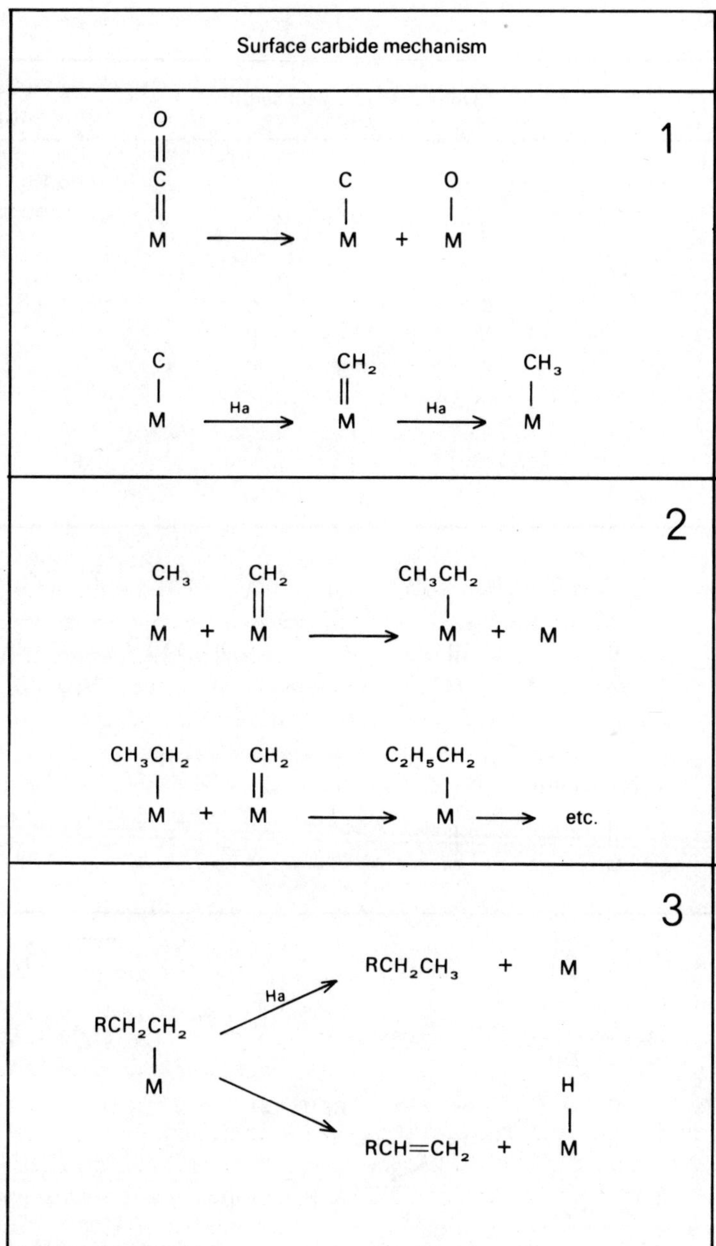

The Fischer–Tropsch reaction

Alternative mechanisms for hydrocarbon and oxygenate synthesis from CO and H_2. 1, initiation; 2, chain propagation; 3, chain termination.
(Ha = hydrogen atoms adsorbed on metal surface, M)

Hydroxycarbene mechanism

1

$$M{=}C{=}O \xrightarrow{H_a} M{=}C(H)(OH)$$

2

$$M{=}C(H)(OH) + M{=}C(H)(OH) \xrightarrow{H_a} M{=}C(CH_3)(OH) + H_2O$$

$$M{=}C(CH_3)(OH) + M{=}C(H)(OH) \xrightarrow{H_a} M{=}C(C_2H_5)(OH) \longrightarrow \text{etc.}$$

3

$$M{=}C(RCH_2)(OH) \xrightarrow{H_a} RCH_2CH_3 + H_2O$$

$$M{=}C(RCH_2)(OH) \xrightarrow{H_a} RCH{=}CH_2 + H_2O$$

$$M{=}C(RCH_2)(OH) \xrightarrow{H_a} RCH_2CH_2OH$$

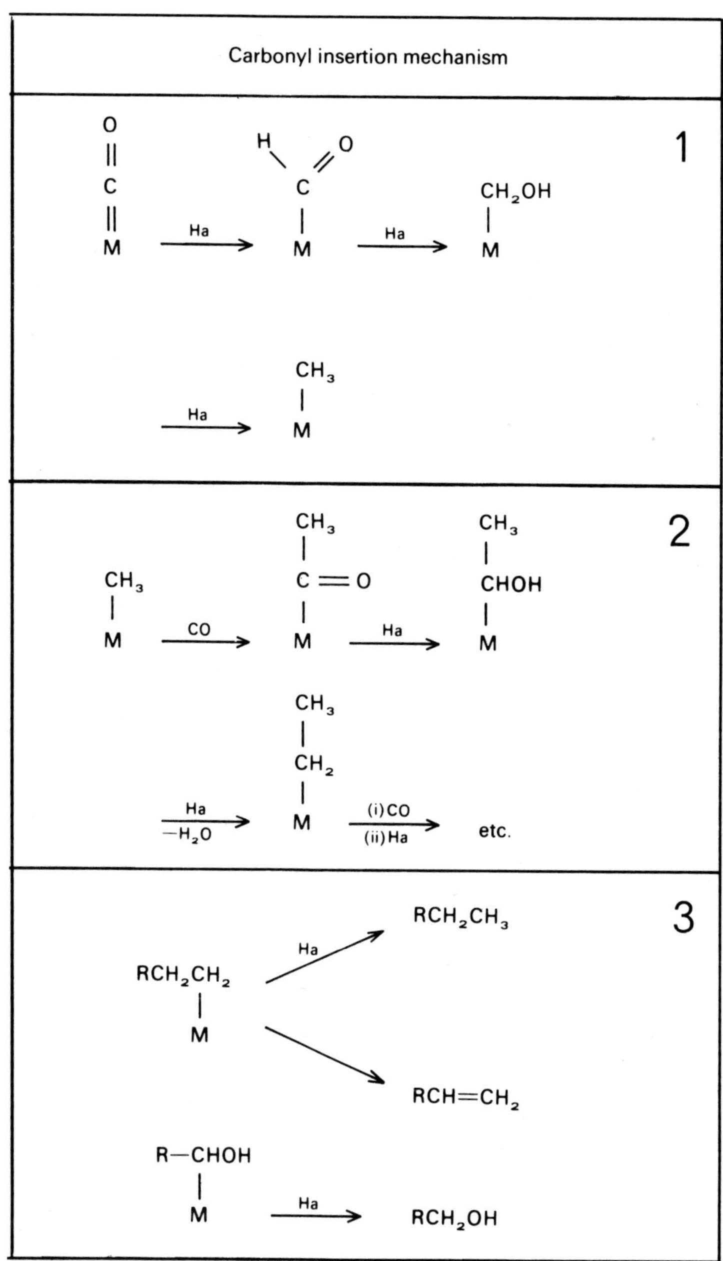
Carbonyl insertion mechanism
1
O
C
M
Ha
H
O
C
M
Ha
CH2OH
M
Ha
CH3
M
2
CH3
M
CO
CH3
C=O
M
Ha
CH3
CHOH
M
Ha
−H2O
CH3
CH2
M
(i) CO
(ii) Ha
etc.
3
RCH2CH2
M
Ha
RCH2CH3
RCH=CH2
R—CHOH
M
Ha
RCH2OH

ever, recent work has amply demonstrated that reasonably stable complexes incorporating these groups can be prepared and can be fully characterised.[4]

The alternatives on pp. 171–3 by no means make a comprehensive mechanistic survey. Other proposals include one involving metal-formyl species in a similar manner to metal-hydroxycarbenes and another incorporating many of the features of all three but requiring the chemistry to occur at a metal oxygen (M–O–C) rather than a metal (M–C) surface.[5]

For the longer term, the very nature of the Fischer–Tropsch process as a "growth" reaction (with heterogeneous catalysts) and the low space-time yields currently achievable would seem to mitigate against using diffusion limited reactions in pores without, at the same time, effecting dramatic improvements in activity. A better approach may well be to change the chemistry entirely and to place the burden of selectivity on a different reaction sequence. The best example of this approach is that adopted by Mobil who have recently shown that methanol, obtainable in very high selectivity from synthesis gas, can be transformed selectively to gasoline using novel shape-selective zeolite catalysts. This reaction is free from the constraints of a "growth" process.

8.2.2 *Methanation*

Although discovered two decades earlier, methanation can best be regarded as a special case of the Fischer–Tropsch reaction. Under a Schulz–Flory kinetic scheme for reaction of C_1 initiating species at the catalyst surface, it is the only product that can be obtained in selectivities approaching 100%. Required conditions are a high rate of hydrogenation of the surface C_1 intermediate and a low rate of chain growth via C—C bond formation.

Methane formation is favoured at low temperatures ($\Delta G = 0$ at around 625°C), high $H_2 : CO$ ratios, and low partial pressures of water.[6] In practice a number of problems are encountered. Sulphur is a potent poison of the nickel catalysts in common usage and must be maintained at very low (ppm) levels. Carbon deposition via the Boudouard reaction ($2CO \rightleftharpoons CO_2 + C$) can lead to catalyst deactivation, but can be controlled by operation at high $H_2 : CO$ ratios (typically 2.5 : 1 or more at 25 bar).

$$CO + 3H_2 \rightleftharpoons CH_4 + H_2O \qquad (\Delta H = -206\ \text{kJ/mol})$$

Heat release in the reaction is the major problem, and one that is not yet fully resolved. Heat must be removed efficiently since increases in temperature not only give lower equilibrium conversions but, most important, can lead to catalyst deactivation via sintering of the dispersed metal particles.

As a methanation catalyst, nickel is pre-eminent. It is relatively cheap, is very active when dispersed in a high surface area form, and is the most selective of all the transition metals. Typical catalyst formulations involve 25 to 75% w/w nickel dispersed on alumina or Kieselguhr, loadings considerably higher than are encountered with other supported catalysts. Nickel is not the most active metal.

This position is held by ruthenium which, while giving a wax-like fraction at high pressures, displays high selectivity to methane at high temperatures, low pressures, and high $H_2:CO$ ratios, conditions typically encountered in commercial operation. However, for the more expensive ruthenium systems to be competitive with nickel, efficient methods of metal recovery and recycle from spent catalysts must be devised.

For such a simple chemical reaction it is perhaps surprising that there are still several areas where improvements could realise significant benefits. There is a need for more thermally stable and less poison-sensitive catalysts and improved methods of heat removal. For the latter, a number of options are under consideration of which perhaps the most promising is the use of catalysts supported on metallic monoliths.[7] Being metallic they have excellent thermal conductivity and with their parallel multichannels coated with a thin layer of catalyst they enable the construction of catalyst beds that give high gas flows with minimal pressure drop across the bed.

Perhaps less surprisingly, selectivity is not a problem. High selectivities to methane are readily achieved and some higher hydrocarbons can be tolerated in the product SNG. Supplies of natural gas are adequate to meet commercial requirements in many parts of the world for the immediate future. Indeed, where supplies are assured, the reverse reaction, steam reforming (6.2.2), is the preferred method for the production of synthesis gas for methanol or ammonia synthesis. In the longer term as supplies dwindle, methanation will become more important and large quantities of SNG are likely to be made available using coal-derived synthesis gas. Methanation also finds use in the final purification of the feedstock gas for ammonia synthesis (6.2.5), purification of coolant in carbon dioxide-cooled nuclear reactors, and, in the laboratory, the determination of carbon oxides in gas chromatography using flame ionisation detectors involves prior conversion to methane.

8.2.3 *Liquid hydrocarbons from methanol*

Methanol has a number of advantages as a fuel or chemical feedstock. It can be obtained from synthesis gas in very high selectivity by well-developed processes, and being a liquid it can be easily stored and transported. As a transportation fuel it has a number of disadvantages, however. Miscibility limits its blending in gasoline to about 5%, and it presents also the associated problems of toxicity, corrosivity, and high affinity for water. Its principle disadvantage is its energy density which is half that of gasoline ($CH_3OH = CH_2 + H_2O$). At similar thermal efficiencies, twice the volume must be transported for the same energy output. It therefore makes sense to seek highly efficient processes for the conversion of methanol to liquid hydrocarbons to provide a cost-effective means of circumventing these problems.

In the mid-1970's Mobil announced the development of a process for the selective conversion of methanol to gasoline range hydrocarbons using their proprietary ZSM-5 zeolite catalysts.[8,9] Methanol to hydrocarbon conversion over a wide range of metal oxides was already well known. The advantage of the Mobil process lies in the high degree of selectivity to low molecular weight ($< C_{10}$) hydrocarbons coupled with useful catalyst lifetimes, features not shown in the earlier work. At high methanol conversion the major portion of the product is in the C_5 to C_{10} range which corresponds closely to petroleum-derived gasoline. A typical percentage breakdown of product is: C_1–C_3, 8 (largely propane), C_4, 16; C_5, 19; C_6, 16; C_7–C_9 aliphatic 13%, and C_6–C_9 aromatic hydrocarbons, 28%. The high aromatic content and the high degree of branching in the C_{5+} aliphatic hydrocarbons assures a high octane number and upgrading via catalytic reforming is unnecessary. This is in contrast to the Fischer–Tropsch process (8.2.1) which gives a predominantly linear aliphatic C_5–C_{10} cut of relatively low octane number.

Water is also a major product and is formed in 56% w/w yield, an inescapable feature of a process which is formally a dehydration reaction; but it is an advantage that the catalyst is able to cope with impure methanol streams containing up to 30% water. In addition, the catalyst shows great versatility and can convert other oxygenated species (e.g. ethanol, acetone) and hydrocarbons to a similar product range and thus has considerable potential for octane uprating of the low molecular weight fraction in Fischer–Tropsch processes.

In the design of the full-scale commercial process a major factor for consideration is heat removal from the reactor. With an overall exotherm of 1700 kJ/kg the adiabatic temperature rise could be as high as 600°C. Two different options have been considered: a two-reactor, fixed-bed process in which 20% of the heat is removed in the first reactor via conversion to dimethyl ether and water, and a fluid-bed reactor.[10] The fluid bed is the more promising but may require greater time for development. The superior heat transfer properties and the ability to continuously regenerate catalyst via oxidation of any laid-down coke would be advantageous. The first commercialisation of this process is taking place in New Zealand where a plant capable of producing 12 500 bbl/year of synthetic gasoline should be on stream in 1985. Factors making the process attractive for New Zealand are its extensive natural gas fields, lack of oil resources, and remote geographical location. Except under such favourable circumstances the Mobil process does not yet appear, however, to be fully competitive with petroleum-based gasoline. Even at such high achieved selectivities, the multistep nature of the overall conversion gives a lower thermal efficiency and a higher capital investment per unit output.

A current view of the mechanistic pathway is given in scheme 3.[8,9] By far the slowest step is the C—C bond formation in which ethylene is made via a C_2 surface species. The basic feature is then that of classical carbonium ion chemistry initiated by the strongly acidic hydroxyl groups residing in the zeolite pores. The sharp cut-off at C_{10} is associated with the unique properties of these pores. Their size (5 to 6 Å) severely constrains transition states leading to larger molecules which, even if formed, meet an additional diffusional constraint and may undergo fragmentation to smaller species before leaving the pore system. These effects are shown in the non-thermodynamic equilibrium distribution of tetramethylbenzenes in the product. The transition state leading to 1,2,4,5,-tetramethylbenzene can be accommodated but those leading to less symmetrical tetramethylbenzenes are of larger effective molecular volume and cannot be accommodated. The Mobil process thus overcomes the limitations of the broad molecular weight distribution associated with the Fischer–Tropsch process via

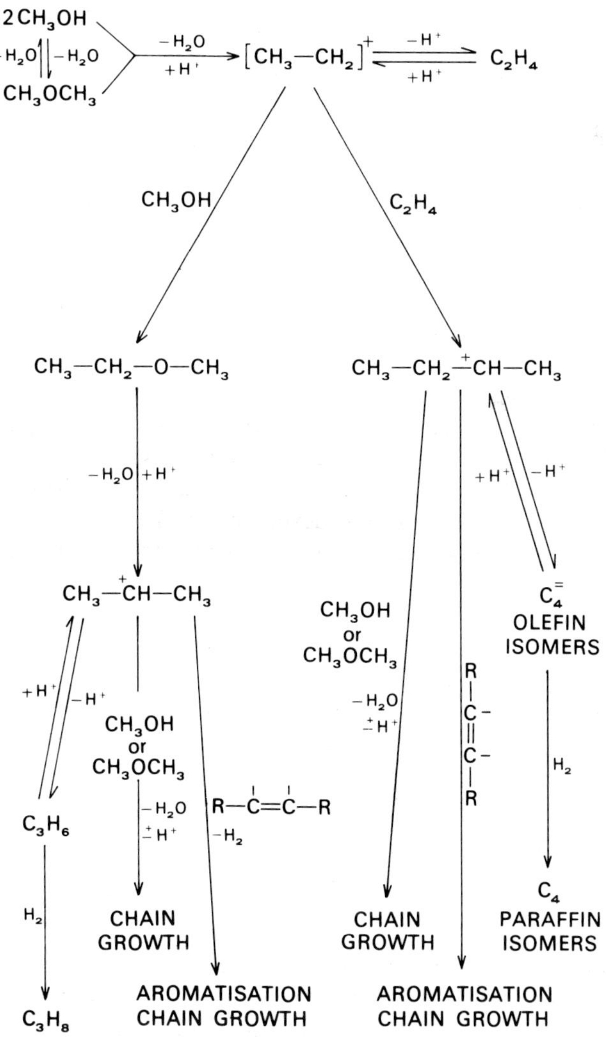

Scheme 3 Mechanism for conversion of methanol to hydrocarbons.

2. FORMATION OF AROMATICS

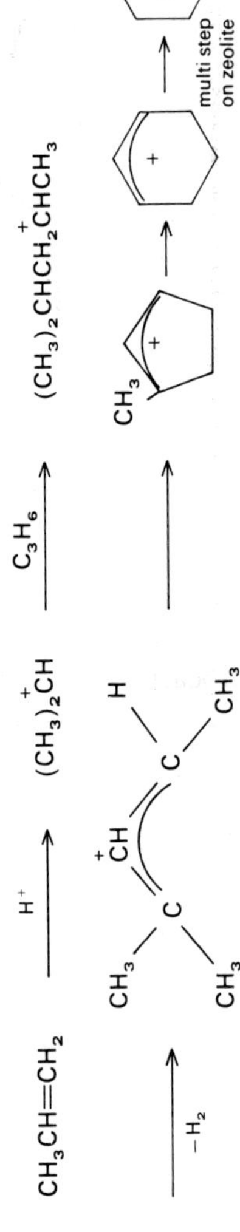

3. ALKYLATION OF AROMATICS

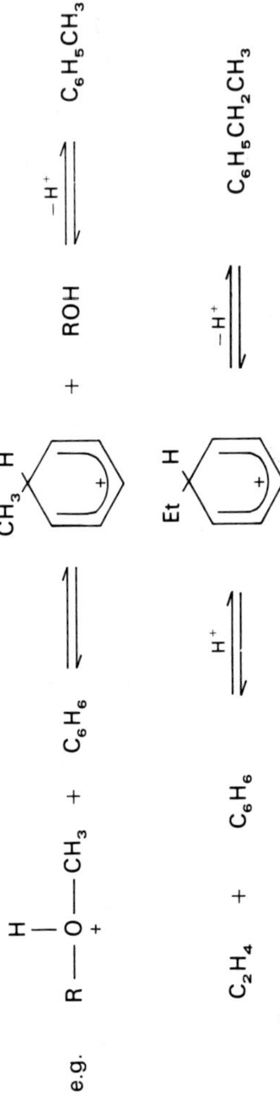

Scheme 3 *(continued)*

the provision of a shape-selective environment. The two-dimensional interconnecting network of pores ensures free passage within the structure and minimises deactivation via blocking of pore mouths (which would lead to coking).

The success of the Mobil process raises the question of the direct conversion of CO/H_2 to hydrocarbons without the need for independent production of methanol. It has already been established that mixtures of H-ZSM-5 and a methanol-forming catalyst or a Fischer–Tropsch catalyst work and give broadly similar product, but a number of associated problems still have to be solved.

The shape-selective properties of the zeolite and the formation of ethylene as an intermediate suggest that new zeolite catalysts could be designed to form the basis for production of lower olefins from methanol. The effective allowable molecular size could be reduced either by partial ion exchange with more bulky cations thereby restricting the pores, or the synthesis of new and smaller pore zeolites which are unable to accommodate transition states leading to C_6-cyclic structures. Other possible developments for these shape-selective zeolites could be for the selective preparation of BTX aromatics or for the preparation of novel heterogenised catalysts stable to metal leaching to replace those in current use in homogeneous systems such as hydroformylation.

8.3 Synthesis of oxygenated products

The potential advantage in the use of Fischer–Tropsch chemistry (8.2) to produce oxygenated products rather than hydrocarbons is that oxygen is included in the product, and the fraction of the synthesis gas feed converted into waste carbon dioxide or water is reduced compared with hydrocarbon formation. Thus, for example, formation of methanol from CO/H_2 involves no loss of water, whereas formation of ethylene involves loss of 56% of water (w/w) on feed, and ethane a little less than this. Formation of ethanol from CO/H_2 leads to a 28% loss (w/w) on feed. The standard free energy change associated with the formation of alcohols increases rapidly with temperature and it would therefore be predicted that, if kinetically possible, alcohol synthesis should be carried out at low temperatures.[11] The conditions which favour alcohol formation at the expense of hydrocarbons are low temperatures, high pressures, high space velocities and high CO/H_2 ratios. Methanol synthesis has been discussed in chapter 6. In this section we discuss ethanol synthesis and the high-pressure synthesis of ethylene glycol. These are the most significant industrial alcohols in addition to methanol—ethanol is more attractive than methanol as a gasoline additive on account of its higher solubility.

8.3.1 *Ethanol production*

Alternatives to making ethanol by fermentation are available via hydration of ethylene or homologation of methanol. In 1951 it was shown that methanol could be converted to ethanol by reaction with syn gas in the presence of dicobalt octacarbonyl in solution:[12]

$$CH_3OH + 2H_2 + CO \xrightarrow[250\ \text{bar}]{185^\circ} CH_3CH_2OH + H_2O$$

The cobalt carbonyl catalyst was not ideal and improved systems using, for example, cobalt salts, iodine and tertiary phosphines have now been developed to give selectivities as high as 70–90%;[13] the reaction proceeds via acetaldehyde as intermediate. Direct synthesis of ethanol from CO/H_2 is possible under high pressures using rhodium catalysts or via Fischer–Tropsch type synthesis over oxide or metal catalysts[14] but both these routes suffer from lack of selectivity associated with high alkane or acetic acid by-product formation. An IFP* catalyst (a modified methanol catalyst) can, however, give reasonable selectivity to ethanol and higher primary alcohols.[15]

8.3.2 *Ethylene glycol*

At present most ethylene glycol is made in a two-step route from ethylene via hydrolysis of ethylene oxide. The cost of this route is increasing as the price of premium oil cuts (naphtha) escalates in real terms relative to a general hydrocarbon base—this will be especially true if the selectivity of ethylene oxidation to ethylene oxide (11.2.1) cannot be markedly increased. A direct one-stage synthesis of ethylene glycol from the more versatile and readily available feedstock CO/H_2 would therefore be a very much superior route which may eventually provide ethylene glycol as cheaply as methanol.

This direct homogeneously-catalysed route has been developed by Union Carbide and on paper this is very attractive. The reaction gives high selectivity to oxygenates (with no methane formation) and the active catalyst is probably a homogeneous rhodium carbonyl metal cluster.[16] The major problem is the use of very high pressures ($\gtrsim$ 1000 bar). The use of rhodium as catalyst is now well established on an industrial scale in the production of acetic acid from methanol (8.5.1) and in propylene hydroformylation (8.4) and should present no problem in the ethylene glycol process provided that it is recovered and recycled. The postulated mechanism of this reaction is given in scheme 4, where the first stage is the formation of a formyl intermediate which is further hydrogenated to a hydroxymethyl species. These two ligands have only recently been identified in stable organometallic complexes.[4] Other metals e.g. cobalt, ruthenium and iridium are active in this reaction but do not work as well.[16a]

The direct route to ethylene glycol from syn gas considered above is the most attractive but if it does not prove to be technically feasible indirect routes may be viable (see scheme 5). Of the reactions indicated in the scheme, all the steps (1)–(5) have been commercialised although they are not all being operated at the

* Institut Francais du Pétrole.

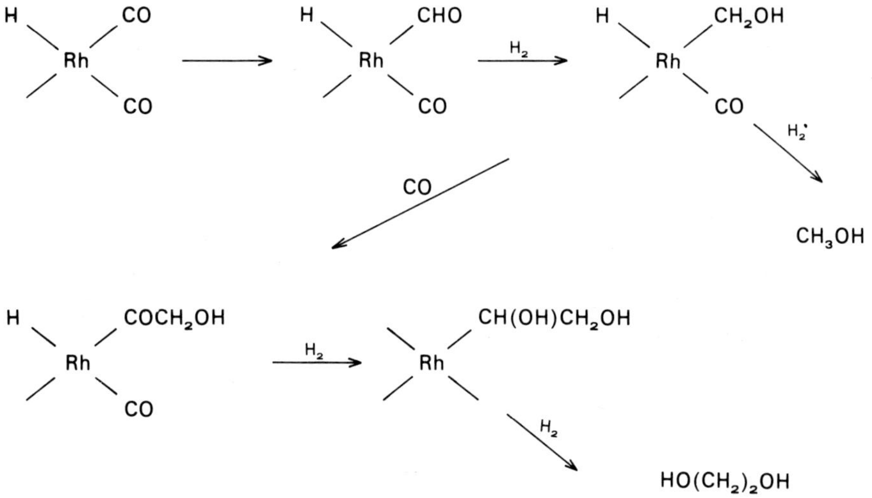

H
CO
Rh
CO
H
CHO
Rh
CO
H2
H
CH2OH
Rh
CO
H2
CH3OH
CO
H
COCH2OH
Rh
CO
H2
CH(OH)CH2OH
Rh
H2
HO(CH2)2OH

Scheme 4 Mechanism for ethylene glycol formation from syn gas.

CO
ROH/O2
Pd
ROOCCO2R
H2/Pt
CO/H2
Rh
HOCH2CH2OH
(1)
CuO/Zn/Al2O3
CH3OH
(2)
Ag
HCHO
CO/H2
Rh
OH−
zeolite
HOCH2CHO
H2/Pt
(5)
H2/Pt
HOCH2CO2R
(4)
CO/H+/H2O
(3)
HOCH2CO2H

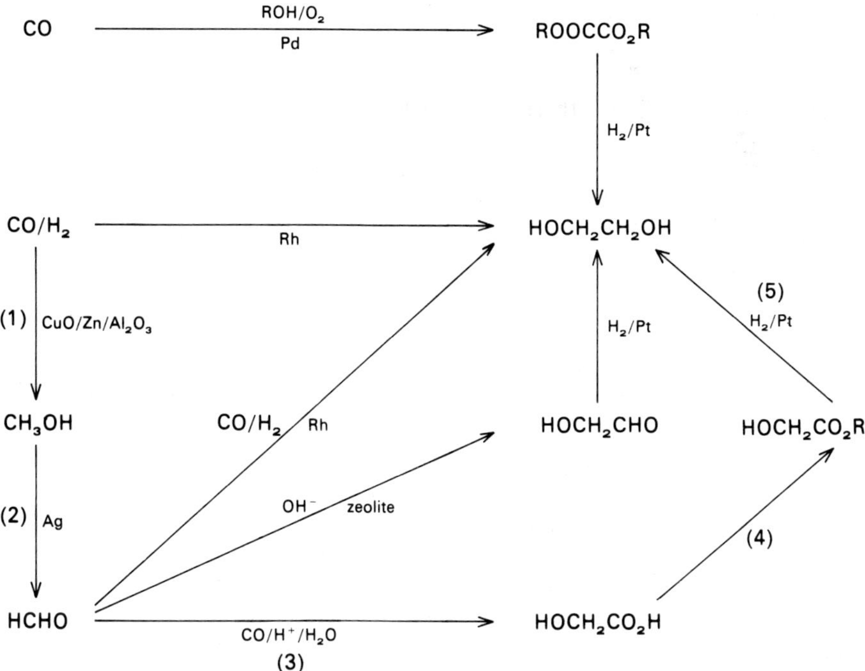

Scheme 5 Interconversions of simple organic molecules produced from syn gas in the preparation of ethylene glycol.

present time. Except for methanol to glycol, all the conversions represented have been reported in the literature. The route to ethylene glycol from formaldehyde via glycollic acid, steps (3), (4), (5), was practised commercially by du Pont until around 1970.

8.4 Hydroformylation

This process, often referred to as the OXO reaction, involves reaction of olefins with carbon monoxide and hydrogen in the presence of a homogeneous cobalt or rhodium catalyst to give isomeric aldehydes.[21,22] Formally, it involves addition of H and CHO across the double bond of the olefin, hence the term hydroformylation. The aldehyde may be further hydrogenated either *in situ* by the same catalyst or in a separate reaction step.

$$RCH{=}CH_2 \xrightarrow{CO/H_2} \underset{(n\text{-product})}{RCH_2CH_2CHO} \xrightarrow{H_2} RCH_2CH_2CH_2OH$$

$$RCH{=}CH_2 \xrightarrow{CO/H_2} \underset{(iso\text{-product})}{RCH(CHO)CH_3} \xrightarrow{H_2} RCH(CH_2OH)CH_3$$

In general, alcohols are the more important products but aldehydes can be useful chemical intermediates. This is well illustrated in the hydroformylation of propylene where *n*-butyraldehyde is both hydrogenated to *n*-butanol and converted, via aldol condensation, to 2-ethyl-1-hexanol.

$$C_3H_7CHO \xrightarrow{H_2} C_3H_7CH_2OH$$

$$C_3H_7CHO \xrightarrow{OH^-} C_3H_7CH(OH)CHEtCHO \xrightarrow[\text{(ii) } H_2]{\text{(i) } -H_2O} C_4H_9CHEtCH_2OH$$

With olefins other than ethylene, a mixture of isomeric aldehydes is produced. From a commercial standpoint, maximum selectivity to the normal, straight-chain product is desirable. The reasons for preferring the *n*- over the *iso*- product relate to chemical utility and improved performance in the end product. Thus, in the hydroformylation of propylene the isobutyraldehyde, readily separable from *n*-butyraldehyde, finds few outlets and, if present in large amount, may need to be disposed of. For higher molecular weight materials, e.g. C_8–C_{10} alcohols used in plasticiser manufacture* and C_{12}–C_{16} alcohols in detergents,† separation is difficult and the product mixture is generally used without purification. For detergents, linear alcohols mimic the natural systems with which they compete, e.g. alcohols derived from coconut oil, and are more easily

* Typically as diethylphthalates.

† Typically as alcohol ethoxylates, $RO(CH_2CH_2O)_xH$.

biodegraded than their branched isomers. Price and performance are however always balancing considerations. Thus, for plasticisers, the branched 2-ethylhexanol (from C_3H_6 via n-C_3H_7CHO) competes with n-octanol (from the more expensive heptene) largely on the basis of cost as a prime consideration.

With olefins higher than propylene, the catalyst may isomerise the double bond along the hydrocarbon chain and this may lead to the production of a number of additional branched products. In general α-olefins, although more expensive than the counterpart internal olefins, are the preferred feedstocks in that they give a much higher proportion of the desired linear product.

The hydroformylation reaction was first operated using supported cobalt catalysts, but it is now known that homogeneous systems are better than heterogeneous ones. The reaction pathway is believed to be of the kind shown in scheme 6 for production of the n-aldehyde. The principal steps involved are activation of the catalyst via carbon monoxide dissociation to give the co-ordinatively unsaturated hydrocarbonyl ($HCo(CO)_3$); alkene co-ordination followed by insertion into Co—H to give an alkyl compound; CO insertion into Co—C to give an acylcobalt carbonyl, and hydrogenolysis to give product aldehyde and regeneration of catalyst. Branched aldehyde results from Markownikoff addition of Co—H to the alkene. *In situ* infrared studies under OXO reaction conditions in the absence of alkene have shown almost complete conversion of $Co_2(CO)_8$ into $HCo(CO)_4$ and of $[Co(CO)_3PBu_3]_2$ into $HCo(CO)_3PBu_3$.[19]

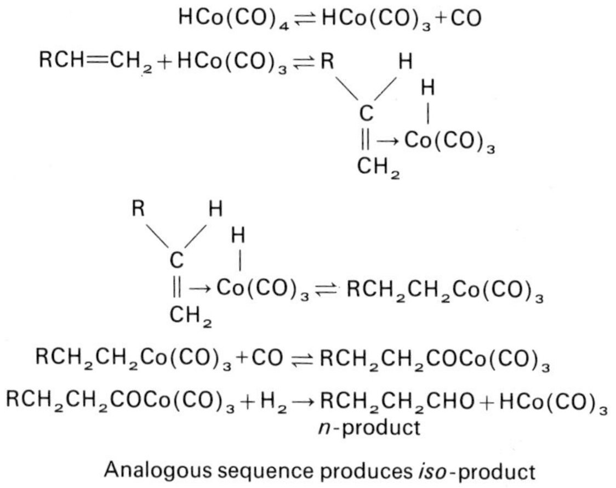

$$HCo(CO)_4 \rightleftharpoons HCo(CO)_3 + CO$$

$$RCH{=}CH_2 + HCo(CO)_3 \rightleftharpoons (RHC{=}CH_2)\rightarrow Co(CO)_3H$$

$$(RHC{=}CH_2)\rightarrow Co(CO)_3H \rightleftharpoons RCH_2CH_2Co(CO)_3$$

$$RCH_2CH_2Co(CO)_3 + CO \rightleftharpoons RCH_2CH_2COCo(CO)_3$$

$$RCH_2CH_2COCo(CO)_3 + H_2 \rightarrow \underset{n\text{-product}}{RCH_2CH_2CHO} + HCo(CO)_3$$

Analogous sequence produces *iso*-product

Scheme 6 Mechanism for cobalt-catalysed hydroformylation.

The addition of phosphines to cobalt systems has a number of effects. (i) It increases the linearity of the product. This improved product ratio relates to both the electronic and the steric effect of the phosphine ligand in favouring the formation of the n-alkylcobalt intermediate. In addition, with internal olefins the phosphine ligand has the beneficial effect of giving a highly linear product. This effect is thought to relate to rapid equilibration among the various alkylcobalt intermediates before CO insertion, the phosphine favouring the less bulky *n*-alkylcobalt species. (ii) It increases the hydrogenating power of the catalyst, and, under comparable conditions, a higher proportion of alcohols are produced. (iii) It stabilises the cobalt catalyst, allowing lower partial pressures of CO to be used without serious thermal degradation. (iv) The benefits in (i)–(iii) are, however, only secured at a penalty in overall reaction rate and it is necessary to employ either more severe conditions or greater catalyst concentration to restore the balance.

The discovery by Wilkinson[20] that rhodium catalyses the hydroformylation of olefins at room temperature and atmospheric pressure opened up the route which has now led to the Johnson Matthey/Union Carbide/Davy McKee low pressure phosphine modified rhodium process for the hydroformylation of propylene.[21] The mechanism is seen to be broadly similar to the cobalt-catalysed reaction (scheme 6) but rhodium is intrinsically much more active than cobalt and may consequently be used under milder conditions. Advantages for the rhodium-based process are flexibility (*n*/*iso* ratios may be controlled from 8 : 1 to 16 : 1) lower plant capital cost, lower running costs and production of a simpler product mix containing no C_4 alcohols. Syn gas can be introduced into the OXO plant without any need to use costly high-pressure compression procedures. The high *n*/*iso* ratios obtained mean that on a plant selectivity of 10 : 1 for propylene, a saving of 20 % on propylene and syn gas costs can be made because less of the product is thrown away. Union Carbide have operated this technology since 1975 and the process is attracting licensees from all over the world. Conditions used in the cobalt- and rhodium-catalysed reactions are compared in Table 8.2, and the layout of the plant is given in figure 8.1.

From the proposed mechanism for propylene hydroformylation given in scheme 7 we can speculate as to why high *n*/*iso* ratios are produced. Initial co-ordination of the olefin is visualised as the first step, followed by insertion into an Rh—H bond to give an alkyl. CO insertion into the Rh—alkyl bond gives an acyl. Oxidative addition of hydrogen gives a dihydroacyl complex and hydrogen transfer to the acyl group gives butyraldehyde along with a catalyst precursor which can react again with CO and olefin to start the next cycle. The presence of excess triphenylphosphine under the low pressure conditions favours high selectivity to normal aldehyde by suppressing loss of phosphine from the

Table 8.2 Comparison of hydroformylation processes based on cobalt and rhodium

	Cobalt process	*Rhodium process*
Pressure	100–400 bar	7–13 bar
Temperature	145–180°C	100°C
n/iso ratio	3 : 1 or 4 : 1	Controllable over wide range normally 8 : 1–16 : 1

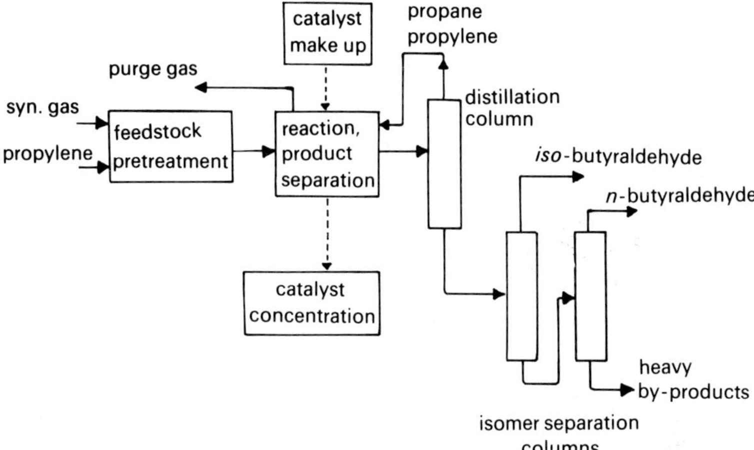

Figure 8.1 Plant flow diagram for rhodium hydroformylation process (JM/UC/Davy McKee). Adapted from Fowler *et al.*, 1976, and Brewster, 1976.

catalyst. When olefin approaches, the steric effect of the two bulky phosphine ligands promotes a high ratio of primary alkyl product formation. In the absence of excess phosphine, *n*/*iso* ratios of 3 : 1 are obtained as for cobalt catalysts.

Both cobalt- and rhodium-catalysed processes run very effectively and are unlikely to be replaced by systems based on other metals. Ruthenium shows some activity in hydroformylation but it is much less active than cobalt or rhodium. Iridium and platinum-tin systems are also active for hydroformylation and the latter can be made active and selective by the addition of diphosphines.[27]

Many attempts have been made to bond hydroformylation catalysts to polymer and inorganic supports in order to provide easy means of separating the catalyst from the products but in general the attempts in this area have been less than satisfactory—the metal leaches off the support, the reaction rates decrease, and the *n*/*iso* ratio tends to decline. These, however, may not be insurmountable problems and work is continuing in this field.[28]

8.5 Carbonylation of organic substrates

Many classes of organic compounds containing the carbonyl group are accessible via reactions with CO. Structural types preparable via carbonylation include: ketones, carboxylic acids and anhydrides, esters, lactones, quinones and hydroquinones, acyl halides, amides, ureas, imides, lactams, phthalimidines, indazolones, quinazolines, and amino acids. In this section we discuss two processes: first, the production of acetic acid via carbonylation of methanol which has already made a major commercial impact, and second, the production of acetic anhydride and vinyl acetate via carbonylation of methyl acetate which

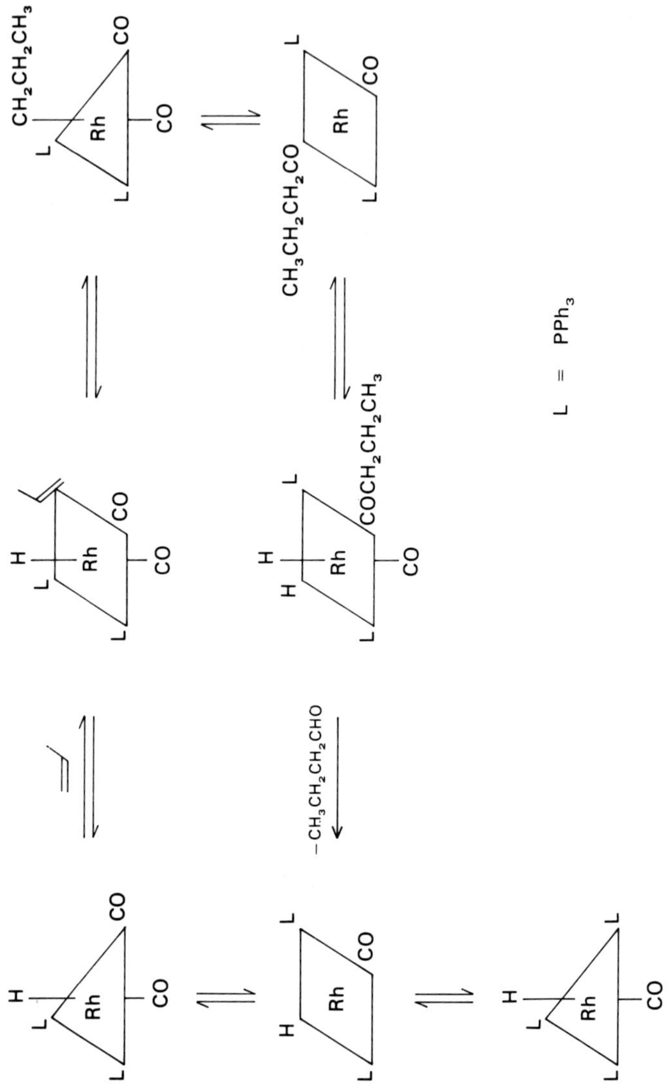

Scheme 7 Mechanism for rhodium-catalysed hydroformylation.

shows great promise for the future production of two lesser, though still significant, tonnage chemicals.

Three other reactions involving carbonylation of olefins that appear to be presently underexploited, but could gain in commercial importance, are the Hydroesterification, Reppe, and Koch reactions.[17,18] The first two of these are formally variations on the hydroformylation reaction. Hydroesterification provides a direct route from olefins to long chain, linear carboxylic acids and derivatives

$$RCH{=}CH_2 + R'OH + CO \rightarrow RCH_2CH_2COOR' + \textit{iso}\text{-product}$$

It also offers a novel route to azelaic acid, an intermediate in the synthesis of the speciality nylon 6,9, from the readily available cyclo-octa-1,5-diene, a dimer of butadiene:[22]

$$\text{cyclo-octa-1,5-diene} \xrightarrow[\text{110°C 100 atm}]{\text{CO/EtOH, [Pd]}} \text{(cyclo-octenyl)}CO_2Et \xrightarrow[\text{(2) } H_2/OH^-\text{, 320°C}]{\text{(1) } H_2O\text{, 100°C}} \underset{\text{azelaic acid}}{HO_2C(CH_2)_7CO_2H}$$

The Reppe reaction is just one of a number of carbonylation reactions that bears the name of one of the pioneers in the field of catalytic carbonylation, Walter Reppe. Homogeneous iron catalysts, in the presence of strong base, effect direct conversion of olefins to alcohols using carbon monoxide and water as reagents. Formally, the reaction involves water gas shift *in situ* and so the reaction has certain potential for the utilisation of hydrogen-lean synthesis gas.

$$RCH{=}CH_2 + CO + H_2O \rightarrow RCH_2CH_2CH_2OH + \textit{iso}\text{-product} + CO_2$$

The Koch reaction is perhaps more limited, producing highly branched carboxylic acids, e.g.

$$RCH{=}CMe_2 + CO + H_2O \xrightarrow{H^+} RCH_2CMe_2COOH$$

This reaction operates under strongly acidic conditions in direct contrast to the mildly basic conditions of the hydroformylation and hydroesterification reactions. Space does not however allow a complete discussion of all the various reactions, and the interested reader is directed to several very useful recent reviews.[17,18]

8.5.1 *Conversion of methanol into acetic acid*

As raw material costs and availability have changed over the past 100 years, preferred industrial routes to acetic acid have changed very significantly. These routes are summarised in Table 8.3 (see also 11.5.1). The current best process is the rhodium catalysed route developed by Monsanto[23] who started up their first plant in 1970. This process, wholly synthesis gas based, was thus fully

Table 8.3 Seven phases of acetic acid process development

Chemical route	*Comments*
Carbohydrate $\xrightarrow{\text{enzyme}}$ CH_3CO_2H aq.	Fermentation route (19th century)
$HC{\equiv}CH \xrightarrow[\text{Hg}]{H_2O} CH_3CHO \xrightarrow{O_2} CH_3CO_2H$	First synthetic route
Butane $\xrightarrow[\text{Mn/Co}]{O_2}$ CH_3CO_2H + by-products	Celanese, BP (ca. 1955)
$CH_2{=}CH_2 \xrightarrow[\text{Pd/Cu}]{O_2} CH_3CHO \xrightarrow{O_2} CH_3CO_2H$	Wacker process (1962)
$CH_3OH \xrightarrow[\text{Co/500 atm}]{CO} CH_3CO_2H$	BASF (1966)
$CH_3OH \xrightarrow[\text{Rh/20 atm}]{CO} CH_3CO_2H$	Monsanto (1970)
$CO/H_2 \xrightarrow[\text{Rh}]{\text{40 atm}} CH_3CO_2H$	Union Carbide (needs further development) (1990?)

competitive with routes based on oil-derived hydrocarbons in the era of cheap oil and is likely to continue to be the most attractive route as oil prices increase and synthesis gas becomes widely available from coal. Even better would be a direct route from CO/H_2 to acetic acid, and indications that this is a possibility have been published by Union Carbide.[14] The Monsanto catalysed methanol conversion process is another very successful homogeneously catalysed reaction based on a precious metal and proceeds with a chemical selectivity of greater than 99%. Iridium is also an active catalyst but not as selective. The lower selectivity is associated with a more complicated catalytic cycle.

The mechanism for the rhodium reaction is given in scheme 8 and the Rh and Co processes are compared in Table 8.4. It can clearly be seen that the substitution

Table 8.4 Comparison of methanol carbonylation processes based on cobalt and rhodium

	Cobalt process	*Rhodium process*
Pressure	500–700 bar	30–40 bar
Temperature	230°C	180°C
Selectivity to HOAc (based on CH_3OH)	90%	> 99%
Metal concentration	10^{-1}M	10^{-3}M

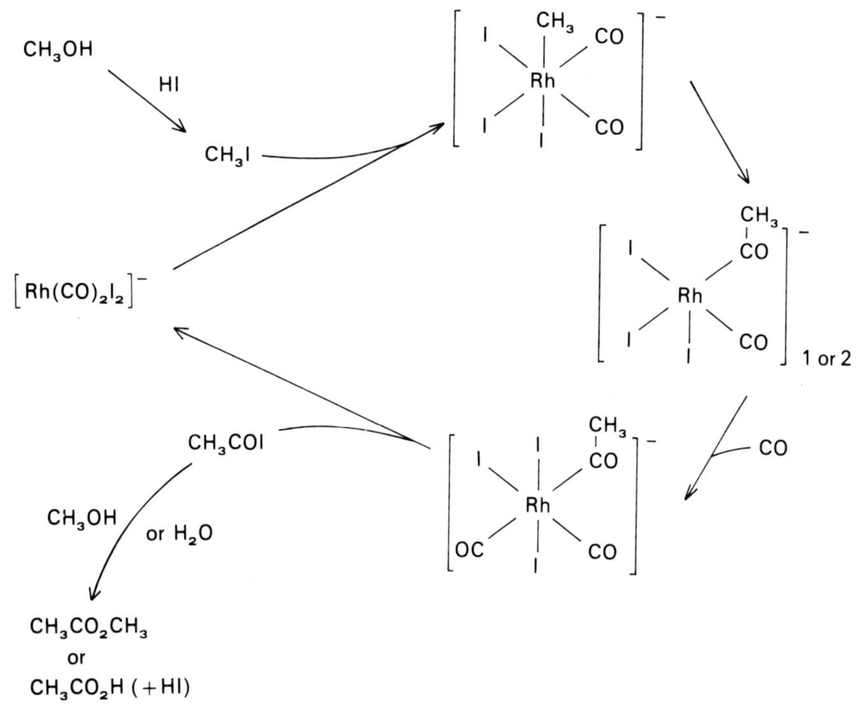

Scheme 8 Mechanism for rhodium-catalysed carbonylation of methanol.

of Rh for Co produces the same sort of advantages as a similar development had in hydroformylation (see 8.4) in lowering the temperature and pressure and increasing the selectivity of the reaction.

Supported rhodium catalysts have been investigated and compared with homogeneous catalyst activities. Similar mechanisms seem to operate in both liquid and vapour phase but to date the supported catalyst approach has suffered from low reactivity (typically 100 times slower per g of Rh) and rhodium loss from the support. Since the rhodium catalyst can be recovered and recycled in the homogeneous reaction the latter is likely to be preferred for some time to come. Gas-phase reactions based on rhodium zeolites have also been studied and may lead to future developments.[24]

8.5.2 *Synthesis of vinyl acetate and acetic anhydride*

Halcon have recently developed routes to acetic anhydride and vinyl acetate based entirely on synthesis gas using rhodium or palladium catalysts. Methyl acetate is a common starting material.[25] Reaction in the presence of carbon monoxide gives acetic anhydride, while with mixtures of carbon monoxide and hydrogen, the major product is 1,1-diacetoxyethane accompanied by smaller amounts of acetic anhydride and acetaldehyde. Vinyl acetate is obtained via the (known) pyrolysis of the 1,1-diacetoxyethane. These reactions are summarised in scheme 9.

$$CO/H_2 \rightarrow CH_3OH \xrightarrow{CO} CH_3COOH \xrightarrow{MeOH} CH_3COOMe$$

$$CH_3OAc \xrightarrow{HI} CH_3I \xrightarrow{[Rh]} CH_3RhI \xrightarrow{CO} CH_3CORhI$$

$$CH_3CORhI \xrightarrow{HOAc} (CH_3CO)_2O + [Rh] + HI$$

$$CH_3CORhI \xrightarrow[H_2]{} CH_3CHO + [Rh] + HI$$

$$CH_3CHO + (CH_3CO)_2O \rightarrow CH_3CH(OAc)_2$$

$$CH_3CH(OAc)_2 \xrightarrow[H^+,\ 170°C]{\Delta} CH_2{=}CHOAc + HOAc$$

Scheme 9 Conversion of methyl acetate to vinyl acetate and acetic anhydride.

Eastman Kodak and Halcon have jointly developed the process for the production of acetic anhydride, a relatively small tonnage chemical. The attraction of this route over the conventional technology (controlled oxidation of acetaldehyde or reaction of ketene with acetic acid) is the high selectivity ($\sim 90\%$) which eliminates the need for by-product separation and disposal.

The vinyl acetate process is less well-developed. To be competitive with the well-developed, ethylene-based route (cf. 11.3.8) high selectivity must be realised. The attractions of the process are its overall simplicity: conceptually it involves conversion of methanol and synthesis gas to vinyl acetate. No acetic acid synthesis step is required since it is internally recycled between the pyrolysis and esterification stages:

$$CH_3OH + 3CO + 2H_2 \rightarrow CH_2{=}CHOAc + 2H_2O$$

Catalyst compositions and conditions are similar to those used in methanol carbonylation to acetic acid.

8.6 Appendix—Product distributions in polymerisation (growth) processes and their implications

Polymerisation, or growth, reactions are frequently encountered in the field of catalysis. In this book examples are discussed in 8.2 (Fischer–Tropsch reactions), 9.2 (ethylene oligomerisation), and in chapter 10 (olefin polymerisation). They have a number of mechanistic features in common which have a direct influence on the primary product distribution. This, in turn, has important implications for the operation of all chemical processes based on growth reactions.

Three key steps characterise growth reactions:

(a) *Initiation*, the generation of active growth centres;
(b) *Propagation*, the sequential addition of fundamental building blocks (i.e. monomer units) to the active centre, coupled with *chain transfer*, the release of product molecules with the regeneration of the active centre;
(c) *Termination*, the destruction of active centres.

The primary product distribution can be obtained by consideration of the simplified scheme below. For the purposes of calculation, only propagation and chain transfer reactions need be considered. Under steady state conditions, initiation and termination reactions occur at equal rates and can be ignored.

$$\begin{array}{ccccccc} \xrightarrow{k_1} & R_{n-1} & \xrightarrow{k_1} & R_n & \xrightarrow{k_1} & R_{n+1} & \xrightarrow{k_1} \\ & \downarrow k_2 & & \downarrow k_2 & & \downarrow k_2 & \\ & P_{n-1} & & P_n & & P_{n+1} & \end{array}$$

R_n and P_n represent active growing centres and product molecules, respectively, each containing n monomer units. k_1 and k_2 are rate constants for propagation and chain transfer which are assumed to be independent of n.

At steady state, the concentration of propagating centres is constant; thus,

$$k_1 R_{n-1} = (k_1 + k_2) R_n$$

if $\dfrac{k_2}{k_1} = \beta$,

$$R_{n-1} = (1+\beta) R_n$$

and hence,

$$R_0 = (1+\beta)^n R_n$$

The mole fraction, F_n, of the product P_n is given by

$$F_n = \frac{P_n}{\sum_{n=1}^{\infty} P_n},$$

which, under steady-state conditions relates directly to the distribution of active centres, i.e.

$$F_n = \frac{R_n}{\sum_{n=1}^{\infty} R_n}$$

$$\therefore F_n = \frac{R_0}{(1+\beta)^n} \bigg/ \sum_{n=1}^{\infty} \frac{R_0}{(1+\beta)^n} = \frac{\beta}{(1+\beta)^n}$$

This relationship is generally expressed graphically as a plot of $\ln(F_n)$ versus n (figure *a*). Plots of product data of this kind can be used to test the conformity of catalytic processes to a simple growth mechanism.

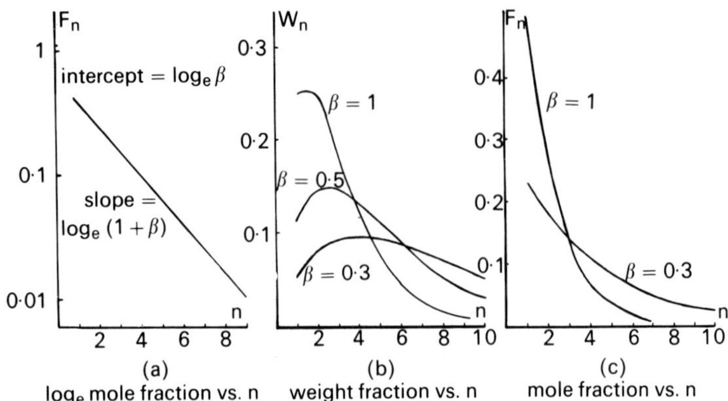

(a) loge mole fraction vs. n (b) weight fraction vs. n (c) mole fraction vs. n

An alternative, and very useful, representation is shown in figure *b* which records the variation in weight fraction, W_n* (the main concern of the industrial chemist) with *n*. Figure *c* gives the equivalent F_n versus *n* relationship for comparison. Figure *b* shows curves of distributions maximised at different values of *n* (i.e. different β).

The implications of growth processes of this kind can readily be seen from these figures.

(i) High selectivity can be only achieved in the production of the first member of the series, e.g. the dimer, butene, in ethylene oligomerisation, and C_1 species in CO/H_2 reactions.† Theoretically, 100% selectivity can be achieved ($\beta \gg 1$).

(ii) High selectivity to higher products is not possible. The second members of the series, hexene and C_2 species in the examples above, can only be obtained in a maximum possible 32% w/w. Optimal selectivity diminishes for products higher in the series.

REFERENCES

1. The Fischer–Tropsch reaction has been the subject of several excellent reviews, and the early work has been well covered: H. H. Storch, N. Golumbic and R. A. Anderson, *The Fischer–Tropsch and Related Synthesis*, Wiley, 1951; *Catalysis*, P. H. Emmett ed., Reinhold, Vol. 4, 1956; M. A. Vannice, *Catalysis Rev.*, 1976, **14**, 153; V. Ponec, *Catalysis Rev.*, 1978, **18**, 151; G. Henrici-Olivé and S. Olivé, *Angew. Chem. Internat. Edn.*, 1976, **15**, 136; H. Pichler and A. Hector in *Kirk–Othmer Encyclopedia of Chemical Technology*, 2nd Edn., 1964, Vol. 4, 446.
2. B. Cornils *et al.*, *Hydrocarbon Processing*, 1976, (Nov.), 105.
3. W. M. H. Sachtler *et al.*, *J. Catalysis*, 1979, **58**, 95; P. Biloen, *Rec. Trav. Chim. Pays-Bas*, 1980, **99**, 33.
4. See e.g. W. A. G. Graham, *J. Organometallic Chem.*, 1979, **173**, C9.
5. R. S. Sapienza *et al.*, *Fundamental Research Homog. Catalysis*, 1979, **3**, 179, and refs. therein.

* $W_n = \dfrac{\beta^2 n}{(1+\beta)^{n+1}}$.

† The monomeric unit in CO/H_2-based reactions is not obvious at first glance: formally, in hydrocarbon synthesis, it can be taken as $(CO + 2H_2) = CH_2(+H_2O)$.

6. G. A. Mills and F. W. Steffgen, *Catalysis Rev.*, 1974, **8**, 159.
7. E. R. Tucci *et al.*, *Hydrocarbon Processing*, 1979, (Feb.), 123.
8. Selected contributions from Mobil Research Laboratories: S. L. Meisel *et al.*, *Chemtech*, 1976, **6**, 86; C. D. Chang, W. H. Lang and A. J. Silvestri, *J. Catalysis*, 1977, **56**, 268; P. D. Caesar *et al.*, *ibid.*, 1979, **56**, 274; N. Y. Chen and W. J. Reagan, *ibid.*, 1979, **59**, 123; W. W. Kaeding and S. A. Butter, *ibid.*, 1980, **61**, 155; D. H. Olson, W. O. Haag and R. M. Lago, *ibid.*, 1980, **61**, 390; and refs. therein.
9. J. R. Anderson *et al.*, *J. Catalysis*, 1979, **58**, 114; 1980, **61**, 477; E. G. Derouane, *ibid.*, 1980, **63**, 331.
10. R. I. Berry, *Chem. Engineering*, 1980, **87** (April 21st), 86.
11. P. J. Denny and D. A. Whan, Chem. Soc. Spec. Period. Rep. *Catalysis*, 1979, **2**, 46.
12. For a recent review, see H. Bahrmann and B. Cornils, *Chem. Zeitung*, 1980, **104**, 39; also ref. 18, p. 226.
13. See e.g. U.K. Pat. 1,546,428 (to Shell); Europ. Pat. 1,936 and 1,937 (to B.P.); U.S. Pat. 4,233,466 (to Union Carbide).
14. M. M. Bhasin *et al.*, *J. Catalysis*, 1978, **54**, 120.
15. U.K. Pat. 1,547,687 (to IFP).
16. R. Pruett, *Ann. N.Y. Acad. Sci.*, 1977, **295**, 239; J. L. Vidal and W. E. Walker, *Inorg. Chem.*, 1980, **19**, 896.
16a. J. F. Knifton, *J. C. S. Chem. Comm.*, 1981, 188; B. D. Dombek, *J. Amer. Chem. Soc.*, 1980, **102**, 6855; D. R. Fahey, *ibid.*, 1981, **103**, 136.
17. P. Pino and I. Wender, *Organic Synthesis via Metal Carbonyls*, Wiley, Vol. 1 1968, Vol. 2 1977; R. Pruett, *Adv. Organometallic Chem.*, 1979, **17**, 1; D. T. Thompson *et al.*, Chem. Soc. Spec. Period. Rep. *Catalysis*, 1977, **1**, 369.
18. *New Syntheses with Carbon Monoxide*, J. Falbe, ed., Springer-Verlag, 1980.
19. M. van Boven *et al.*, *Ind. Eng. Chem. Prod. Res. Dev.*, 1975, **14**, 259.
20. G. Wilkinson *et al.*, *J. C. S. Chem. Comm.*, 1965, 17; *Chem. Ind.*, 1965, 560.
21. R. Fowler *et al.*, *Chemtech*, 1976, **6**, 772; *Hydrocarbon Processing*, 1976, (Sept.), 247; E. A. V. Brewster, *Chem. Engineering*, 1976, **83** (Nov. 8th), 90.
22. D. T. Thompson, *Platinum Met. Rev.*, 1975, **19**, 88.
23. D. Forster, *Adv. Organometallic Chem.*, 1979, **17**, 255.
24. M. S. Scurrell and R. F. Howe, *J. Molec. Catalysis*, 1980, **7**, 535.
25. *European Chem. News*, 1980, (Mar. 3rd), 24; Ger. Pat. 2,610,035 (to Halcon); U.S. Pat. 4,115,444 (to Halcon).
26. Examples of several reactions in Scheme 5 are: (a) oxalate route, K. Nishimura *et al.*, Preprints A.C.S. Div. Petrol Chem., 1979, **24**, 355; F. Rivetti and U. Romano, *Chem. Ind.* (*Milan*), 1980, **62**, 7; (b) glycollic acid route, U.S. Pat. 4,136,112 and U.K. Pat. 1,452,258 (to Chevron); (c) formaldehyde to glycolaldehyde using zeolites, A. H. Weiss *et al.*, *Ind. Eng. Chem. Prod. Des. Dev.*, 1979, **18**, 522, U.S. Pat. 4,238,418; (d) formaldehyde to glycolaldehyde using Rh, A. Spencer, *J. Organometallic Chem.*, 1980, **194**, 113; J. A. Roth and M. Orchin, *ibid.*, 1979, **172**, C27; (e) formaldehyde to glycol, U.K. Pat. 1,585,604 (to National Distillers).

CHAPTER NINE

CARBON–CARBON BOND FORMATION. II: OLIGOMERISATION, ISOMERISATION, METATHESIS AND HYDROCYANATION

R. PEARCE

9.1 Introduction

This chapter considers the conversion of olefinic and aromatic building blocks to end products in more detail. Attention is focused on the processes of ethylene oligomerisation to α-olefins, olefin metathesis, and xylenes isomerisation. These reactions are treated in some detail since they illustrate a number of important features that confront designers of catalytic processes. A problem common to all growth processes is that of product distribution. Mechanistic constraints imposed by the basic growth reaction give a distribution of products that is often not in line with the demands of the market. Ethylene oligomerisation to α-olefins illustrates this nicely. A somewhat narrow spectrum of products is required (C_8 to C_{16}) and to be successful the overall process must be capable of effective utilisation of product olefins outside this range. One of the most elegant solutions involves the marrying of three catalytic reactions: ethylene oligomerisation, olefin isomerisation and metathesis. Similar problems are involved in xylenes utilisation: the main market demand is for the *p*-xylene-isomer for conversion to terephthalic acid. At thermodynamic equilibrium this comprises around 20% of C_8-aromatics. Selective extraction of *p*-xylene and isomerisation of the residual C_8-species are the key to their effective utilisation.

9.2 α-Olefins via ethylene oligomerisation

Higher olefins, like ethylene, are chemical intermediates and need further processing to give end products. The main end uses of C_6–C_{10} olefins are as alcohols in plasticisers (e.g. dialkylphthalates) and the C_{12}–C_{16} olefins as the hydrophobic functional group in surface active agents (e.g. alkyl sulphates, alkyl sulphonates, alkylbenzene sulphonates and alcohol ethoxylates, $RO(CH_2CH_2O)_xH$). Among the many chemical combinations that have surfactant properties, those based on modified linear primary alcohols are among the most important. "Hard" detergents, branched alkyl benzene sulphonates, used in the 1950's, caused the problem of uncontrollable foam in many inland waterways, and this created a demand for more readily biodegraded products. Alkyl chains with minimal branching show the best degradability and are therefore preferred in surfactant formulations. Linearity in plasticiser alcohols is less important, but premium performance is associated with straight-chain products.

These alcohols are produced synthetically by hydroformylation of olefins (cf. 8.4); equally effective are some natural materials, e.g. coconut or tallow alcohols. As we have already noted, the main problem in all the processes leading to higher olefins is that the primary product distribution and the market demands for the end products do not match. They all produce a wide spread of molecular weights, from C_4 to C_{20+}, whereas the market demands a much narrower range, largely restricted to the C_8–C_{16} range. Making a better match has been a preoccupation of research workers for many years. Of the various options, the pathway via ethylene oligomerisation can give premium quality α-olefins in a product distribution that can most closely match market demand. Modification of the primary product distribution of this reaction to give a final distribution closer to the demands of the market can be effected in a number of novel ways. These are considered in the sections that follow.

9.2.1 *Primary reaction steps*

Two main reactions form the basis of all commercialised ethylene oligomerisation processes. These are:

(a) The *growth reaction*, in which ethylene inserts into a metal–alkyl bond extending the chain by two carbon atoms,

$$L_xM{-}R + C_2H_4 \rightarrow L_xMCH_2CH_2R \qquad (1)$$

This reaction is dominant at low temperatures.

(b) The *transalkylation* or *chain transfer reaction* in which the alkyl group attached to the metal centre is converted into an α-olefin,

$$L_xMCH_2CH_2R + CH_2{=}CHR' \rightarrow L_xMCH_2CH_2R' + CH_2{=}CHR \qquad (2)$$

A lower olefin, e.g. ethylene, displaces a higher alkyl chain with formation of a metal ethyl and the higher α-olefin. This reaction assumes greater importance with increasing temperature. Side reactions also occur, but these can be kept to an acceptable level by a careful control of reaction conditions. These are dealt with later.

9.2.2 *Stoichiometric alkyl growth*

At low temperature (90–120°C with aluminium alkyls) reaction (1) is very much faster than reaction (2). Ethylene is consumed, lengthening ethyl groups to butyl, hexyl, octyl etc., but no α-olefin is produced. A distribution of metal alkyls of varying chain lengths is produced, the average length of which depends on the number of ethylene molecules added per metal centre. Since rates of insertion are roughly independent of chain length the distribution is of the Poisson type, i.e. the mole fraction (X_p) of $M(CH_2CH_2)_pCH_2CH_3$ is given by $X_p = n^p e^{-n}/p!$ (where n is the average number of molecules added to the initial aluminium ethyl). The maximum in this distribution increases with n.

In order to obtain α-olefins from the metal alkyls it is necessary to effect reaction (2). The most obvious choice is to transalkylate with ethylene, regenerating fresh triethylaluminium. Separation problems can arise, however, as for example in the separation of $AlEt_3$ (b.p. 207°C) and 1-dodecene (b.p. 213°C). Such a process has not been commercialised. Oxidation of the aluminium alkyls to alkoxides and their hydrolysis overcomes these difficulties. This is the basis of the Alfol Process[1] which produces linear primary alcohols in a Poisson-type distribution (figure 9.1a). The aluminium alkyl is destroyed in the conversion to alcohols and is not available for recycle, but the higher value of the alcohol products compensates for this loss.

9.2.3. *Catalytic ethylene oligomerisation*

If metal alkyls contact excess ethylene at higher temperatures (180°C for aluminium alkyls) then both reactions (1) and (2) occur at comparable rates. The result is the *catalytic* production of α-olefins whose distribution depends on the relative reaction rates k_1 and k_2. This distribution is of the geometric type and the mole fraction (X_p) of $CH_2{=}CH(CH_2CH_2)_pH$ is generally expressed as:

$$X_p = \frac{\beta}{(1+\beta)^n} \qquad (\text{where } \beta = k_2/k_1)$$

The mole fraction decreases with increasing molecular weight for any value of β (see 8.6). Neglecting side reactions, it is independent of conversion and its shape depends mainly on temperature since this determines the magnitude of the parameter β.

The catalytic oligomerisation reaction is the source of α-olefins for all the commercialised processes which use either aluminium, titanium, or nickel-based catalysts. Different values of β are used in each of these, reflecting either the desire for maximisation of a particular molecular fraction in the product or the different methods of upgrading the less valuable portions of the product distribution.

The Gulf Process[2] is the simplest: the olefins are separated and used without further modification. Values of β are chosen therefore to maximise the total value of the whole product distribution. A typical value of $\beta = 0.44$ gives the distribution in figure 9.1b which shows that significant fractions of the less desirable butenes and C_{18+}-olefins are produced.

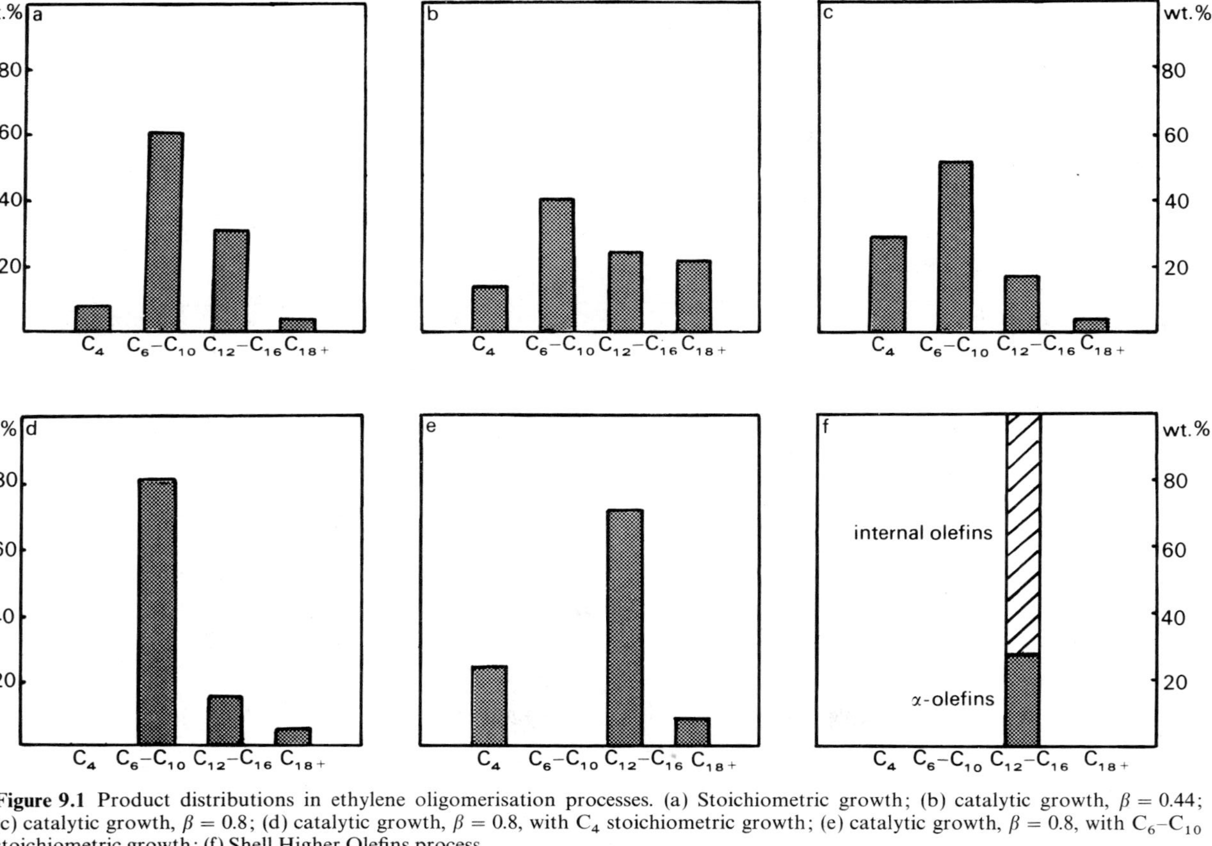

Figure 9.1 Product distributions in ethylene oligomerisation processes. (a) Stoichiometric growth; (b) catalytic growth, $\beta = 0.44$; (c) catalytic growth, $\beta = 0.8$; (d) catalytic growth, $\beta = 0.8$, with C_4 stoichiometric growth; (e) catalytic growth, $\beta = 0.8$, with C_6–C_{10} stoichiometric growth; (f) Shell Higher Olefins process.

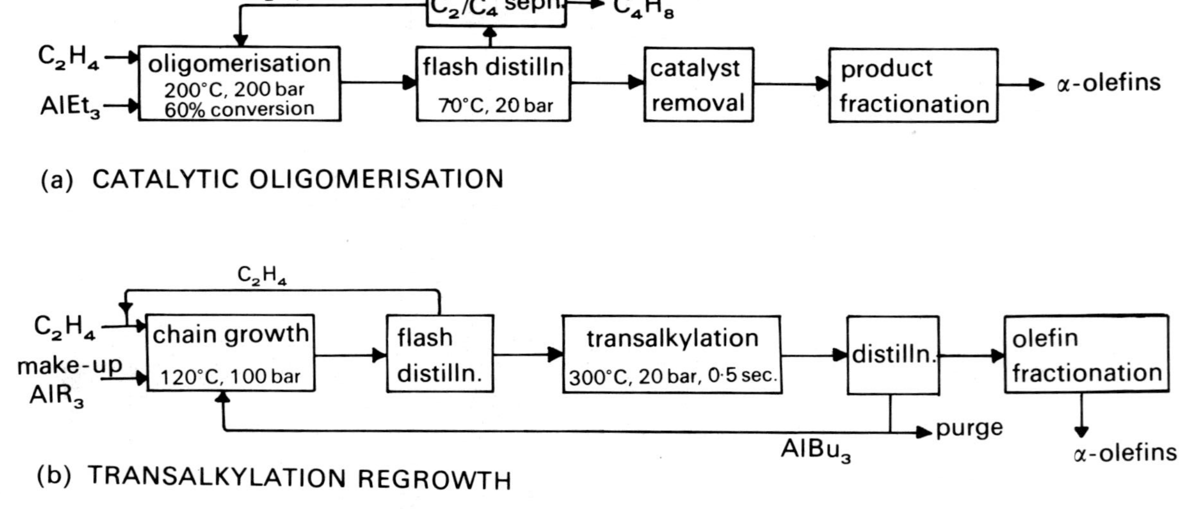

Figure 9.2 (a) Catalytic oligomerisation; (b) transalkylation regrowth. Reaction conditions are best estimates from patent examples.

Ethyl Corporation and Shell have shown that this distribution can be modified by secondary processing of the less desirable fractions (9.2.4 and 9.2.6).[3]

9.2.4 *Modifying the molecular weight distribution* (*Ethyl Corporation Process*)

Transalkylation of a metal alkyl with ethylene to produce an α-olefin and a metal ethyl occurs during the catalytic ethylene oligomerisation steps. It is also possible to carry out this reaction with higher α-olefins in place of ethylene with the net result of alkyl-olefin exchange (reaction (1), $R' =$ ethyl, butyl, hexyl, etc.). This reaction can be carried out in a separate stage in one of the modifications described by Ethyl Corporation as a means of increasing the chain length of the lower, less desirable olefins.[4]

Recycling aluminium alkyls are mixed with butene at high temperatures and short contact times (e.g. 300°C, 0.5 sec) in order to minimise side reactions (see below). Tributyl aluminium is separated from the excess butene and product α-olefins and then undergoes stoichiometric growth with ethylene in a separate reactor. Growth should be restricted to approximately one ethylene per alkyl chain to minimise formation of C_{14+}-olefins which are difficult to separate from the tributylaluminium after transalkylation. The "grown" aluminium alkyls are stripped of ethylene and recycled to the transalkylation reactor. The product α-olefins have a Poisson distribution peaking at C_6–C_{10} when pure butene is used.

A combination of catalytic growth ($\beta = 0.8$ shown in figure 9.1c), followed by transalkylation and stoichiometric growth of the butene fraction, gives a typical final product distribution as shown in figure 9.1d. The majority of the product is in the plasticiser olefin (C_6–C_{10}) range. For maximisation of the detergent range (C_{12}–C_{16}) product, the C_6–C_{10} fraction must be "grown". The result of the operation of this option is shown in figure 9.1e. It will be noted that only the C_6–C_{10} fraction has been "grown" and a substantial fraction of butene remains. This butene may be "grown" as well but limitations imposed by the closeness of boiling points of tributylaluminium and C_{14+} olefins necessitate separate growth sections for butene and C_6–C_{10} olefins.

9.2.5 *Side reactions*

The main side reaction to be avoided is a variant of reaction (1) in which α-olefins insert into the metal-alkyl bond (3):

$$L_xM{-}R + CH_2{=}CHR' \rightarrow L_xM{-}CH_2CHRR' \qquad (3)$$

β-Branched aluminium alkyls are produced which readily undergo transalkylation with ethylene to form 1,1-disubstituted olefins, $CH_2{=}CRR'$. This side reaction can occur during catalytic growth and during transalkylation with butene and higher olefins in the regrowth section. It can be minimised by a careful choice of reaction conditions. In the catalytic growth section this is achieved by:

(a) maintaining excess ethylene throughout; ethylene conversions are therefore limited to ca. 60%;
(b) avoidance of backmixing by the use of long tubular reactors (these have the added benefit of ensuring isothermal operation);
(c) deactivation of the remaining aluminium alkyl catalyst via hydrolysis before distillation of the product olefins. This also minimises olefin isomerisation, another undesired but less important side reaction, but leads to contamination of the product with small quantities of alkanes.

In the transalkylation/regrowth section the side reaction is minimised by operating the transalkylation reaction at high temperatures and short times and by using a low temperature work-up for the separation of the α-olefins from the aluminium alkyls in the transalkylation reactor product stream.

9.2.6 *Modifying the molecular weight distribution* (*Shell Higher Olefins Process*)

The Shell (SHOP) Process, the most modern of the ethylene oligomerisation processes, has a number of novel features.[5] Firstly, they have identified a transition-metal-based catalyst that is considerably more active than the aluminium alkyls employed in the Gulf or Ethyl processes. This catalyst is separated from the α-olefin products by phase separation in a polar solvent and is available for recycle to the catalytic oligomerisation reactor. Secondly, the manner of dealing with the higher and lower fractions is unique. Both the higher and lower α-olefins are isomerised to internal olefins and then undergo combined metathesis. These steps are shown schematically in figure 9.3.

In the catalytic oligomerisation step, three well-mixed phases are rapidly circulated through a long tubular reactor. These phases are: a phosphine-liganded transition-metal catalyst in the polar solvent, product α-olefins, and

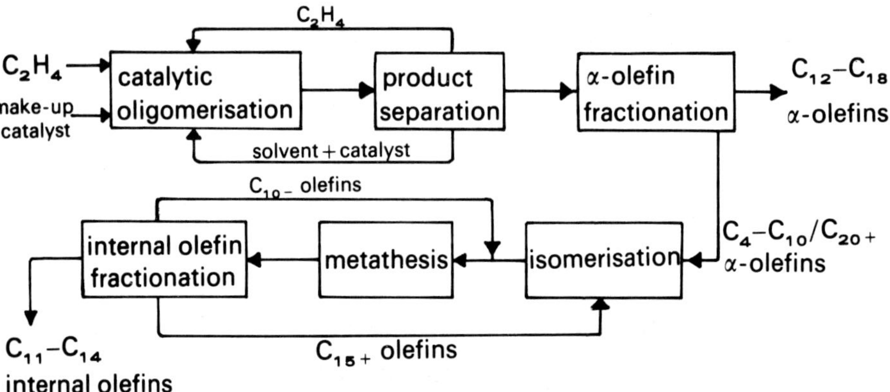

Figure 9.3 The Shell Higher Olefins process.

gaseous ethylene. Ethylene is also dissolved in the product and solvent phases. After reaction, ethylene is stripped, and the catalyst/solvent phase separated from the product phase. By a careful choice of catalyst and solvent almost complete separation can be achieved. A number of catalyst–solvent combinations have been exemplified in patents, and it is believed that phosphines of the type $R_2PCH_2CH_2COO^-$ ($R = C_6H_5$ or cyclo-C_6H_{11}) in combination with 1,4-butanediol as solvent are used, with nickel as the active metal. The chelating P, O functionality of these special phosphines gives excellent activity and ensures solubility in the polar solvent. The high activity of the catalyst keeps metal concentrations low.

Several advantages accrue from this method of separation. The catalyst is available for direct recycle to the oligomerisation reactor, an option not available with aluminium alkyls. It keeps the concentration of α-olefins in the catalyst phase at very low concentrations, minimising branched olefin formation, and means that adventitious recycle of some α-olefins to the oligomerisation reactor is not detrimental to product quality.

The product stream, after clean-up to remove traces of entrained catalyst and solvent, is distilled into the various α-olefin fractions. The C_{10}–C_{18} α-olefins form less than 30% of the total product. The remainder (the C_{10-} and C_{20+} fractions) is further processed via isomerisation and metathesis. In the isomerisation reactor the α-olefins are isomerised to a near-equilibrium distribution of straight-chain olefins. The stream then proceeds to the metathesis step. This reaction, considered in more detail in 9.3, involves exchange of alkylidene units among the various olefins. Thus, for example, a combination of but-2-ene (C_4) and docos-11-ene (C_{22}) will give tridec-2-ene (C_{13}). Mechanistic details of this reaction are discussed in detail in 9.3.3.

All possible metathetical combinations take place, giving a statistical distribution of products. Only 10 to 15% of this is in the desired C_{11}–C_{14} range, necessitating extensive recycle of the lighter and heavier fractions. Some additional features of the metathesis step are:

(a) Optimally, equal weights of light and heavy olefins are fed to this reaction stage. This is advantageous in allowing a greater proportion of detergent range products to be generated in the catalytic growth step. By comparison, the Ethyl Process cannot handle the higher molecular weight olefins and must operate at higher values of β to minimise their production.
(b) Although the double bond distribution in the feed is almost uniform, it is predominantly located towards the end of the chains in the desired C_{11}–C_{14} products. *Only internal olefins are produced.*
(c) Both even *and* odd number olefins are obtained.
(d) Since any non-isomerised olefins form products outside the desired C_{11}–C_{14} range, the isomerisation reaction must be taken as close as possible to equilibrium.

Shell have described an alternative ethenolysis step for dealing with the higher molecular weight fractions.[6] This involves metathesis of the isomerised olefins with ethylene generating α-olefin products. This alternative has not been commercialised. It is believed that the lower olefin purity (90% α-olefin) and lower overall ethylene utilisation are the main problems.

It therefore seems that the SHOP Process solves the basic problem of a mismatch between product distribution in the oligomerisation process and the demands of the market. What incentives are there for improvement? The SHOP Process effects the solution with certain limitations: energy consumption is increased by the extensive recycles required and the premium α-olefins constitute only 30% of the total product. Due to the complex separations required in all the ethylene growth processes, there are economies to be gained in operating at the large scale, e.g. 200 000 te/yr for SHOP.

Table 9.1 Examples of other olefin and diene dimerisation and oligomerisation reactions

Catalyst	*Reaction*	*Comments and applications*
Aluminium alkyl	$C_3H_6 \rightarrow$ 2-Me-pent-1-ene	Isoprene precursor
Transition metal (Ziegler system)		
M=Ni[a,b]	$C_3H_6 + C_4H_8 \rightarrow$ branched C_6–C_8 olefins	hydroformylation feedstocks for plasticiser alcohols
M=Co, Rh, or Ni[a,c]	$C_2H_4 + C_4H_6 \rightarrow$ hexa-1,4-diene	termonomer for ethylene/propylene elastomers
M=Ni or Ti[a,d]	$C_4H_6 \rightarrow$ cyclo-dodecatriene	precursor to nylon-12
Acids		
SiO_2-supported H_3PO_4	$C_3H_6 \rightarrow$ branched C_9 and C_{12} olefins	gasoline octane improver; feedstock for alkyl benzene sulphonates[e]
Supported H_3PO_4 or ion-exchange resin	*iso*-$C_4H_8 \rightarrow$ di-isobutene	gasoline octane improver; hydroformylation feedstock for plasticiser alcohols
Alkali metals		
Na/K_2CO_3	$C_3H_6 \rightarrow$ 4-Me-pent-1-ene	Monomer for speciality polymer

[a] P. W. Jolly and G. Wilke, *The Organic Chemistry of Nickel*, Academic Press, Vol. 2, 1975.
[b] B. Bogdanovic, *Adv. Organometallic Chem.*, 1979, **17**, 105; Y. Chauvin *et al.*, *Chem. Ind.*, 1974, 335.
[c] A. C. L. Su, *Adv. Organometallic Chem.*, 1979, **17**, 269.
[d] P. Heimbach, *Aspects Homog. Catalysis*, 1974, **2**, 79; G. P. Chuisoli, *ibid*, 1970, **1**, 77; P. Heimbach, P. W. Jolly and G. Wilke, *Adv. Organometallic Chem.*, 1970, **8**, 29; J. Tsuji, *ibid.*, 1979, **17**, 141.
[e] Branched structure of alkyl chain gives poor biodegradability. Application now largely displayed by products based on linear olefins.

Future directions could take several courses. Better control of the primary production distribution could be effected, e.g. by using a shape-selective catalyst that cuts the distribution above a certain carbon number fraction. A cheaper feedstock could be substituted, e.g. oligomerisation of butenes or butadiene. Wholesale changes could be made in the routes leading to higher alcohols (the major end use of α-olefins at present), e.g. selective partial oxidation of linear alkanes. None of these various options will however be easy to achieve, and major new catalyst discoveries will be needed before commercialisation.

9.3 Olefin metathesis

First discovered by workers at Phillips Petroleum as recently as the late 1950's,[6] it now seems remarkable that the general class of metal-catalysed reactions known as olefin metathesis had eluded research workers in catalysis for so long. Development of the first commercial process quickly followed the initial discovery; only later did a more detailed understanding of the reaction pathway emerge. This reaction, also variously termed *olefin dismutation* or *disproportionation*, involves an approximately thermoneutral and statistical redistribution of alkylidene entities between olefins (4).

$$2(R^1R^2C{=}CR^3R^4) \rightleftharpoons R^1R^2C{=}CR^1R^2 + R^3R^4C{=}CR^3R^4 \qquad (4)$$

9.3.1 *Scope of the reaction*

Reported applications of metathesis are now numerous, many showing great promise (Table 9.2). Development of several of these has proceeded as far as the pilot plant scale, but in only two cases have they gone into large-scale commercial operation.

Equation (4) is written in a very general form—with one or two exceptions, the reaction applies to a broad range of cyclic and acyclic olefins.

Applications fall under four headings:

(a) *Increase or decrease in chain length.* Metathesis is a useful vehicle for altering the chain length of acyclic olefins. The first commercial application, the now defunct Triolefin process, was adopted by Shawinigan near Montreal to deal with a local surplus of propylene and a deficiency of polymerisation grade ethylene. The second, and more enduring, is in the Shell Higher Olefins process (9.2.6) which uses metathesis to generate C_{12}–C_{16} olefins from higher and lower molecular weight fractions.

An interesting, although as yet unexploited, alternative route to either plasticiser (C_6–C_8) or detergent (C_{12}–C_{16}) range olefins is via metathetical upgrading of propylene and/or butenes. These cheap lower olefins are "grown" by a combination of metathesis and isomerisation in an integrated multistage operation. Overall success of the process depends on the provision of a dual function catalyst, i.e. one effecting both metathesis and isomerisation in the same reactor. MgO is most effective, promoting selective positional isomerisation of the double bond without catalysing unwanted side reactions, e.g. skeletal isomerisation, oligomerisation, or dehydrogenation. Reaction proceeds along the lines of equation (5):

$$\begin{array}{l} CH_3CH{=}CH_2 \rightleftharpoons CH_3CH{=}CHCH_3 + C_2H_4 \\ \qquad\qquad\qquad\quad \downarrow\uparrow \\ \qquad\qquad\qquad\quad CH_3CH_2CH{=}CH_2 \rightleftharpoons CH_3CH_2CH{=}CHCH_2CH_3 + C_2H_4 \\ \qquad\qquad\qquad\quad \downarrow\uparrow \; C_3H_6 \text{ or 2-butene} \\ \qquad\qquad\qquad\quad CH_3CH_2CH{=}CHCH_3 + C_2H_4 (\text{or } C_3H_6) \end{array} \tag{5}$$

(b) *Specific diolefins.* α,ω-dienes, useful comonomers for the formation of cross-linkable polymers and precursors for α,ω-difunctional compounds, can be prepared via metathetical ring opening of cyclo-olefins with ethylene (6a). Cyclopentane (from selective hydrogenation of cyclopentadiene), cyclo-octene (via cyclodimerisation/hydrogenation of butadiene) and cyclododecene (via cyclotrimerisation/hydrogenation of butadiene) are readily available cyclo-olefins. Other carbon number cyclo-olefins are more difficult to prepare. Cyclohexene is unfortunately peculiarly unreactive in this and all other metathesis reactions.

$$(CH_2)_n\text{-ring with } CH{=}CH \xrightarrow{C_2H_4} CH_2{=}CH(CH_2)_nCH{=}CH_2 \tag{6a}$$

$$(CH_2)_n\text{-ring with } CH{=}CH \longrightarrow \left[(CH_2)_nCH{=}CH \right]_m \tag{6b}$$

(c) *Ring opening polymerisation of cis-cyclo-olefins.* A variety of cyclo-olefins, except cyclohexene, undergo ring opening polymerisation (6b) to give high-molecular polymers. The most significant of these are the polypentenamers which find application as speciality cross-linkable rubbers. Materials of predominantly *cis*- or *trans*-microstructure can be prepared.

(d) *Functionally substituted olefins.* The lack of success in the metathesis of olefins bearing polar substituents, particularly those containing nitrogen and oxygen, has been a serious disappointment. Those reactions that do take place with any facility are the exceptions rather than the rule. In some respects, this lack of reaction is none too surprising: polar molecules are known to be potent catalyst poisons (see below). So far success has been achieved only where the functional group is relatively remote from the double bond or where it has been deactivated, e.g. for amine functions *via* quaternisation.

9.3.2 *Catalyst systems for olefin metathesis*

(a) *Supported oxide catalysts (heterogeneous catalysts).* Catalysts of this type have generated the most interest for commercial application. Oxides, MoO_3, WO_3, or Re_2O_7, or mixed oxides, e.g. $CoO.MoO_3$ supported on silica or alumina, typically 5–10% w/w, have provided the majority of examples. In commercial practice, silica is preferred over alumina, giving catalysts that are active and stable at higher temperatures. Operation at high temperature greatly simplifies periodic regeneration procedures. In use, the catalysts are slowly deactivated by the accumulation of coke, presumably derived from carbonisation of polymeric byproducts. This is removed by controlled combustion. The support must not promote side reactions. To this end, it is necessary to suppress any residual acidity which would otherwise catalyse side reactions such as isomerisation and alkylation. This is achieved via neutralisation with small amounts of alkali metal salts.

There is some ambiguity as to the state of the active catalyst species. For Mo, oxidation states of V and VI have been variously estimated, which are surprisingly high for a reaction occurring at high temperatures under (formally) reducing conditions.

Certain potent catalyst poisons must be scrupulously removed from the olefin feed. Polar species containing N or O are the main poisons, especially water, amines, and (to a somewhat lesser extent) oxygen. It is believed that they preferentially co-ordinate to the catalyst sites, either destroying propagating centres or blocking access of olefin. In addition, some unsaturated hydrocarbons, e.g alkynes and conjugated diolefins, which occur as common contaminants in olefinic streams, strongly inhibit the reaction. They readily form polymer which is converted to "coke".

(b) *Supported metal complexes (heterogeneous catalysts).* Metal carbonyls, allyls, and alkyls all form active catalysts when supported on silica or alumina. On the whole they show no real advantages over those of type (a): in fact, there are positive disadvantages in that many, although activated by stoichiometric quantities of oxygen, are deactivated in air, thus complicating preparation and handling procedures.

(c) *Modified Ziegler–Natta catalysts (homogeneous catalysis).* These refer to catalyst systems composed of a transition metal salt, usually the halide, and a Main Group metal alkyl or hydride cocatalyst. Tungsten, molybdenum and rhenium are the most active transition metals, and compounds such as $EtAlCl_2$, BuLi, $SnEt_4$, and $LiAlH_4$ or $NaBH_4$ are commonly used cocatalysts. There has been some dispute as to the truly homogeneous nature of many of these systems and it is now recognised that there are both homogeneous and heterogeneous active components in many of the solutions. Such a debate is somewhat academic to the catalyst user, however, when both components show similar activity and selectivity. Both halide, which is present in either the transition metal component (e.g. WCl_6), or the cocatalyst (e.g. $EtAlCl_2$), and traces of oxygen are considered essential for the generation of highly active and long-lived catalysts. The extreme sensitivity of the active centres to oxygen impurities once generated makes assignment of the oxidation state difficult, but it is thought that one of the key roles of the cocatalyst is to effect reduction of the metal precursor to give the active catalyst species, e.g. W^{VI} to W^{IV}.

(d) *Cocatalyst-free systems (homogeneous catalysis).* A small number of stable metal–carbene complexes form active catalysts e.g. $(OC)_5W{=}CPh_2$. They do not require the Main Group metal cocatalyst of the Ziegler-type systems above. It was the recognition that the propagating centre closely resembled metal-carbene complexes that opened the way for a full and detailed understanding of the reaction mechanism (see below). The observation of catalytic activity with a preformed carbene complex was therefore key to this understanding.

9.3.3 *Reaction mechanism*

Propagation. Detailed studies have convincingly demonstrated that the catalytic chain is propagated by two different odd-carbon-number species, metal-carbenes (A) and metallocyclobutanes (B) as shown below. It is now clear that the reaction does not proceed via pairwise exchange of alkylidene units at the catalytic centre (C), as had been originally proposed. This relatively simple scheme is able to accommodate most of the subtleties observed experimentally. A detailed treatment of the finer points is well beyond the scope of this book, but fortunately, a thorough coverage is to be found in the many excellent review articles now available.[7]

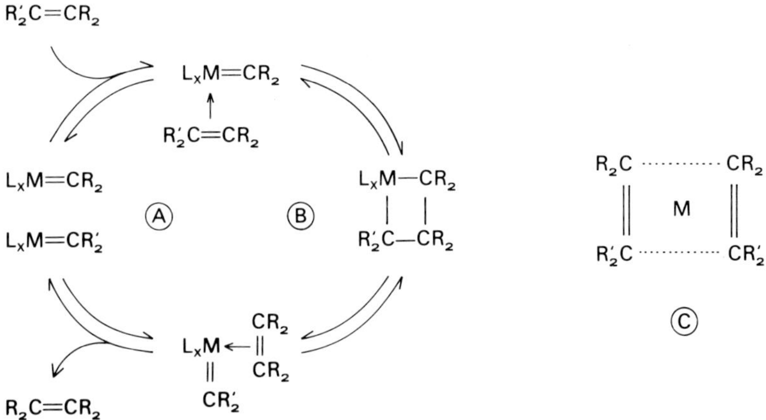

The olefin metathesis reaction is unique in that there are such striking similarities between homogeneous and heterogeneous catalysts. The vast majority of known catalysts, both homogeneous and heterogeneous, are based on molybdenum, tungsten or rhenium in the form of their oxides, halides or carbonyls. Oligomerisation and polymerisation are side reactions with those metals, but with chromium, the other Group VI metal, olefin polymerisation is dominant. These relationships among the Group VI metals, coupled with a number of other observations, have suggested a unifying mechanism for metathesis and polymerisation.[8] The polymerisation pathway is outlined below (R = growing chain).

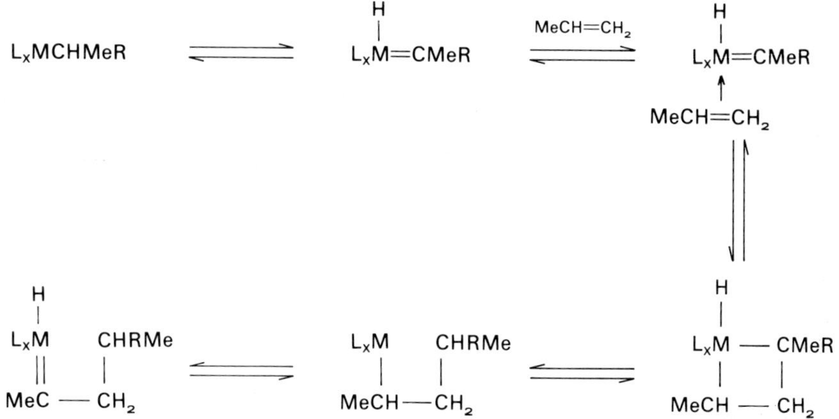

Such a proposal is by no means universally accepted, but it does not discredit the conventional insertion mechanism for metal-catalysed olefin polymerisation (10.7.2).

Initiation of active centres. In the rare examples using preformed carbene complexes an initiation step is not necessary. For the others, the generation of metal-carbenes or metallocyclobutanes requires explanation. Two main routes have been proposed

(a) *α-elimination*

$$LMCl_2 \xrightarrow{MeLi} LMMe_2 \rightarrow \underset{Me}{\overset{H}{LM}}=CH_2 \rightarrow LM=CH_2 + CH_4 \tag{7}$$

$$\underset{OX}{LM}-H \xrightarrow{CH_2=CHR} \underset{OX}{LM}CH_2CH_2R \rightarrow \underset{OX}{\overset{H}{LM}}=CHCH_2R \tag{8}$$

Reaction (7) is appropriate to homogeneous systems containing Main Group cocatalysts and goes via consecutive alkylation, α-elimination, and reductive elimination. The α-elimination step, a key feature also of the scheme above, involves transfer of a hydrogen atom from the α-carbon to the metal with formation of a hydrido carbene complex. At first considered implausible, it has been amply demonstrated in the chemistry of some model niobium and tantalum complexes.[9] In equation (8), appropriate to heterogeneous catalysts, the hydrido species arises via interaction of the metal with surface hydroxyl groups of the support (X = support surface).

(b) *η^3-allyl-metallocyclobutane interconversion*

This process (9) provides an initiation pathway in situations where there is no obvious route via metal alkyls or hydrides.

$$L_xM \xrightarrow{CH_2{=}CHCH_2R} L_xM \leftarrow \begin{matrix} CH_2 \\ \| \\ CH \\ | \\ CH_2R \end{matrix} \longrightarrow \begin{matrix} L_xM \\ | \\ H \end{matrix} - \begin{matrix} CH_2 \\ CH \\ CHR \end{matrix} \longrightarrow L_xM \begin{matrix} \diagup CH_2 \diagdown \\ \diagdown CHR \diagup \end{matrix} CH_2 \qquad (9)$$

Termination and side reactions. Termination refers to all processes that lead to destruction or deactivation of active centres. Poisoning is the major factor in loss of active sites. Polar compounds, e.g. H_2O or O_2, which are activators in small quantities in many cases, strongly poison the centres in larger amounts. Their action may be via destruction of propagating centres (e.g. oxidation of carbenes to carbonyl compounds) or by bonding strongly to the metal centre, preventing co-ordination of the olefin.

Other possible termination reactions are: olefin formation via bimolecular coupling of metal carbenes (for homogeneous catalysts); olefin formation via the inverse of equations (8) and (9); and cyclopropane formation via reductive elimination from the metallocyclobutane intermediate. Side reactions include cyclopropane formation, isomerisation and olefin oligomerisation/polymerisation. Extensive polymerisation will eventually lead to a build-up of a coke-like material and this too will deactivate the catalyst by preventing access of the olefin.

9.4 Xylenes separation and isomerisation

9.4.1 *Separation processes*

The main source of xylenes is a narrow boiling-range fraction obtained from the catalytic reforming of naphtha (cf. 5.2). This consists of the four possible C_8 aromatic hydrocarbons: ethylbenzene (EB) and *o*-, *m*- and *p*-xylene.

Table 9.3 gives details of their physical properties and major end uses. From this we can discern two of the main problems associated with their utilisation:

(a) For uses as fuels or as solvents no further separation is needed, but, when used as intermediates in chemical synthesis, the individual isomers are required in purities often exceeding 99%. The similarity in their physical properties makes separation into individual components difficult.
(b) The most convenient sources of C_8's contain all the isomers in proportions near to thermodynamic equilibrium. Market requirements are quite different, however. The most abundant isomer, *m*-xylene, is the least used, and *p*-xylene, present in only ca. 20%, is the most in demand as a precursor to terephthalic

Table 9.2 Some applications of olefin metathesis (see ref. 6)

Starting material	*Product*	*Catalyst*	*Comments*
C_3H_6	$C_2H_4 + 2{-}C_4H_8$	$WO_3 . SiO_2$	Triolefin Process, commercialised 1966 abandoned 1975
$C_nH_{2n} + C_mH_{2m}$ ($n \leqslant 10$) ($m \geqslant 18$)	C_QH_{2Q} ($Q = 2$–$40+$)	e.g. $MoO_3 . Al_2O_3$, $CoO . MoO_3 . Al_2O_3$	used in Shell Higher Olefins Process for generation of C_{12}–C_{16} olefins from fractions outside this range
$C_8H_{17}CH{=}CH_2 + C_{13}H_{27}CH{=}CH_2$	$C_8H_{17}CH{=}CHC_{13}H_{27}$ (25%)	$(Ph_3P)_2Mo(NO)_2Cl_2/EtAlCl_2$	*cis*-9-tricosene (muscalure) pheromone of housefly
$Me_2C{=}CH_2 + C_3H_6$ (or $2{-}C_4H_8$)	$Me_2C{=}CHMe$	$WO_3 . SiO_2$	isoprene precursor
$Me_3CCH{=}CMe_2{}^* + C_2H_4$	$Me_3CCH{=}CH_2$	$WO_3 . SiO_2/MgO$	perfumery intermediate
cyclo-$C_nH_{2n-2} + C_2H_4$ ($n = 5, 8, 12$)	$CH_2{=}CH(CH_2)_{n-2}CH{=}CH_2$	e.g. $Mo(CO)_6 . Al_2O_3$	specific α,ω-diene synthesis
cyclo-C_nH_{2n-2}	*cis*- or *trans*-linear polymer	e.g. $WF_6/EtAlCl_2$	linear polypentenamer, cross-linkable rubber
1-R-cyclo-octa-1,5-diene	novel polymers	e.g. $WCl_6/EtAlCl_2$	formally, novel alternating copolymers of butadiene and isoprene (R=Me) or chloroprene (R=Cl)
$Me(CH_2)_7CH{=}CH(CH_2)_7COOMe$	$MeOOC(CH_2)_7CH{=}CH{-}(CH_2)_7COOMe$	$WCl_6/SnMe_4$	α,ω-diacid synthesis; functionalised olefin
$C_nH_{2n} + {}^{14}C_2H_4(C_2D_4)$	$>C{=}{}^{14}CH_2(D_2)$	various	specifically labelled alkene synthesis
$PhCH{=}CHPh + C_2H_4$	$PhCH{=}CH_2$		route to styrene from toluene via oxidative coupling to stilbene

* isobutene dimer

Table 9.3 The C_8 aromatic fraction

Component	*Major end use*	*% in typical reformer feedstock*	*Composition at thermodynamic equilibrium; 600 K (%)*	*B.p. (°C)*	*M.p. (°C)*
Ethylbenzene (EB)	Styrene	20	7	136.2	−95.0
o-xylene	Phthalic anhydride	20	21	144.4	−25.2
m-xylene	Isophthalic acid	42	50	139.1	−40.5
p-xylene	Terephthalic acid (TA)	18	22	138.4	13.3

acid, the key constituent of polyesters. EB, a precursor of styrene, is needed in quantities that exceed its availability in the C_8 fraction. Due to problems associated with its separation (see below), it is mainly produced via reaction of benzene with ethylene over an acid catalyst ($AlCl_3$, or, more preferably, zeolites such as H-ZSM-5) in a separate process.

Various options exist for xylenes separation and utilisation. The closeness of boiling points limits distillation as a method: *o*-xylene, the highest b.p. component, can be separated fairly easily, but EB, the lightest, can be separated only by the use of very efficient fractionating columns (these are costly to construct and also have high energy demands), and *m*- and *p*-xylene cannot be separated by distillation. The highest melting component, *p*-xylene can be isolated via crystallisation, but the relatively small initial quantity of *p*-xylene in the feedstock restricts the extent of extraction to 60–70%. An alternative and highly efficient means of separation is by continuous solid-liquid extraction using zeolites as adsorbents (UOP Parex Process).[11]

It will be appreciated that the selective demand for the individual isomers means that few, if any, plants do a complete separation of all the components. The unwanted material is most effectively utilised by "recycle to extinction". The C_8 fraction, depleted in one or more isomers after separation, is isomerised to a position closer to thermodynamic equilibrium and returned to the separation stage. The majority of plants thus consist of two interdependent sections: separation and isomerisation. The remainder of this section is concerned with the latter and the catalysts used. Firstly, the reactions that can take place over the various catalysts are considered and then the various catalysts themselves.

9.4.2 *Reactions over isomerisation catalysts*

A number of reactions occur over the isomerisation catalysts—some desirable, some less so. The most important are outlined below.

(a) *Isomerisation.* The most important of all the possible reactions over the catalysts is simple interconversion among the isomeric xylenes. This proceeds according to the pattern in eqn. (10):

$$o \rightleftarrows m \rightleftarrows p \tag{10}$$

The *o* to *p* interconversion does not take place. These isomerisations are catalysed by Brönsted acids, the most convenient source being the acidic hydroxyl groups that occur on the surface of silica-aluminas, amorphous or crystalline (section 2.7). Reaction takes place via adsorption of the aromatic on to the acid site, protonation to give a Wheland-type intermediate, and 1,2-migration of a methyl group, e.g.

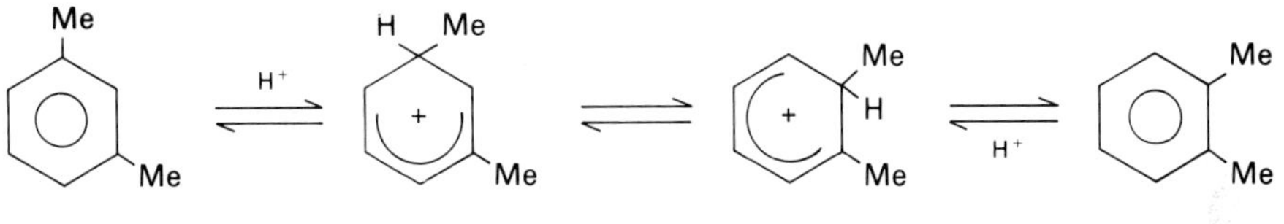

(11)

EB does not enter into this acid-catalysed pathway. It is separately converted to xylenes in the presence of H_2 via a hydrogenation-isomerisation-dehydrogenation sequence, however, using catalysts having the dual functions of hydrogenation/dehydrogenation and Brönsted acidity.[12]

The most important sequence produces *o*-xylene as the primary aromatic product:

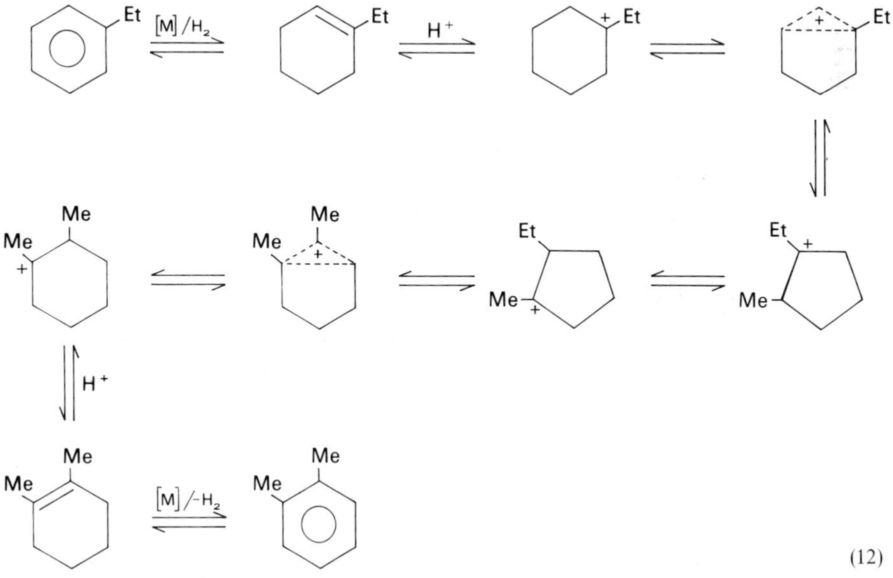

(12)

There are alternative pathways that lead to *m*-xylene but, involving secondary rather than tertiary carbenium ions, their overall contribution is smaller.

(b) *Disproportionation/transalkylation.* Formally, these reactions involve alkyl group transfer between aromatic molecules: e.g. xylenes give toluene and trimethyl benzenes, EB gives benzene and diethylbenzene, and cross combinations also occur. These reactions are acid-catalysed and are thought to proceed as in (13).

$$PhCH_2R \rightleftharpoons Ph\overset{+}{C}HR \xrightleftharpoons{PhCH_2R} PhCHR\text{–}[C_6H_5(H)(CH_2R)]^+ \rightleftharpoons C_6H_6 + {}^+CHR\text{–}C_6H_4\text{–}CH_2R \rightleftharpoons RCH_2\text{–}C_6H_4\text{–}CH_2R \quad (13)$$

With EB, this reaction can take place via an alternative pathway:

$$PhEt \xrightleftharpoons{H^+} C_6H_6 + C_2H_4 \quad (14)$$

$$PhEt + C_2H_4 \xrightleftharpoons{H^+} C_6H_4Et_2 \quad (15)$$

Reactions of this type are well-known; the reverse of reaction (14) is the main preparative route to EB.

(c) *Hydrogenation and hydrogenolysis.* In the presence of H_2 and a hydrogen activation function on the catalyst, alkyl aromatics can undergo either hydrogenation to cyclic aliphatic hydrocarbons or hydrogenolysis of the alkyl group(s). Hydrogenation has already been mentioned in the context of EB isomerisation above. Hydrogenolysis reactions are typified by the conversion of xylenes to toluene and methane and EB to benzene and ethane. With EB, a two-step reaction is possible: i.e. cracking (14), followed by hydrogenation of the product ethylene.

(d) *Coke formation.* Coke formation is thought to proceed via the formation of fused aromatic ring systems which condense further to a highly carbonaceous "coke-like" material. This can either poison the catalytic sites or severely hinder access of molecules to those sites still active. Its formation is minimised by operation in the presence of H_2 (where precursors can be removed via hydrogenation) or in the liquid phase at low temperatures (where the precursors either do not form or are removed by dissolution in the liquid phase).

9.4.3 *Isomerisation catalyst systems*

Before considering the various types of catalyst in detail, it is useful to comment on their diversity. Why are there so many different types? The complexity of possible reaction pathways, coupled with the need to promote only the desirable ones and to suppress those less desired, has meant that the ideal catalyst system has been slow to emerge. Each catalyst has features that are appropriate to the separation process with which it is associated.

In general, there is a need to promote (a) isomerisation among xylene isomers, and (b) pathways that reduce levels of EB. The main pathways to be suppressed are those resulting in loss of xylenes to other products. EB, occurring in all feedstocks, presents a number of problems. It is usually present in levels above those at thermodynamic equilibrium but is difficult to transform to the isomeric xylenes. It is both difficult and costly to separate from other C_8's, and yet it can be prepared readily in high purity by alternative chemistry (e.g. alkylation of benzene with ethylene).

Catalyst systems can be grouped into three general types according to the conditions under which they operate most effectively:

Vapour phase in the absence of H_2 at 300–500°C, ca. 5 bar. Several types of acidic catalysts, e.g. fluorided aluminas, silica-aluminas, and zeolites, may be used to isomerise xylenes. Catalysts based on amorphous silica-alumina were among the first to be used, and some preferred versions were highly selective. However, EB reacts partly by eliminating ethylene and partly by transalkylation with xylenes. Since the latter reaction results in the downgrading of xylenes, this type of catalyst is appropriate when the EB concentration in the feedstock has been reduced to low levels, e.g. by distillation. These catalysts are particularly susceptible to coking and require frequent regeneration.

Much research has been reported on the use of zeolites for xylenes isomerisation, but in general, although they show high activities, their selectivities are poor and they coke-up very rapidly. In recent years, however, certain zeolite frameworks (e.g. ZSM-5) have been discovered which have overcome these difficulties, and Mobil and others claim that they have advantages over amorphous silica-aluminas. In particular, they rapidly and selectively catalyse EB disproportionation, probably through ethylene elimination and alkylation—see (14) and (15). The ethylene tends to alkylate EB rather than xylenes, probably because the transition state for the former is more easily accommodated in the constrained environment of the zeolite pores. For the same reason, the formation of coke precursors is also restricted, resulting in long periods between regeneration.

Vapour phase in the presence of H_2 at 400–500°C, 10–25 bar. Catalysts operating under these conditions have dual functionality; hydrogenation/dehydrogenation provided by a metal component and isomerisation by Brönsted acids. They are produced by incorporation of a metal into catalysts described in the category above. Two main types can be identified.

(a) *Platinum supported on amorphous silica-alumina* has been the most widely used of all of the various types of isomerisation catalysts. Examples are to be found in the Octafining Process (Atlantic Richfield/Engelhard) and the Isomar Process (Universal Oil Products).[14] Both use a catalyst containing ca. 1% Pt prepared by impregnation. The type of catalyst and reaction conditions are very similar to those used in catalytic reforming (cf. 5.2).

The main advantages of this type of catalyst are the long lifetime, and the ability to isomerise EB to xylenes, equation (12). The main drawbacks are that relatively high levels of hydrogen are needed (5–20 mole/mole C_8) which must be recovered and recycled, and most serious of all, hydrogenation to saturated cyclic aliphatic hydrocarbons is an important side reaction. Hydrogenation, essential in the pathway for EB conversion, consumes expensive hydrogen and degrades a portion of the feed-stock. Aromatic by-products are useful in gasoline blending, having a high octane number, whereas cyclic aliphatics have particularly low octane ratings.

(b) *Zeolites*, especially those containing a hydrogen activating metal are very effective under these conditions.[15] Metals of choice are platinum or, especially, the less expensive nickel which are incorporated via ion-exchange or via cocrystallisation during zeolite preparation. Their main advantage lies in their ability to effect high rates of EB conversion while minimising side reactions such as hydrogenation. This permits the use of feedstocks containing high levels of EB without the need for an EB separation column in the system or the drawback of significant xylenes degradation. Side reactions are at such low levels that ratios of xylenes isomerisation to disproportionation of 10^3 are achievable.

The high rates of EB conversion appear to be associated with a change in the reaction pathway. The hydrogenation-isomerisation-dehydrogenation route is insignificant. Hydrogenolysis to benzene and ethane is the main reaction. It is most likely that this proceeds in two stages: firstly via equation (14) with acid-catalysed cracking to benzene and ethylene, and, secondly, hydrogenation of ethylene to ethane by the metal component, making it unavailable for alkylation of xylenes or EB. Zeolites are considerably more active than catalysts of type (a) above (about 10 times) and so advantage may be taken of either a reduction in reactor size, or, if charged to an existing plant, the expensive zeolite may be diluted in an inert matrix, e.g. alumina. Since licensing by Mobil in the mid-1970's, they have been widely adopted and have largely displaced catalysts of type (a) (cf. 4.4.1).

Liquid phase in the absence of H_2. Two quite different catalyst systems have been described:

(a) *HF/BF_3 at 100°C.* This forms part of the Mitsubishi Gas Chem. Co. Process[10] which uses the same HF/BF_3 combination for selective liquid-liquid extraction of *m*-xylene. At temperatures above those used for extraction, the HF/BF_3 effects isomerisation of xylene isomers, equation (10). Mild conditions ensure that other reactions are insignificant. Thus, EB is unaffected and its removal forms an integral part of the whole process which separates EB, *o*-, and *p*-xylene by distillation.
(b) *Zeolites.* Zeolites, e.g. H-ZSM-5, which operate satisfactorily in the gas phase appear to work equally well in the liquid phase.[16] Differences are essentially ones of detail.

9.4.4 *The alternatives; routes to end products not involving C_8 separation processes*

The problematical C_8 separation processes are often wholly directed towards *p*-xylene and hence to terephthalic acid or dimethyl terephthalate. Other isomers are isomerised and recycled to extinction. There must exist, therefore, incentives for the preparation of pure *p*-xylene, terephthalic acid, or related species which are readily convertible to terephthalic acid, that avoid the C_8 separation route. To this end, toluene is the starting material of choice: it is freely available and cheap, and, finding few chemical outlets, is either hydrodealkylated to benzene or blended in gasoline. Four interesting options are outlined in Table 9.4.

9.5 Adiponitrile via hydrocyanation of butadiene

The du Pont nickel-catalysed hydrocyanation[20] of butadiene is included in this chapter for two main reasons. Firstly, it plays a significant part in the preparation of the industrially important nylon 6,6 (cf. 7.8). Secondly, it represents a good example of the benefits to be derived from the use of homogeneous catalysts. Indeed, there are no heterogeneous catalysts that will accomplish these transformations satisfactorily.

Despite successful commercial operation for more than ten years, there are scant details on process conditions and reaction mechanism. However, it is generally considered that the reaction is carried out in three stages ($L{=}(ArO)_3P$):

(i) *Preparation of mononitriles*[21]

$$C_4H_6 + HCN \xrightarrow[NiL_4/L]{100°C} \begin{cases} CH_3CH{=}CHCH_2CN & (70\%) \\ CH_2{=}CHCH(CN)CH_3 & (30\%) \end{cases}$$

(ii) *Isomerisation*[22]

$$CH_2{=}CHCH(CN)CH_3 \underset{NiL_4/ZnCl_2}{\overset{120°C}{\rightleftharpoons}} CH_3CH{=}CHCH_2CN$$

The undesired branched mononitrile is transformed into the linear, 3-pentene-nitrile via equilibration using a similar NiL_4 catalyst system. At this stage

Table 9.4 Terephthalic acid and intermediates from toluene

Route	*Catalyst*	*Notes*
(i) *Disproportionation*[17] 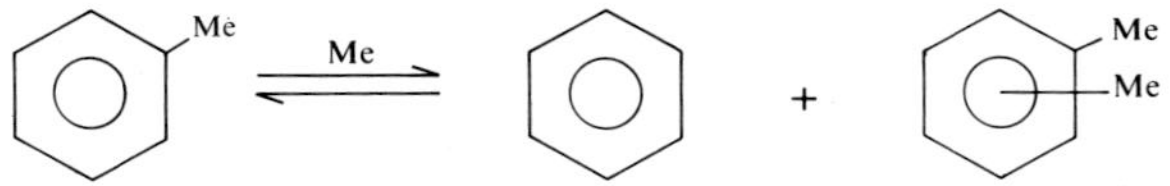 Me ⇌ (Me) + Me, Me	Zeolite in acid form	High selectivity to *p*-xylene over modified H-ZSM-5; these are examples of product-selective reactions: i.e. diffusitivity of *o*- and *m*- 10^{-3} to 10^{-4} that of *p*-isomer.
(ii) *Methylation*[17] 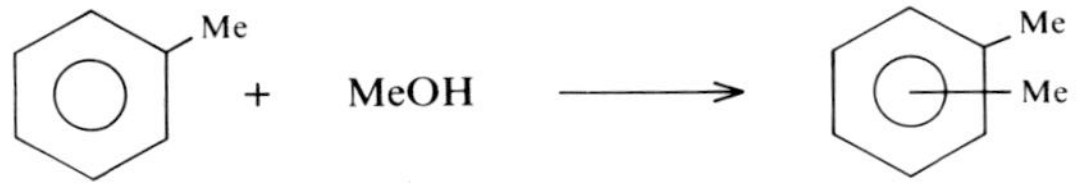 Me + MeOH ⟶ Me, Me	Zeolite in acid form	Unselective disproportionation used commercially to convert toluene to benzene and mixed xylenes
(iii) *Carbonylation*[18] 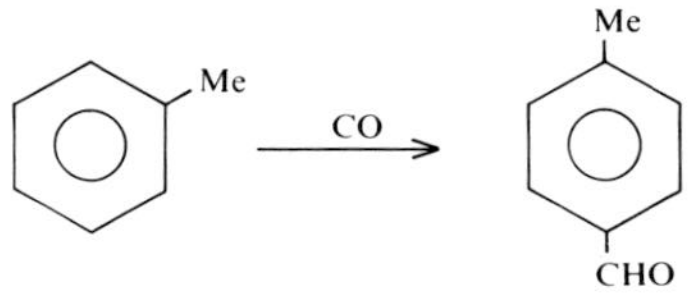 Me —CO⟶ Me, CHO	HF/BF_3	*o*-aldehyde by-product separated by distillation
(iv) *Henkel Process*[19] 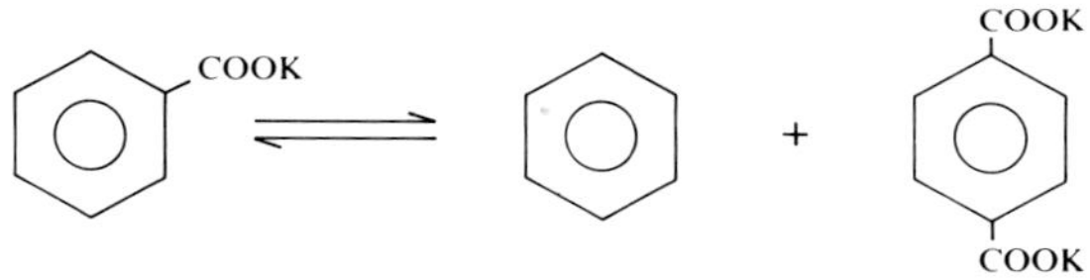 COOK ⇌ + COOK, COOK	Cd^{2+} or Zn^{2+} in melt	Extreme insolubility of terephthalate gives high selectivity; process commercialised, but no longer operated. Benzoic acid produced via selective oxidation of toluene

formation of other isomeric products must be avoided: the most stable conjugated nitrile $CH_3CH_2CH{=}CHCN$, is reported to be a catalyst inhibitor in stage (iii).

(iii) *Preparation of dinitriles*[23]

$$CH_3CH{=}CHCH_2CN + HCN \xrightarrow[\substack{ZnCl_2 \\ 80°C}]{NiL_4/L} \begin{cases} NC(CH_2)_4CN & (83\%) \\ CH_3CH(CN)CH_2CH_2CN & (15\%) \\ CH_3CH_2CH(CN)CH_2CN & (2\%) \end{cases}$$

Selectivity at this stage is vital. The major byproducts, 2-methyl-glutaronitrile and ethylsuccinonitrile, must be rigorously separated from the desired adiponitrile and, finding no satisfactory commercial outlets, must be disposed of.

A reasonable mechanistic scheme can be proposed, as follows:

Stage (i)

$$NiL_4 + HCN \rightleftharpoons HNiL_4(CN) \rightleftharpoons HNiL_3(CN) + L$$

$$HNiL_3(CN) + C_4H_6 \rightleftharpoons (\eta^3\text{-}CH_2CHCH{-}CH_3)NiL_2(CN) + L$$

$$(\eta^3\text{-}CH_2CHCH{-}CH_3)NiL_2(CN) \begin{cases} \overset{a}{\rightleftharpoons} CH_3CH{=}CHCH_2CN \\ \overset{b}{\rightleftharpoons} CH_2{=}CHCH(CN)CH_3 \end{cases}$$

(An alternative formulation of $HNiL_x(CN)$ is $(HNiL_x)^+CN^-$)

Stage (ii). This proceeds via the operation of the last step in stage (i) above.

Stage (iii)

$$CH_3CH{=}CHCH_2CN \xrightleftharpoons{HNiL_3(CN)} CH_3CH{=}CHCH_2CN \cdot HNiL_3(CN) \rightleftharpoons \begin{cases} CH_3CH_2CH(NiL_3(CN))CH_2CN \;\text{(A)} \rightarrow CH_3CH_2CH(CN)CH_2CN \\ CH_3CH(NiL_3(CN))CH_2CH_2CN \;\text{(B)} \rightarrow CH_3CH(CN)CH_2CH_2CN \end{cases}$$

$$\text{(B)} \rightleftharpoons CH_2{=}CHCH_2CH_2CN \cdot HNiL_3(CN) \rightleftharpoons CH_2(CH_2)_3CN\text{–}NiL_3(CN) \rightarrow NC(CH_2)_4CN$$

This sequence of events proceeds via formation of olefin-nickel π-complexes, insertion into the Ni—H bond to give a nickel alkyl, and reductive elimination to give the dinitrile products. Further $HNiL_3(CN)$ is generated by reaction of the product (NiL_3) with HCN. The high selectivity to adiponitrile is presumably associated with (i) the greater stability of the linear C over the branched A and B intermediates which are in equilibrium if (ii) reductive elimination to give the dinitrile products is rate-determining. The role of the ligand is crucial. Among phosphorus compounds only phosphites produce active catalysts (electronic effect) and among phosphites only the relatively bulky triarylphosphites give high selectivity in the final stage (steric effect in favouring the less crowded intermediate C). Clearly, catalysts that give higher selectivities to the desired products in stage (i) and, particularly, stage (iii) are very desirable. To this end, copper catalysts have been developed that can give 90% selectivity to 3-pentene-nitrile in stage (i).[24]

REFERENCES

1. P. A. Lobo, D. C. Coldiron, L. N. Vernon and A. T. Ashton, *Chem. Eng. Progr.*, **58**, 85, (May) 1962.
2. E.g. U.K. 1,020,563.
3. See also K. Ziegler, *Angew. Chem.*, **72**, 829, 1960.
4. E.g. U.K. 990, 748; 1,019,411; U.S. 3,384,651; 3,487,097.
5. (a) Patents relating to overall process, e.g. U.S. 3,647,915; 3,726,938; 3,776,975 and U.K. 1,416,317; (b) patents describing oligomerisation catalysts, e.g. U.S. 3,635,937; 3,637,636; 3,644,563; 3,661,803; 3,676,523; 3,686,159; 3,686,351; and 3,737,475.
6. R. L. Banks, *Chemtech*, **9**, 496, 1979; *ACS Preprints, Div. Petrol Chem.*, **24**, 399, 1979.
7. R. J. Haines and G. J. Leigh, *Chem. Soc. Rev.*, **4**, 155, 1975; N. Calderon *et al.*, *Adv. Organometallic Chem.*, **17**, 449, 1979; *Angew. Chem. Int. Ed. Engl.*, **15**, 401, 1976; R. H. Grubbs, *Progr. Inorg. Chem.*, **24**, 1, 1978; T. J. Katz, *Adv. Organometallic Chem.*, **16**, 283, 1977; J. J. Rooney and A. Stewart, *Chem. Soc. Spec. Period Rep. Catalysis*, **1**, 227, 1977.
8. K. J. Ivin, J. J. Rooney, C. D. Stewart, M. L. H. Green and R. Mahtab, *JCS Chem. Comm.*, 604, 1978.
9. R. R. Schrock, *Accounts Chem. Research*, **12**, 98, 1979.
10. J. J. H. Masseling, *Chemtech*, **6**, 714, 1976.
11. D. P. Thornton Jr., *Hydrocarbon Proc.*, 151, Nov., 1970; H. J. Bieser and G. R. Winter, *Oil Gas J.*, **73**, 74, Aug. 11th, 1975.
12. Cf. K. H. Robschlager and E. G. Christoffel, *Ind. Eng. Chem. Prod. Res. Devel.*, **18**, 347, 1979.
13. E.g. U.K. 1,544,691, U.S. 3,856,873 (to Mobil).
14. P. M. Pitts, J. E. Connor and L. N. Leun, *Ind. Eng. Chem.*, **47**, 770, 1955; Petroleum Refinery, **38**, 278, Nov., 1959.
15. E.g. U.S. 4,163,028; 3,856,872 (to Mobil).
16. E.g. U.S. 3,856,871 (to Mobil).
17. N. Y. Chen, W. W. Kaeding and F. G. Dwyer, *J. Amer. Chem. Soc.*, **101**, 6783, 1979; W. W. Kaeding *et al.*, *J. Catalysis*, **67**, 159, 1981.
18. S. Fujiyama and T. Kasahara, *Hydrocarbon Proc.*, 147, Nov., 1978.
19. Cf B. Raecke, *Angew. Chem.*, **70**, 1, 1958; J. Ratusky, *Coll. Czech. Chem. Comm.*, **32**, 2504, 1967; J. Szammer and L. Otvos, *Chem. Ind.*, 38, 1978.
20. For a general review of metal catalysed hydrocyanation see: F. S. Brown in *Organic Synthesis via Metal Carbonyls*, Vol. 2, I. Wender and P. Pino eds., Wiley-Interscience, 1977, p. 655; *Aspects Homog. Catalysis*, **2**, 57, 1974.
21. See e.g. U.K. 1,104,140 (to du Pont).
22. See e.g. U.S. 3,676,481 (to du Pont).
23. See e.g. U.K. 1,377,228; U.K. 1,178,950 (to du Pont).
24. U.K. 1,429,169 (to ICI).

CHAPTER TEN

CARBON–CARBON BOND FORMATION. III: POLYMERISATION

J. P. CANDLIN

10.1 Introduction

The polymer industry consumes well over half the output of the organic chemical industry if we include in this total materials such as stabilisers, plasticisers and solvents (for surface coatings) used to make the finished articles. The majority of polymers are derived from oil; natural materials, e.g. natural rubber and cellulose acetate, make only a minor contribution. Many polymers are now regarded as commodity chemicals, e.g. low and high density polyethylene (LDPE and HDPE). They are however better described as speciality (effect) chemicals, in that there are many grades of individual polymers, each having different properties and made for a particular application. For example, polyethylene grades made for injection or blow moulding, films, wire and cable coating are not interchangeable and these applications depend on properties like the melt viscosity, toughness and brittleness of the polymer. These properties are, in turn, dictated by chemical and physical parameters such as molecular weight and molecular weight distribution and the varying amounts of crystalline or amorphous polymer.

Earlier, polymers were obtained more by chance than designed synthesis. Today, with our increased knowledge of the relationship between polymer structure and properties such as toughness and strength at various temperatures, it is possible to predict the properties of new (co)polymers or blends and to set about their preparation in a planned manner.

This chapter outlines selected aspects of catalysts in the polymer industry, concentrating on those aspects that are particular to polymerisation reactions, with emphasis on olefin polymerisation. We exclude examples in which the formation of polymers is a matter of simple organic chemistry. For example, the same catalyst is used for the formation of poly(ethylene terephthalate) from ethylene glycol and terephthalic acid or ester (namely antimony and germanium oxides) as for the preparation of ethyl benzoate from ethanol and benzoic acid. Examples of Lewis acid and base catalysts which do not involve carbon-carbon bond formation have also been omitted despite their use in the preparation of many polymers, e.g. polyacetal (from formaldehyde) and poly(ethylene oxide).

Table 10.1 Comparison of thermosets and thermoplastics

Thermoset	*Thermoplastic*
Cross-linked 3D structures	Long chain molecules, often without branching
Solvent insoluble	Soluble in some solvents
Amorphous	Crystalline and/or amorphous
Very high MW (cannot be determined)	MW determined by solution techniques
Decompose on heating without melting	Melt on heating (some materials depolymerise)
Rigid brittle structures	Can be flexible, soft or brittle
Will not flow under pressure	Can flow under pressure
Only moulded articles obtained	Can be moulded or fabricated
All thermosets are self-extinguishing	Many highly inflammable
Have good dimensional stability or fail slowly under load at elevated temperatures	Sometimes fail catastrophically at high temperatures
Impossible to remove cross-links	Can be converted into thermoset by post-treatments

10.2 The nature of polymers and polymerisation processes

Of the two categories of polymers—thermoset and thermoplastic (Table 10.1)—this chapter concerns the latter. Thermoplastics comprise about 80% of the total polymer market although thermosets have been produced for much longer. The lower growth rate of thermoset use is due to their limitations in modern

Table 10.2 Comparison of step growth and chain growth polymerisation

Step growth	*Chain growth*
Monomer concentration decreases in a 2nd or higher order manner	Monomer decreases in 1st order manner
Viscosity of medium increases rapidly at end of reaction	Viscosity increases monotonously
Polymerisation can proceed in absence of catalyst	Catalyst or initiator required
Product molecules (oligomers) converted to higher MW product	Product molecules generally do not undergo further reaction
At partial conversion, mixture of monomer and low MW oligomers	At incomplete conversion only monomer and high MW product
Intermediates isolatable at any stage	Intermediates usually not isolatable—low concentration of growing chains
MW increases with time	MW independent of time
MW independent of temperature (at complete conversion)	MW generally decreases with increasing temperature
MW distribution decreases on completion of reaction	MW distribution very dependent on nature and concentration of catalyst
Empirical composition of polymer generally differs from monomer	Empirical composition identical to monomer

production processes. Elastomers, e.g. vulcanised rubber, form an intermediate between these two groups.

Polymerisation reactions form two categories (Table 10.2)—chain growth (formerly addition polymerisation) and step growth (condensation polymerisation). Step growth polymerisations are simple reactions which occur repeatedly: the reaction may be stopped at any stage and the intermediates isolated. An example is the formation of polyesters via condensation of diacids (or their esters) with diols. A catalyst accelerates the reaction, and removal of product water (or alcohol) permits formation of high molecular weight polyesters.

Chain growth is the rapid formation of individual polymers via transient intermediates. Three steps are identified—initiation of the growing chain, propagation and chain transfer, and termination. The remainder of this chapter will be concerned with chain growth polymerisations, but it is first appropriate to consider polymer structures. Homopolymers are formed by the linking together of monomer units (A) in a simple chain: XAAAAA . . . AAAY, where X and Y are end groups which may be derived from the monomer (e.g. vinyl or ethyl groups in polyethylene), the initiator or catalyst, or chain transfer agent. These end groups comprise only a minor part of the polymer molecule and do not affect overall mechanical properties. They may, however, provide reactive sites and can influence properties such as thermal stability and can be used in cross-linking and chain extension reactions.

With symmetrical monomers such as ethylene the chain structure is simple, but with less symmetrical α-olefins (e.g. propylene, vinyl chloride or styrene) and dienes (e.g. butadiene), several possibilities exist. For example, the monomers can join in a head-to-tail or head-to-head arrangement, although the latter is generally rare. In regular head-to-tail arrangements there can be stereochemical configurational differences between the secondary carbon atoms along the polymer backbone (D or L). This is best illustrated for polypropylene for which two stereoregular structures, called *isotactic* (which has configurations, DDD . . . or LLLL . . .) and *syndiotactic* (DLDLDLD . . .), are known. This can be represented as follows.

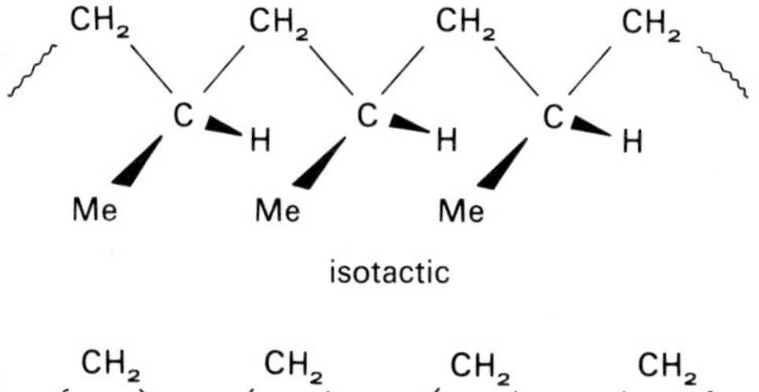

isotactic

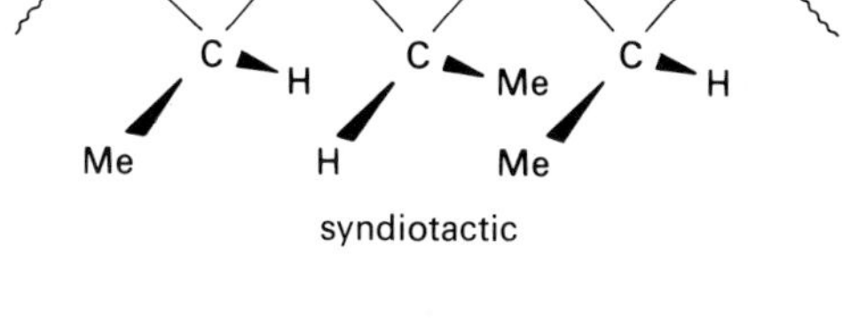

syndiotactic

When the regular stereochemical arrangement is not present and a random configuration of the methyl and hydrogen groups occurs (DDDLDDLLL...) the material is called atactic. Isotactic (and syndiotactic) polypropylene can crystallise forming a stiff polymer with a high softening point (≃150°C) whereas the non-crystalline atactic polypropylene has the properties of a viscous gum with no strength at all.

Copolymers are as important industrially as homopolymers.[1,2] The polymer structure is now more complicated and the arrangement of the different monomer units within the polymer chain does have a considerable influence on polymer properties. Four main types can be identified.

```
—AABAAABBBBAA—        random
—ABABABABAB—          alternating
—AAAAABBBBB—          block
—AAAAAAAAAA—          branched or graft
   B     B
   B     B
   B     B
         B
         B
```

Random copolymers have a number of important uses. Thus, a small amount of a comonomer may be included to act as an internal plasticiser and improve the properties of otherwise brittle homopolymers, e.g. 2–15% vinyl acetate copolymerised with vinyl chloride gives a tough, flexible copolymer. The incorporation of a small amount of comonomer can give reactive sites within the polymer, e.g. 2% isoprene in polyisobutene (butyl rubber) gives free double bonds for cross-linking (vulcanisation) during processing. Random copolymers usually have low crystallinity and are therefore less stiff. Thus, the inclusion of small amounts (5–10%) of α-olefin into polyethylene significantly reduces crystallinity (converting HDPE into LDPE), while copolymers of ethylene and propylene in equal amounts have elastomeric properties.

Block and graft copolymers are valued for their often novel physical properties. The homopolymer blocks within the polymer molecules are generally not miscible and there can exist domains in the polymer matrix rich in one homopolymer phase. This phenomenon can be utilised in styrene–butadiene and propylene–ethylene/propylene polymers which have the properties of thermoplastic elastomers. At high temperatures (>100°C) the material exhibits the melt flow properties of a typical thermoplastic and can be fabricated as such. At room temperature, which is below the glass transition temperature* of polystyrene but above that of polybutadiene, the styrene-rich regions become glass-like and provide solid anchor sites for the still rubbery butadiene-rich regions. The material then has the typical properties of a cross-linked rubber. (Cross-linked natural or synthetic rubber has the properties of a thermoset polymer and cannot be fabricated via thermoplastic methods.)

*The glass transition temperature, T_g, is the temperature below which an amorphous polymer becomes a glassy, brittle solid.

The features usually associated with thermoplastics, e.g. toughness, strength, elasticity, and fibre-forming, are found when sufficient monomer units are joined together to manifest properties such as crystallinity, inter-chain attraction and entanglement. These features are best-developed for molecular weights in the region 5×10^3 to 10^6. Some improvements in physical properties take place with increasing molecular weight, but beyond values of $\simeq 10^6$ the disadvantages outweigh any advantages. At very high molecular weights, melt viscosity becomes so high that extrusion methods for fabrication are not possible and sintering or forging techniques have to be used (e.g. ultra-high molecular weight polyethylene (UHMWPE), and PTFE).

It is known that parameters such as chain flexibility, interchain attraction and regularity of the polymer dictate the polymer properties. These parameters, in turn, affect the glass transition temperature and the ability of the polymer to crystallise. The molecular weight and molecular weight distribution and the size of the crystalline domains (spherulites) also determine the properties and therefore applications of polymers. Further details are available in a number of excellent text books.[3–7]

10.3 Industrial polymerisation techniques

10.3.1 *General considerations*

Catalysts are necessary for the production of useful polymers. Without them either the reaction (e.g. polymerisation of ethylene or propylene) does not take place at all, or an uncontrollable, sometimes explosive polymerisation can occur giving a virtually useless product. Two roles of the catalyst can be discerned. Firstly, it must initiate the reaction, and secondly, it must control the polymerisation process to give the desired end product. This subtle control is nowhere more important than in the polymerisation of propylene (10.8.2).

Polymerisation catalysts do not strictly conform to the generally accepted definition. They are often incorporated into the product molecules and are rarely, if ever, recoverable unchanged at the end of the reaction. Nevertheless, since one mole of a polymerisation catalyst is able to convert 10^6 or more moles of monomer into polymer, the classification is justified. An alternative classification particularly appropriate to free radical reactions (10.5), is that of "initiator".

It is not usually desirable that catalyst residues should be recovered from the product. Separation can be costly and the catalyst components in present use are not of sufficient value to warrant recovery and recycle. However, catalyst residues can adversely affect polymer properties and must be removed (or reduced) when this effect is unacceptable. Strong incentives exist, therefore, for the preparations of catalysts that leave harmless residues or are of sufficient activity to leave residues at an acceptably low level.

Polymerisation processes[8–10] can be operated in continuous or batch modes. The continuous method is preferred since it generally lends itself to smoother operations, more uniform products and lower operating costs. However, in a large number of processes, a simple continuous operation is not practical. In a simple back-mixed reactor the polymer when withdrawn will be composed of material with a range of residence times within the reactor. The disadvantages of this kind of operation are apparent where:

(a) minimisation of catalyst residues is an important consideration—the material of short residence time will contain large amounts of catalyst residues, and the product may be unacceptable if catalyst activity decays with residence time.
(b) catalyst properties deteriorate or change with reaction time. This is particularly important in the polymerisation of propylene. As the reaction proceeds, the catalyst makes increasing amounts of the undesired non-stereoregular (atactic) polymer. Polymer properties therefore also deteriorate with increasing residence time.
(c) the polymers may degrade at reaction temperature. Here it is important that the residence time be as short as possible.
(d) polymer is deposited on the reactor walls and associated equipment and requires regular removal. This is a particular problem in the production of PVC.
(e) frequent changes of grade are required. During the change-over period, the majority of the product will be a mixture of grades and will be unsaleable.

Continuous polymerisations are therefore operated in a manner that more closely approximates plug-flow. This is achieved either by the use of long tubular reactors or a series of reactors in a cascade. Fluidised bed reactors also suffer less from the problems listed above. As product is withdrawn from the bottom of the reactor, it is generally composed of larger, heavier particles.

Polymerisation processes, batch or continuous, can be carried out in two main ways: in bulk, either with liquid or gaseous monomer, or in diluent either as a slurry (suspension or emulsion) or in solution. These are considered in more detail below.

10.3.2 *Bulk polymerisation*

Bulk polymerisation involves reaction between the monomer (either in the liquid or gas phase) and the catalyst in the absence of a diluent or solvent. This procedure is simplest in concept but may not always be the best. There are problems of heat transfer and hence control of reaction temperature is required to prevent runaway reactions which can lead to catalyst degradation and inferior polymer. All polymerisations are exothermic—some typical heats of polymerisation are 66, 104, 112 and 154 kJ/mole for styrene, ethylene, vinyl chloride and tetrafluoroethylene respectively. This corresponds approximately to 500 kWh per tonne of polymer. Heat removal is made difficult by the low thermal conductivities and low heat capacities of the monomers and, in liquid-phase reactions as conversion increases, by the increasing viscosity of the solution (when the polymer is soluble in the monomer) or slurry (when the polymer phase separates). These difficulties have been largely overcome in three different ways—firstly, by the use of specially designed reactors, e.g. long tubular reactors of high surface/volume ratio, secondly by taking the reaction to a low conversion and using the unconverted monomer as a heat transfer medium, and thirdly, by performing an initial polymerisation in one reactor and completing it in another. The first two techniques are used in the high pressure processes for the production of LDPE. The third method has advantages when the final stages of polymerisation are carried out in a mould, e.g. the production of finished articles from poly(methylmethacrylate). With this polymerisation large volume contractions ($\simeq 20\%$) occur on conversion of monomer to polymer. Using the prepolymerisation technique, the majority of the shrinkage takes place in the first stage and that occurring in the mould is thus minimised.

Polymerisation with gaseous monomer using a heterogeneous catalyst is particularly suitable for the production of polyethylene and polypropylene. Figure 10.1 compares liquid-phase and gas-phase processes for the production of LDPE. Processes using either fluidised beds[11] (Union Carbide process for PE), horizontal reactors with weirs and rotating paddles[12] (Amoco process for PE) or helically-stirred reactors[13] (BASF process for PE and/or PP) are coming into use. The procedure involves adding the heterogeneous catalyst, either as a powder or slurry, and the gaseous monomer to the polymer powder. The newly-formed polymer then becomes the carrier for the incoming catalyst. The temperature of the reaction is controlled by operating at low conversion per pass through the reactor and removing heat from the monomer (the heat transfer medium) in an external heat exchanger before recycle to the reactor. In addition, internal cooling coils may be incorporated in the fluidised bed for finer control.

Temperature control is crucial. The polymerisation should be operated at as high a temperature as possible to minimise the costs associated with heat removal at the relatively low reaction temperatures ($<100°C$). It must not, however, be allowed to rise above a critical level, since LDPE and HDPE soften above 90° and 110° respectively and the particles may become tacky and aggregate. With propylene and current catalyst systems, the polymer quality deteriorates at temperatures above 75°C. Hence there are strong incentives for the production of polypropylene catalysts that can be used at significantly higher temperatures.

The main advantage of the gas phase method is that the polymer is easily separated from the monomer, and with modern high activity and stereospecific catalysts no further processing is required for the removal of either catalyst residues or undesirable polymeric product, e.g. low molecular weight or atactic material in polypropylene.

Modern high-speed polymer processing machines require polymer of high packing density and with good flow properties. Traditionally this has been achieved by granulation in which the polymer is melted, combined with stabilisers, and extruded into laces which are chopped into short ($\simeq 5$ mm)

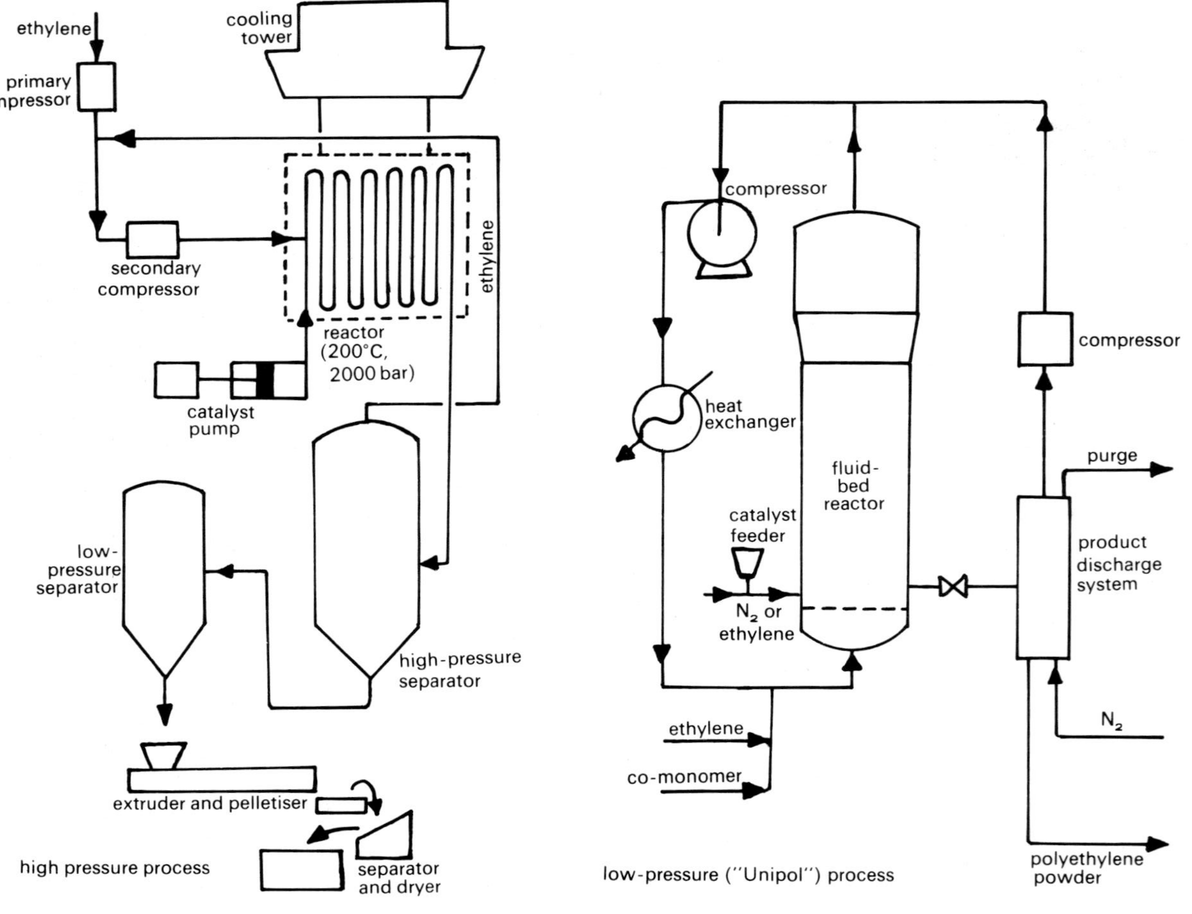

Figure 10.1 Liquid-phase and gas-phase bulk polymerisation processes for LDPE.

lengths. Significant savings can be made if this step can be dispensed with. To do so requires a tight control on the ex-reactor polymer powder. In the production of polyolefins the gas phase method is one in which this can readily be achieved. Requirements are for a narrow distribution of particle sizes with very few fines ($<100\,\mu$) or lumps ($>5\,mm$) and an absence of tackiness. The correct choice of catalyst and catalyst formulations is vital for achieving these goals. With heterogeneous catalysts the catalyst becomes incorporated within the polymer particle and so the shape and size of the polymer particle will be an enlarged version of the catalyst particle ($\simeq$ thirty-fold increase in diameter for highly active catalysts). Thus, a control of the particle size and distribution of the catalyst will govern the properties of the final powder (see section 10.7.1).

10.3.3 *Diluent polymerisations*

The problems encountered in liquid-phase bulk polymerisation could be overcome by carrying out the reaction in an inert diluent. This approach reduces the viscosity of the reaction medium and helps remove heat. However, difficulties do arise, the principal one being that removal and purification of the diluent can be difficult and expensive. Also, many diluents can take part in the reactions by acting as chain transfer agents and can change the nature of the polymer. There is only a restricted range of diluents that are truly inert.

Solution polymerisation. This is not a widely practised method but it is used for the low-pressure production of LDPE (10.8.1). Advantages include the easy removal of heterogeneous catalyst residues in applications where solutions of the polymer are used, e.g. film casting, fibre and wire coating and in adhesives. Because of the fire hazards associated with organic solvents, attention is turning to materials prepared by water-borne suspension or emulsion polymerisation methods.

Suspension polymerisation. This method, widely used for the production of PVC, polystyrene and poly(vinyl acetate), is also referred to as bead, pearl or granular polymerisation on the basis of the shape and size of the final polymer particles. Industrial processes are conducted in aqueous suspensions with the monomer dispersed as droplets of about 0.01 to 0.5 cm diameter. Heat transfer problems are minimised by the high surface-to-volume ratio of the droplets and the high heat capacity and low viscosity of the aqueous medium. Suspension is maintained by agitation, and coalescence of the droplets is prevented by stabilisers, e.g. poly(vinyl alcohol) or finely divided inorganic solids such as kaolin or alumina. Monomer-soluble initiators are used and the process is effectively one of a large number of bulk polymerisations on the micro scale. Some disadvantages are the presence of the suspending agents, which can affect product properties, and also the need for filtration and drying which adds to the capital and process costs. This process is usually operated on a batch scale because of the difficulties of polymer particle shape and size control when using continuous reactors.

Organic diluents, instead of water, have recently been used for "reverse suspension" polymerisations. A commercial process involves the use of a water-soluble monomer, e.g. quaternary ammonium acrylate, and a water-soluble initiator as a suspension in a continuous hydrophobic organic phase. The possibility of a process using monomer plus catalyst droplets or aerosol suspended in an inert gas has been considered, but has not yet been used in a commercial manner.

Emulsion polymerisation. Many of the major industrial polymers are manufactured by emulsion polymerisation techniques. A number of synthetic rubbers and plastics, e.g. polybutadiene and PVC, are made from coagulated or spray-dried emulsions or lattices, and emulsion polymers are used in water-based paints, adhesives and coatings.

Emulsion polymerisation is superficially similar to suspension polymerisation, but the mechanism of polymerisation and the form of the polymer are quite different. It involves the polymerisation of the monomer dispersed in water using a dispersing agent, e.g. an anionic or non-ionic surfactant or soap. These agents stabilise the monomer as micelles (typically 5 nm) and droplets (typically 1500 nm). Water-soluble initiators, e.g. persulphates, are used, which generate free radicals ($S_2O_8{}^{2-} \rightarrow 2SO_4\cdot$) in the aqueous phase. These diffuse through the solution until they encounter a micelle or droplet. Since the number and surface area of the micelles are so much greater than those of the droplets, most of the polymerisation takes place in the micelles. The droplets serve as a reservoir of monomer for the growing polymer chains by diffusion through the aqueous phase.

The rate of initiation and the number of micelles is such that, on average, there are relatively long periods between successive encounters of the radicals by the micelles. This means that termination reactions are suppressed and one important feature of emulsion polymerisations is that very high molecular weight polymers can be obtained at high polymerisation rates (cf. 10.5). Other benefits include the use of water, which is an ideal heat transfer medium; also, the viscosity of the polymer latex is low and is independent of the molecular weight of the polymer. "Sticky" polymers can also be prepared and handled with ease since the polymer in the micelles cannot coagulate. Further, the diffusion path length between the droplets and micelles is small, so it is possible to effect some control in copolymerisation and prepare reasonably "homogeneous" random or block copolymers. The products from emulsion polymerisations are more uniform than those obtained by the previous methods, and have improved properties such as clarity and mechanical strength. One disadvantage is that the polymer lattices contain contaminants, e.g. dispersing agents and initiator residues, which may need to be removed. As with suspension polymerisation this process is carried out batch-wise. An interesting development is non-aqueous emulsion polymerisation in which the catalyst and monomer are soluble in the diluent but the polymer is not, e.g. combinations of organic peroxide, methyl methacrylate and hexane. Dispersing agents are used to stabilise the growing polymer particle. Formulations of products prepared this way are used in oil-based paints and coatings.

10.4 Chain growth polymerisations

Chain growth polymerisations[14–17] are conveniently classified according to the nature of the chain propagating species—free radical, ionic (including cationic and anionic) and organometallic. All chain growth reactions involve the common steps of initiation, propagation and termination. Chain propagation includes chain transfer which can have an important influence on the structure and properties of the final polymer. The majority of monomers involved in chain growth polymerisation are olefinically unsaturated, e.g. ethylene, styrene, vinyl chloride, methyl methacrylate and butadiene. Other species that polymerise in this way include the cyclic ethers, e.g. ethylene oxide and tetrahydrofuran.

10.5 Free radical polymerisation

The widest range of monomers can be polymerised by a free radical pathway—roughly half of the total output of the polymer industry is obtained this way. The key reaction steps are summarised in the scheme on page 230.

The initiation step includes a pre-initiation stage where free radicals are generated from a suitable precursor. These add to a monomer to give the growing chain, and the initiating radical becomes part of the final polymer molecule. Its concentration in the product is so small that it rarely affects the mechanical properties of the polymer; these "catalyst residues" have to be minimised, however, since they do affect the stability to heat and light of the polymer.

The most commonly used initiators are organic peroxides, azo-compounds and redox systems, e.g. Fe^{2+}/H_2O_2. The rate of generation of free-radical initiators by heating the peroxide or parent azo-compound has been thoroughly studied, since it is important to know the monomer/initiator concentration ratio—this determines the molecular weight and the extent of branching in the polymer (high ratios give high molecular weight polymer). When peroxides and azo-compounds are used as initiators, those compounds with a bond dissociation energy of 100–160 kJ/mole are most suitable, and are used in a temperature range where the decomposition rate is 10^{-4} to 10^{-6} sec^{-1}. Polymerisation rates are usually dependent on the square root of the initiator concentration. The choice of the initiator depends on the monomer and polymerisation method used. Thus for bulk or suspension polymerisation (polystyrene and PVC), monomer-soluble peroxides are used, whereas for emulsion polymerisation, water-soluble peracids or redox systems are most suitable.

Propagation involves sequential addition of monomer to the radical centre.

(1) Initiation

(a) Initiator formation ($I \rightarrow R\cdot$)

$$\text{e.g. } (PhCOO)_2 \rightarrow 2\,PhCOO\cdot \rightarrow Ph\cdot + CO_2$$

(b) Chain initiation

$$R\cdot + CH_2{=}CHX \rightarrow RCH_2\dot{C}HX$$

(2) Propagation

(a) Chain propagation

$$RCH_2\dot{C}HX \xrightarrow{CH_2=CHX} RCH_2CHXCH_2\dot{C}HX \xrightarrow{n(CH_2=CHX)} R(CH_2CHX)_nCH_2\dot{C}HX$$

(b) Chain transfer

$$\underset{\text{polymer molecule}}{\sim CH_2CHX \sim} + \underset{\text{growing chain}}{R(CH_2CHX)_nCH_2\dot{C}HX} \rightarrow \sim CH_2\dot{C}X \sim + R(CH_2CHX)_nCH_2CH_2X$$

$$\sim CH_2\dot{C}X \sim \xrightarrow{CH_2=CHX} \sim CH_2\underset{\displaystyle\underset{}{CH_2\dot{C}HX}}{\overset{|}{C}X} \sim$$

branched polymer chain

(3) Termination

$$R(CH_2CHX)_nCH_2\dot{C}HX + R(CH_2CHX)_mCH_2\dot{C}HX \xrightarrow{\text{coupling}} R(CH_2CHX)_nCH_2CHXCHXCH_2(CHXCH_2)_mR$$

$$R(CH_2CHX)_nCH_2\dot{C}HX + R(CH_2CHX)_mCH_2\dot{C}HX \xrightarrow{\text{disproportionation}} R(CH_2CHX)_nCH{=}CHX + R(CH_2CHX)_mCH_2CH_2X$$

With unsymmetrical monomers, $CH_2{=}CHX$, propagation proceeds in a head-to-tail manner with few, if any, head-to-head placements. This relates to the relative stabilities of the radicals $RCH_2\dot{C}HX$ and $R\dot{C}HXCH_2$. These exceptions can occur when growing chain ends combine in a termination reaction. When this occurs, the physical and mechanical properties of the polymer are drastically altered since the polymer no longer crystallises in a regular manner.

Termination is a bimolecular process involving coupling or disproportionation of radical centres, the relative contributions depending on the nature of the radicals and reaction temperature. For example, above 60°C polystyrene terminates largely by coupling, whereas poly(methyl methacrylate) terminates entirely by disproportionation.

The molecular weight of the polymer is determined by the relative rates of initiation, propagation and termination. Termination reactions take place with approximately zero activation energy and are therefore diffusion-controlled. They are favoured at high concentrations of growing chains, i.e. at high rates of initiation, and become less significant as the viscosity of the medium increases. Molecular weight can thus be controlled by a simple variation of rates of initiation and monomer/initiator ratio. However, in bulk and solution polymerisations, molecular weight can increase as the reaction proceeds due to a phenomenon known as the Trommsdorf or gel-effect. With increasing conversion the solution becomes more viscous and although this has little effect on the rates of initiation or propagation, it does hinder termination reactions, which are diffusion-controlled. With increasing viscosity, growing chains diffuse less rapidly through the solutions. The polymers produced in this way have a molecular weight distribution broadened to high molecular weights. When this is undesirable it can be controlled by the addition of chain transfer agents (see below).

In some cases mixtures of initiators are used, usually a small concentration of a very active initiator in combination with one less active. This can be used to shorten polymerisation times without significant reduction of molecular weight and is also useful where polymerisations are carried out in two stages, e.g. in the production of poly(methyl methacrylate). In this example, the monomer-swollen polymer is fabricated into the final article, often in a mould, using the more stable, less active, radical generator as initiator. Mixtures of organic peroxide and water-soluble initiators can be used to prepare graft copolymers in emulsion polymerisation techniques. For example, butadiene can be polymerised with the peroxide to give predominantly 1,2-polybutadiene. The pendant vinyl groups, which are more prone to attack with the redox initiator, can form growth centres for polymerisation with a second monomer, e.g. methyl methacrylate.

Free radical initiators do not usually result in stereochemical control with substituted olefinic monomers and any effect is quickly lost on further monomer additions giving an atactic, random, configuration. Most of the commercial polymers prepared at elevated temperatures with free radical catalysts are atactic. When control does occur it is a result of interaction between the incoming monomer and either the growing end of the polymer chain or the penultimate monomer in the chain. This tends to yield an alternating, syndiotactic configuration which is the dominant form at low temperatures.

Chain transfer reactions, in which the propagating centre is transferred from the growing polymer chain to another molecule, occur readily, and have important influences on molecular weight and molecular structure. Chains can be transferred to polymer, to monomer or to other compounds in the reaction medium. In chain transfer to polymer, the growing chain abstracts an atom, usually hydrogen, from another polymer molecule. The new radical centre can then grow and form a branch on the polymer molecule. In chain transfer to monomer the growing centre is transferred to a monomer molecule which, in turn, grows to form a new polymer molecule. This results in a reduction of the molecular weight of the final polymer. Monomers differ in their susceptibility to chain transfer; those that are formed by hydrogen abstraction and give stable radicals are particularly prone to chain transfer. For example, extensive chain transfer to monomer takes place during the polymerisation of vinyl acetate where hydrogen abstraction gives the stabilised radical, $CH_2{=}CHOC(O)\dot{C}H_2 \leftrightarrow CH_2{=}CHOC({=}CH_2)\dot{O}$. Chain transfer with propylene occurs so readily that the formation of high molecular weight polymer by a free radical pathway is not possible—an extreme case.

Intramolecular chain transfer is also possible and has an important influence in determining the structure of LDPE. Butyl branches, a characteristic feature of LDPE, are generated by a chain transfer reaction via a pseudo six-membered ring. Chain branches have a profound influence on polymer properties in that they prevent close association of polymer chains, thus reducing the ability to crystallise and hence lower the density of the polymer.

$$\sim CH(H)-CH_2-CH_2-CH_2-\dot{C}H_2 \longrightarrow \sim \dot{C}H-CH_2-CH_2-CH_2-CH_3 \xrightarrow{C_2H_4} \sim CH(Bu)\cdot CH_2\,\dot{C}H_2$$

Chain transfer is often deliberately induced, both to limit molecular weight and to minimise chain branching. Chain branching cannot be suppressed without at the same time influencing molecular weight. Transfer agents, added in controlled amounts, are molecules with easily abstractable hydrogen atoms, e.g. mercaptans, which give radicals capable of initiating new chains. Molecules that give very stable free radicals incapable of initiating a new chain, e.g. high-hindered phenols, are used as stabilisers and chain transfer then becomes a termination reaction.

10.6 Ionic polymerisations

As a general rule, olefinic monomers containing withdrawing groups are more easily polymerised anionically, while those containing electron donating groups are more easily polymerised cationically. Some groups, such as phenyl, are ambivalent—styrene is thus polymerised by cationic, anionic and also free radical mechanisms. There are differences in the overall reaction pathway depending on whether the propagating centre is an anion or cation.

10.6.1 *Anionic polymerisation*

Although less commonly used industrially than free radical polymerisation and demanding a purer reaction system and a greater degree of control of temperature, anionic polymerisation has the beneficial features of closer control of molecular weight and molecular weight distribution and the minimisation of chain branching. It is the only general method available for the preparation of block copolymers, but is wholly unsuitable for the preparation of random or alternating copolymers.

The initiation step involves the production of an anion by reaction of the monomer with a base, the initiator.

$$A^-M^+ + CH_2{=}CHX \rightarrow [ACH_2CHX]^-M^+$$

The choice of initiator is determined by the relative basicities of the initiating and propagating ions: the initiator must be the stronger base or reaction will not occur. With the monomers of most commercial interest such as styrene and butadiene, strong bases, e.g. alkali metals, or metal alkyls, e.g. BuLi, are required. These initiators are relatively expensive and both they and the propagating centres are very sensitive to destruction by common impurities such as oxygen and water. It is largely for these reasons that anionic polymerisations are not widely utilised. When the basicity of alkali metals is increased by coordination, e.g. lithium with tetramethylethylenediamine, ethylene can be oligomerised to form polyethylene of low molecular weight (MW $\simeq$ 2000).

The propagation step is formally similar to that in free radical polymerisation. The main difference relates to the effect of solvent on reaction rate. Polar solvents such as diethylether, tetrahydro- and dimethylfuran, enhance the rate by assisting ion-pair separation through solvation of the counter cation. With monomers containing electron-withdrawing groups, e.g. α-cyanoacrylates, anionic polymerisations can take place in the presence of very weak bases such as water. These materials are used for adhesives ("super glue") because of the strong polar attraction between the polymer and the substrate to be bonded.

In ultra-pure systems, termination reactions do not take place. Hydrogen transfer reactions between chains, and hydride ion transfer to the counter ion, are very unlikely, and also direct combination of the growing chain is precluded by the charge. For the same reasons, chain transfer to polymers or monomers are rare events. These features set anionic polymerisations apart from free radical and other polymerisation pathways.

A very close control of molecular weight and molecular weight distribution is possible. Initiation takes place only in the early stages of the reaction, in contrast to free radical pathways, and the molecular weight can thus be controlled and calculated from the monomer/initiator ratio. The absence of natural termination reactions gives rise to the phenomenon of "living" polymers. After exhaustion of the original monomer charge, the chain will continue to grow if supplied with fresh monomer even after several days.

The "living" nature of the polymer chains can be used to good effect in the production of block copolymers and terminally functionalised polymers. A–B block copolymers can be prepared by addition of a second, different monomer after consumption of the initial charge. A minor limitation is that the propagating anion of the second monomer must be of similar or lower basicity to that derived from the first monomer. When the basicities are favourable, higher order blocks (A–B–C, etc.) can be made. For example, a range of thermoplastic elastomers (see section 10.2), which are block copolymers of styrene and butadiene, are prepared this way. Terminally functionalised polymers are prepared by chain termination with an added reagent. A wide range of functional groups can be introduced, e.g. —H using H^+, —OH with oxygen, $—CO_2H$ with carbon dioxide, $—CH_2CH_2OH$ with ethylene oxide, etc. Block copolymers can be prepared from these functional groups either by condensation reactions or by using them as centres for initiation of a new polymer chain. Telechelic polymers (α,ω-difunctional polymers) can be obtained by polymerisations initiated by alkali metals or alkali metal complexes, e.g. sodium naphthalene. In the polymerisation of styrene for instance the initiation step involves the formation of an anionic radical which dimerises to give a dianion containing two propagating centres.

$$\mathrm{PhCH{=}CH_2 + Na^+\,Naph^{\bullet -} \rightarrow [PhCH{=}CH_2]^{\bullet -}Na^+ + Naph}$$

$$\downarrow$$

$$\mathrm{Ph\bar{C}HCH_2CH_2\bar{C}HPh} \quad (\mathrm{Na^+}\ \ \mathrm{Na^+})$$

Reaction with further styrene monomer yields a polymer which can be terminally functionalised at both ends. Instead of termination, reaction with a different monomer, e.g. butadiene, yields a triblock copolymer, A–B–A. These triblocks can also be made by chain extension reactions.

$$2 \sim \mathrm{AAAAA^-Li^+ + Br(CH_2)_nBr \rightarrow \sim AAAA{-}(CH_2)_n{-}AAAA} \sim$$

Star-shaped polymers can be prepared by using multi-functional anionic initiators, e.g. four-arm polymer from 1,2,4,5-tetrachloromethylbenzene.

Chain transfer using an added chain transfer agent can be used to limit molecular weight. In the metal alkyl initiated polymerisation of butadiene, significant reduction in molecular weight can be obtained using toluene as the chain transfer agent. The α-hydrogens on the toluene are abstracted by the growing polymer chain and the benzyl anion formed, being of similar basicity, is able to initiate a new chain. In anionic polymerisation, basicity requirements severely restrict the range of chain transfer agents that can be used.

10.6.2 *Cationic polymerisation*

Ionic polymerisations can also proceed with a cation as the growing centre. A wide range of intermediates is possible: carbon-centred ions, carbenium ions (R_3C^+), in the polymerisations of olefins; heteroatom centred ions, e.g. carboxonium ions ($RO^+{=}CR_2$) in the polymerisation of aldehydes, and oxonium ions (R_3O^+), sulphonium ions (R_3S^+) and immonium ions ($R_2N^+{=}CR_2$) in the polymerisation of cyclic oxides, sulphides and imides respectively.

Initiators consist of a Lewis acid catalyst, e.g. BF_3 or $AlCl_3$, in the presence of a cocatalyst, e.g. water or alkyl halides. The active initiator is generated in a pre-equilibrium,

$$BF_3 + H_2O \rightleftharpoons H^+[BF_3OH]^-$$

$$AlCl_3 + t\text{-}BuCl \rightleftharpoons t\text{-}Bu^+[AlCl_4]^-$$

the establishment of which may be slow compared with the initiating and propagating steps. A characteristic of many cationic polymerisations is their rapid rates of propagation. If water is used as cocatalyst, careful control is required, since in excess it acts as a catalyst poison.

With olefinic monomers, cationic polymerisation is relatively restricted. Problems are encountered in the initiation and propagation steps. Monomers that form unstable carbenium ions, e.g. ethylene, cannot be polymerised. The double bond must be the site of highest electron density in the monomer. If the molecule contains electron withdrawing groups, e.g. CN, CO_2R, OCOR, NR_2, reaction will take place at these sites and stable non-propagating ions will be produced, e.g. $CH_2{=}CHN^+R_3$. Large bulky groups, e.g. $CH_2{=}CPh_2$, inhibit polymerisation and only dimers and trimers are formed. Chain transfer reactions occur easily. The growing chain can transfer a proton to the counter ion which in turn can initiate a new chain, e.g.

$$\sim CH_2\overset{+}{C}HXA^- \rightarrow \sim CH{=}CHX + H^+A^-$$

When the growing cation is very unstable and can isomerise by hydride or alkyl transfer, e.g. with propylene, only low molecular weight branched oils are formed.

These restrictions have limited the widespread use of cationic polymerisation to poly(isobutene). Butyl rubber, a copolymer of isobutene with 2–5 % isoprene (to provide sites of unsaturation in the polymer for subsequent cross-linking), is produced using $AlCl_3$ or BF_3 as catalyst. The problem of heat removal is particularly acute in this polymerisation where the reaction proceeds rapidly at temperatures as low as $-100°C$. At higher temperatures the growing chain is very prone to proton elimination, giving low molecular weight products. Heat removal is controlled by carrying out the reaction in a low-boiling solvent, such as liquid ethylene, and removing heat by the latent heat of vaporisation of the solvent.

As with anionic polymerisation, termination reactions do not take place with highly purified systems when carried out at low temperatures. Termination can take place by destruction of associated anion, e.g. $[AlCl_4]^-$ catalysed reactions yield chloro-ended polymers.

Cationic polymerisations show promise for the preparation of block copolymers (though not from olefinic monomers, only cyclic ethers) and also graft copolymers. These latter compounds are made by starting with a preformed polymer (such as PVC or chlorinated EP rubber) which, in the presence of aluminium halides forms initiating centres on the backbone, thereby allowing other monomers (e.g. isobutene or styrene) to form the graft copolymer.

10.7 Coordination polymerisation

10.7.1 *Introduction*

This is the most recently-discovered class of polymerisation, but it has rapidly overtaken both anionic and cationic methods in industrial applications. With it are associated the names of Phillips Petroleum Company, Standard Oil of Indiana, and Ziegler/Natta, and the discovery of high density polyethylene and polypropylene.[32]

In this type of polymerisation, transition metal coordination compounds in the presence of main group metal alkyls are the initiating and propagating centres. It can be classified as a special type of anionic polymerisation in that the charge on the intermediates can be written (polymer$^-$—M^+). The metal centre, M, however, can be modified with coordinating ligands and this can alter both the activity of the catalyst and the nature of the polymer obtained. The unique feature about the catalysts used is that they polymerise ethylene and propylene to HDPE and PP respectively. Other monomers include 1-butene,4-methyl-1-pentene and butadiene, and combinations thereof have all been polymerised on a commercial scale. LDPE can ordinarily be prepared at high pressures and temperatures, but coordination catalysts can bring about such polymerisations under very mild conditions coupled with stereochemical control, e.g. syndiotactic polypropylene can be made at atmospheric pressure at $-70°C$.

All the catalysts used for the industrial production of PE and PP are heterogeneous, and, although highly active homogeneous systems (such as $CP_2TiR_2/AlR'_3/H_2O$) have been developed for laboratory-scale preparation of PE, even these systems may contain colloidal dispersions or clusters of aluminoxanes formed from aluminium alkyls and water. Ethylene/propylene rubbers are made via soluble vanadium catalysts; indeed, it appears that good

EP rubbers (random, non-crystalline, soluble in hydrocarbon) cannot be made using heterogeneous catalysts.

The term "Ziegler systems" is nowadays used very broadly and has been used to describe metathesis catalysts (section 9.3.2). These catalysts usually contain a Group VI transition metal and an aluminium compound cocatalyst. Polymeric materials, polypentamers, have been made on a development scale using cyclopentenes as the starting monomer.

Catalysts that have been used on an industrial scale for the production of polyolefins can be classified as follows:

(a) *Ziegler–Natta catalysts.* Catalysts of this type are produced by a combination of a transition metal complex and a cocatalyst which is a Main Group metal alkyl or hydride. Transition metal halides or alkoxides and aluminium alkyls, e.g. $AlEt_3$, $AlEt_2Cl$ are most commonly used. The transition metal compounds used are almost exclusively those of titanium, vanadium and chromium. As stated above, they can be either heterogeneous or homogeneous for the preparation of PE and PP or EP rubbers respectively.

(b) *Supported catalysts.* These can be of two types—supported Ziegler–Natta and supported metal complex catalysts. The former are a development from heterogeneous Ziegler–Natta but the latter can be considered as a class of catalysts in their own right. They are obtained by the interaction of a discrete metal complex, e.g. metal alkyl or oxo-complex, with a solid metal oxide, e.g. SiO_2 or Al_2O_3, and do not always need a cocatalyst for the generation of an active system.

There have been several thousand papers, patents, reviews and books written on the subject of polyolefin polymerisation, especially dealing with Ziegler–Natta catalysts,[18–21] but in this section the industrial aspects will be emphasised.[22–25] A key feature of heterogeneous catalysts is their mechanical behaviour during the reaction. To work successfully the polymerisation catalyst must undergo extensive fragmentation, in contrast to other types of heterogeneous catalyst, such as cracking catalysts, where mechanical stability is most desirable. This is important when the polymer is insoluble and is deposited as a discrete phase. If fragmentation did not take place the catalyst would be rapidly deactivated as it became encapsulated in a layer of polymer. In the process of fragmentation new catalytically active sites are generated from hitherto inaccessible areas, and from consideration of "active site" concentrations and propagation rates of polymerisation, it has been suggested that fragmentation dictates the activity and mileage (the maximum yield of polymer obtained in a given time) of a catalyst. Many catalysts fragment to the fundamental crystallites, ranging in size from 10–100 nm, which become evenly distributed throughout the polymer particle. This effect, known as "replication", is important in achieving control of the physical properties of the final polymer particle since the shape and size of the individual polymer particles fully reflect the characteristics of the catalyst particle from which it was produced. With the better catalyst systems, the diameter of the final polymer particles is approximately thirty times that of the original catalyst particle.

Catalyst systems for the commercial polymerisation of ethylene and propylene are often divided into "generations" reflecting their degree of development. The first generation catalysts gave relatively low yields of polymer and hence unacceptably high levels of catalyst residues in the polymer, necessitating their deactivation and removal (called de-ashing). Figure 10.2 gives an outline of a first generation polypropylene process. The polymer work-up sections were a major part of the plant, and significant cost reductions were realisable if these steps could be simplified until ultimately only one polymer work-up step, drying to remove entrained monomer (and diluent in diluent-borne processes), was required. The second generation, non-de-ashing catalysts and their associated processes incorporate these improvements, as can be seen by comparing the complexity of figure 10.2 with the simplicity of modern gas phase processes exemplified in figure 10.1. For the development of second generation catalysts, polymer yield was an important, but not sole, aspect of catalyst performance that required improvement. Performance criteria for preferred olefin polymerisation catalysts are detailed in section 10.8.2.

Finally, while the catalysts have proved themselves extremely useful in the polymerisation of ethylene and propylene, it must be stressed that only a relatively restricted range of monomers can be polymerised. Monomers with polar functional groups, e.g. methyl methacrylate or acrylonitrile, interact strongly with the metal centres and cause catalyst deactivation or destruction.

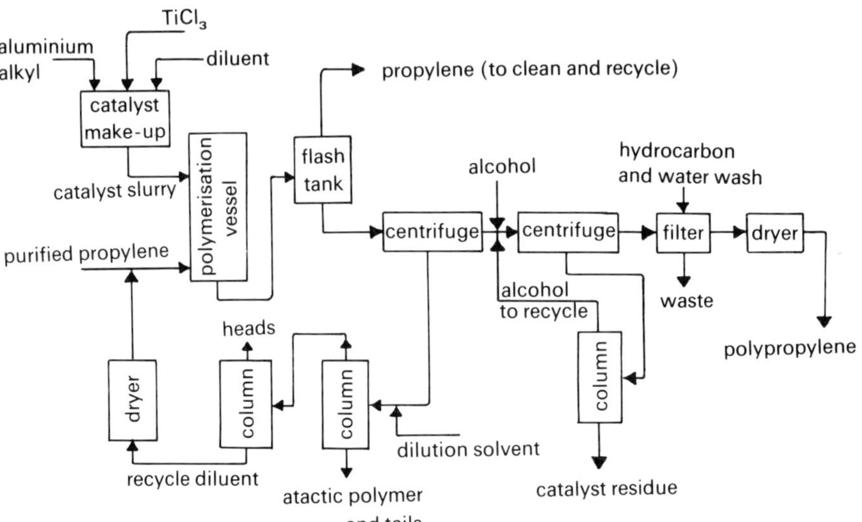

Figure 10.2 First generation continuous process for the Ziegler–Natta polymerisation of propylene.

Those catalyst systems that have been reported to polymerise vinyl chloride have been shown to be sources of free radicals; in other words, the polymerisation follows conventional free radical pathways.

10.7.2 *Initiation and propagation pathways*

Despite the considerable research effort mounted in both academic and industrial laboratories, there are still important uncertainties concerning the nature of the active catalyst centres and the detailed mechanism of initiation and chain propagation.

(a) *Nature of the catalytically active centre.* The growing polymer chain is bound to the metal centre by a covalent transition metal–carbon bond. The initiator is a metal–alkyl or metal–hydride compound which can be generated *in situ* by reaction with the cocatalyst. For example, with $TiCl_3$ a Ti—Et bond is formed via alkyl/halogen exchange with $AlEt_xCl_{3-x}$. The metal polymer bond is prone to cleavage through hydrogen migration (see below) thereby forming a new catalyst centre. The transition metal is not chemically attached to the bulk of the polymer in contrast to other polymerisation initiators which become incorporated as end groups. Thus coordination catalysts are closely related to oligomerisation (cf. 9.2.6), hydrogenation and metathesis catalysts.

In heterogeneous catalysts, relatively few of the metal atoms form active centres. This can be expected for materials such as titanium trichloride where the majority of the titanium atoms will not be exposed at favourable sites on the surface. In the case of solid $TiCl_3$ catalysts the active centre concentration is only 0.05 to 20% of the available surface sites. As a general rule, the activity of polypropylene catalysts is roughly proportional to the number of active sites, although this does not hold true for polyethylene catalysts where factors like diffusion of monomer, breakdown of catalyst during polymerisation, play a role.

(b) *Nature of propagating step.* Chain propagation involves two distinct steps: coordination of the monomer to the metal to give a π-complex and insertion of the complexed olefin into an adjacent metal–carbon bond, possibly via intra-nucleophilic attack on the olefin.

M(R)(□) $\xrightarrow{CH_2=CH_2}$ M(R)(π-CH_2=CH_2) $\longrightarrow$ M($CH_2 \cdot CH_2R$)(□)

□ = vacant coordination site

However, this widely-held view of the propagation step has recently been challenged. Attempts have been made to unify the mechanistic pathways in polymerisation (and oligomerisation) and metathesis via a common metal-carbene or metallocyclobutane intermediate (cf. 9.3.3).

One feature of transition metal catalysts is their ability to form highly isotactic polymers from α-olefins, particularly propylene. This must reflect a very high degree of stereochemical control at each insertion step. Thus, in the production of isotactic polymer the catalyst centre must incorporate each molecule of propylene in a regular head-to tail fashion with the same stereochemical configuration at each secondary carbon atom. The origin of this high degree of stereochemical control is generally believed to be asymmetry in the coordination sphere of the catalyst centre, leading to a favoured complexation of the α-olefin. Interactions from previously added units in the growing polymer chain may also influence this control.

(c) *Role of the cocatalyst.* The cocatalyst, when present, can play a number of roles. The most important is the generation of a metal–carbon bond via alkylation, e.g.

$$L_xMCl + AlR_3 \rightarrow L_xM{-}R + AlR_2Cl$$

The cocatalyst can also act as a reducing agent, as in the reduction of vanadium(V) to vanadium(III) in homogeneous systems and chromate esters to a reduced chromium species in supported catalysts. Stabilisation of the active centre by the formation of halogen or alkyl bridges may also occur. This stabilises the metal centre by completing the coordination sphere of the metal, but allows a ready pathway for the generation of a vacant coordination position via dissociation

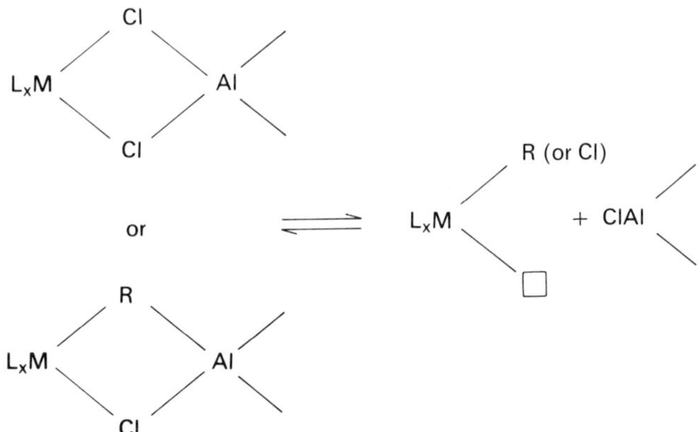

Similar bridged intermediates have been isolated in the case of ethylene polymerisation catalysts based on rare earth complexes and aluminium alkyls.

(d) *Chain transfer reactions.* These reactions control the molecular weight of the polyolefin obtained. They are responsible for the fact that it is possible to obtain polymers of molecular weight greater than 10^6 and yet, with similar catalysts, to make only low molecular weight oligomers (cf. 9.2).

Chain transfer reactions can occur spontaneously either by β-hydrogen elimination to give a metal hydride which can initiate a new chain, or by transfer to monomer.

$$M{-}CH_2CH_2R \longrightarrow M{-}H \;+\; CH_2{=}CHR$$

$$M{-}CH_2CH_2R \xrightarrow{C_2H_4} \begin{array}{c} M{-}CH_2{-}CHR \\ \quad\quad\quad\; | \\ CH_2{=}CH_2 \quad\; H \end{array} \longrightarrow \begin{array}{c} M \\ | \\ CH_2CH_3 \end{array} \;+\; CH_2{=}CHR$$

Chain transfer can also be induced by added reagents. Hydrogen is most commonly used (this is often called "hydrogen modification") since it is easily maintained at a desired level (1–20% depending on catalyst, monomer and temperature) and it can be recycled with the monomer. Other chain transfer agents include zinc alkyls (by alkyl/growing polymer exchange) and cyclopentadiene (for chromium-based catalysts).

The relative proportions of these reactions, which can be determined by end-group analysis (e.g. of vinyl groups by infrared analysis) depends on temperature (high temperatures favour spontaneous chain transfer), monomer (hydrogenolysis occurs more readily with propylene than ethylene) and the metal centre (zirconium-based catalysts are more easily hydrogen modified than titanium-based ones).

10.7.3 *Heterogeneous Ziegler–Natta catalysts*

The early polymerisation catalysts for ethylene were based on titanium tetrachloride and aluminium trialkyls. The polymeric products of this reaction, however, are dependent on many factors such as the rate and sequence of adding the reagents, temperature, the degree of agitation and the purity of the reagents and solvents. These factors were all found to affect the molecular weight and unsaturation present in the polymer, which, coupled with the low activity of the system and poor yields for stereoregular polypropylene, gave considerable scope for improvement (see 10.7.5 and 10.8.2).

10.7.4 *Homogeneous Ziegler–Natta catalysts*

At the present time these catalysts are not used for the homopolymerisation of ethylene and propylene, but they are applied in the production of elastomers based on the copolymerisation of ethylene and propylene and the stereospecific polymerisation of butadiene to give *cis*-1,4-polybutadiene, a material with properties closely matching those of natural rubber.

Elastomers are non-crystalline amorphous materials. For polyolefins, these properties are achieved by random copolymerisation so that homopolymer blocks are not formed. Cross-linking can be achieved using peroxides. Alternatively, elastomers are made by copolymerising a third diene monomer with the ethylene and propylene. The diene monomer contains double bonds of differing reactivity, e.g. 1,4-hexadiene or ethylidene*nor*bornene, with the result that the copolymer contains unsaturation and can be cross-linked (or vulcanised) with similar reagents to those used for natural rubber, such as sulphur.

The homogeneous catalysts yielding these products are based on vanadium compounds, e.g. $VOCl_3$, VCl_4 or $VO(OR)_3$, and aluminium alkyls, $RAlCl_2$. The catalyst lifetime is very short in contrast to heterogeneous catalysts. The active species is thought to be an alkyl vanadium(III) complex stabilised via halogen bridges to aluminium which decomposes to an inactive vanadium(II) state. Continuous regeneration of the catalyst is achieved by the addition of trace amounts of organochlorine "oxidising" agents, e.g. C_2Cl_6, which do not destroy the organo-aluminium component.

10.7.5 *Supported Ziegler–Natta catalysts*

These catalysts have been developed in the pursuit of a more active catalyst system. The unusual feature of the supported Phillips catalyst (10.7.6) is that the support is not inert and plays an intimate role in controlling the activity of the catalyst. The early Ziegler–Natta catalyst based upon titanium trichloride had low surface areas ($< 10\,m^2/g$) and interest was focused on increasing the activity by increasing the surface area. A series of catalysts was therefore made, usually based upon the reaction of titanium tetrachloride with the surface hydroxyls of inorganic oxides (SiO_2, Al_2O_3) followed by reaction with aluminium alkyls. Although improvements were obtained it was not until the use of titanium tetrachloride with magnesium compounds (MgO, $Mg(OH)_2$, $Mg(OR)_2$, $MgCl_2$, magnesium alkyls, etc.) together with aluminium cocatalysts, that polymerisation activities of greater than a thousand times that of the original Ziegler systems were achieved. It is not yet known why magnesium derivatives have this effect; similar calcium derivatives are less effective. One possible explanation may be favourable electron balance in intermediate surface species containing Mg–O–Ti linkages. When solid magnesium chloride supports are used highly isotactic polypropylene can be obtained (10.8.2).

10.7.6 *Supported metal complex catalysts*

Catalysts of this type are formed by interaction of a discrete transition metal complex and an oxide support such as silica, alumina or silica–alumina.

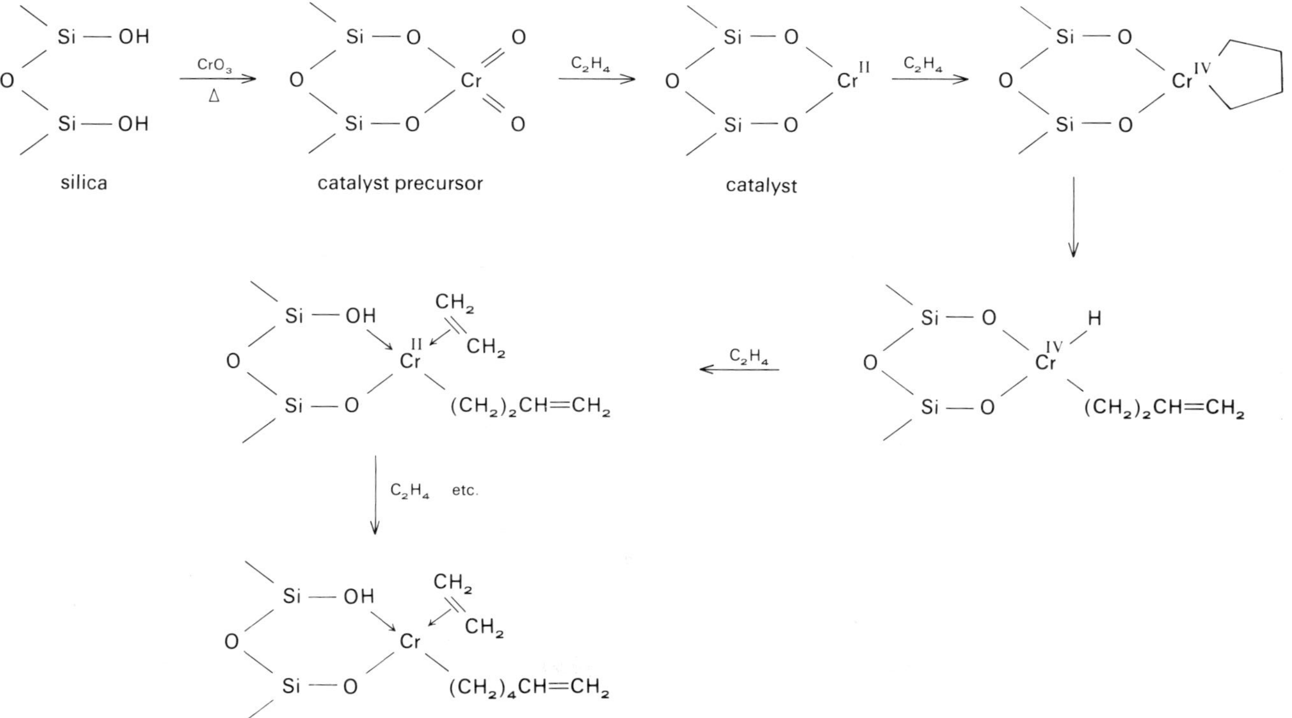

silica
catalyst precursor
catalyst
CrO_3
Δ
C_2H_4
C_2H_4 etc.
$(CH_2)_2CH{=}CH_2$
$(CH_2)_4CH{=}CH_2$

Cocatalysts, such as aluminium alkyls, which are necessary for the generation of active centres for Ziegler systems, are generally not required.

The earliest catalyst for ethylene polymerisation—basically MoO_3 supported on alumina—was discovered in the early 1950's by Standard Oil of Indiana. However, the first commercial process was achieved by Phillips Petroleum Company at approximately the same time as the discovery of the Ziegler-type catalysts. The Phillips catalyst is prepared by impregnation of silica or silica-alumina with an aqueous solution of a chromium salt, e.g. CrO_3, followed by calcination in air at 400°–1000°C. Cooling under nitrogen gives a free-flowing powder containing 0.5–5% chromium. This, the catalyst precursor, becomes active on contact with ethylene, changing in colour from orange to indigo, and in this activated state is very susceptible to poisons, such as oxygen or water (indeed, it has been used as a scavenger for oxygen in feed gas streams). The role of the support is not simply that of inert carrier. It stabilises the chromium centres, possibly as Cr(II) or Cr(IV), via formation of Si–O–Cr linkages (CrO_3 when heated alone above 200°C is converted to the catalytically inactive Cr_2O_3). A possible mechanism for polymerisation is via repetitive insertion into the chromium alkyl or hydride bond (see page 243).

Further chromium-based catalysts have been developed by the Union Carbide Corporation. For example, organochromate esters $(Ph_3SiO)_2CrO_2$ when reacted with a silica surface, can be activated with aluminium alkyls to form ethylene polymerisation catalysts. Other catalysts were developed which did not require cocatalysts, e.g. by the interaction of chromium tris-allyl or chromocene with silica surfaces.

$$\mathrm{O}\begin{matrix}\diagup \mathrm{Si-OH} \\ \diagdown \mathrm{Si-OH}\end{matrix} + (C_5H_5)_2Cr \xrightarrow{-C_5H_6} \mathrm{O}\begin{matrix}\diagup \mathrm{Si-O}\diagdown \\ \diagdown \mathrm{Si-O}\diagup\end{matrix}\overset{IV}{Cr}\begin{matrix}\diagup H \\ \diagdown C_5H_5\end{matrix}$$

Propagation proceeds via insertion of ethylene into the Cr—H bond, not the chromium-cyclopentadienyl bond, which is retained to influence the chromium centre during polymerisation.

In the late 1960's it became clear that if organotitanium species were the true catalysts formed by reaction of titanium compounds and aluminium alkyls, and if these compounds could be prepared and reacted with supports as described above, then active polymerisation catalysts could be obtained. This led to a series of catalysts based on tetrabenzyls, tetraallyls, tetraalkyls (e.g. Me_3SiCH_2) or di-arenemetal(O) complexes of the Group IV metals (Ti, Zr and Hf) in combination with high surface area alumina as support.

With many of these systems the non-supported complex is not active and it has also been found that one particular combination of support and complex is most effective, e.g. SiO_2/Cr, Al_2O_3/Zr or Ti, MgO/Ti. The reasons for this are not understood. Although all of these second generation catalysts are highly active for ethylene polymerisation, they are unsuitable for propylene, giving a polymer of low stereoregularity.

10.8 Selected polymerisation reactions

Three processes which provide an insight into applications of catalysts will be discussed. Both LDPE and polypropylene (PP) are very large tonnage materials and each provides a good illustration of the polymerisation control that can be achieved with catalysts.

A catalyst for LDPE and PP must satisfy the following criteria before it can be considered for use in a commercial process.

(a) It must be of a sufficiently high "mileage", so that catalyst residues need not be removed from the polymer and, for preference, no catalyst deactivation step should be necessary. The ideal catalyst should be active and not decay appreciably with time. Typical acceptable levels of catalyst residues are Ti <50, V or Cr <10, Al <150 and Cl <100 ppm. Ash (from the support) should also represent less than 0.1 %. These values are dictated by UV and thermal stability, electrical insulation, corrosion and toxicity considerations.
(b) For preference, it should be applicable in a diluent-free bulk phase process; either liquid (for PP) or gas phase.
(c) The polymer particles should be in a form that can be used directly in processing equipment and should not require granulation. This is coupled with the need to produce a polymer with a high packing density.
(d) Control of molecular weight and molecular weight distribution (for LDPE) should be achievable simply and reliably, so that a wide range of polymer grades can be produced.
(e) For PP, it must be capable of producing polymer of high stereoregularity and should require no treatment for separation of atactic material. High stereoregularity leads to high crystallinity which, in turn, means that a stiff polymer is obtained.
(f) It should produce either a random ethylene/α-olefin copolymer (for LDPE) or a block copolymer (for PP).
(g) It must be sufficiently robust to withstand reasonable temperature fluctuations (10° to 20°C) within the polymerisation reactor without serious deterioration of polymer properties. It should not physically break down by mechanical means to yield fines, either in catalyst handling or in the polymerisation reactor.
(h) Finally, and perhaps the most important, it must be possible to produce the catalyst reliably on a large scale (typically 1 to 5 tonnes for a plant of capacity 50 000 tonnes/yr), and it should be possible to handle it in a safe and convenient manner. For preference, it should be stable to long term storage.

10.8.1 *Low density polyethylene (LDPE)*

LDPE prepared via free radical-initiated polymerisation of ethylene in a high pressure process has a low degree of crystallinity (approximately 40%), because of the presence of long and short chain branching (typically 20–30 branches per 1000 carbon atoms). HDPE, on the other hand, prepared by the Phillips or Ziegler low-pressure process, is a more crystalline (80–90%) and stiff polymer with a higher softening point, because of the absence of chain branches. LDPE and HDPE do not normally compete for the same applications.

It was soon appreciated that it was possible to mimic the structure of LDPE by making a random copolymer of ethylene containing a small quantity (5–10%) of an α-olefin, e.g. 1-butene or 1-hexene, using a low-pressure process. The economics of this new route, bearing in mind the use of expensive α-olefin as comonomer, had to show distinct advantages over the well-established high-pressure process. The main shortcomings of the high-pressure process are:

(a) the need for expensive heavy-duty equipment designed to withstand pressures up to 2000–3000 bar;
(b) high energy consumption in compressing gases to the high pressures;
(c) the use of dedicated processing equipment for the removal of monomer dissolved in the polymer and for granulation.

The catalyst required for the low-pressure process must satisfy some of the criteria of section 10.8, and also show high efficiency of incorporation of the α-olefin into the polymer chain. Further, since LDPE is used as a packaging material, the inclusion of low molecular weight material in the product must be avoided. Two processes have recently been commercialised using diluent and gas phase techniques. The precise details of the catalysts used are never normally disclosed but possible candidates can be observed from patents. The diluent processes, as described by du Pont[26] and Dow[27] are continuous solution processes (see 10.3.3) carried out at a temperature (>120°C) where the polymer is soluble in the hydrocarbon diluent, using either vanadium complexes and aluminium alkyls, or magnesium-alkyl reduced titanium catalysts, respectively. Union Carbide, on the other hand, claim that the gas-phase fluid bed process, called the Unipol process (figure 10.1) is the most economical (see 10.3.2). Two series of catalysts have been described, their choice depending on the molecular weight distribution of polymer required. Polymers of broad molecular weight distribution are generated by a Phillips-type chromium catalyst[28] whereas polymers of narrow molecular weight distribution are made by physically depositing a magnesium chloride/titanium tetrachloride/THF complex onto silica and activating with aluminium alkyls.[29] Typical operating conditions are 20 bar at 75°–100°C with an average residence time in the reactor of 3–5 hours. Polymers produced in this way differ in detail from the high pressure polymers and have been called "linear LDPE". Advantages of the new processes over the original high-pressure technology are claimed: a 50% reduction in capital and operating costs, while energy consumption is reduced by 75%. It therefore appears likely that no new high-pressure plants will be built.

10.8.2 *Polypropylene*

In the early days of polypropylene manufacture (the 1960's), catalysts which gave adequate yields of reasonably stereoregular (90%) polymer were satisfactory. The polymerisations were carried out in hydrocarbon diluent and the 10% non-stereoregular (atactic) polymer, being soluble in the diluent, was removed by filtration (see figure 10.2). However, second generation catalysts have to fulfil the more stringent needs outlined in 10.8.

Titanium trichloride catalysts, together with aluminium alkyls as activators, have been used extensively for polymerising propylene. Four crystalline modifications of $TiCl_3$ are known. The α-form is obtained by reduction of $TiCl_4$ at 800°C using H_2 or metallic titanium and has a low surface area ($\simeq 1\,m^2/g$). The β-form is prepared by aluminium alkyl reduction of $TiCl_4$ at low temperatures ($\simeq 0$°C). Heating the β-form above 150°C or treating at 60°C with $TiCl_4$ gives the γ-form, whilst prolonged grinding of the α- or γ-forms in a ball mill yields the δ-modification. The α-form, having low surface area is not particularly active and the β-form gives polymer of low stereoregularity.

The majority of commercial catalysts are made by the reduction of $TiCl_4$ with either aluminium metal or aluminium alkyls, giving an approximate composition of $TiCl_3 : 0.33\ AlCl_3$, containing cocrystallised aluminium chloride. Much effort has gone into developing post-treatments of this material to give a highly active and stereoregulating catalyst. Ball-milling with Lewis bases has been found to be particularly successful. Among the wide range of Lewis bases examined, certain phosphines (e.g. PBu_3), ketones (e.g. camphor), esters (e.g. methylmethacrylate and ethyl phenyl acetate) and sulphones (e.g. diphenylsulphone) have been found to be effective. The exact mechanism for the improvement is not known. Certainly the surface area of the catalyst is increased, but whether the improvement is due to facile fragmentation or chemical modification of the titanium centre or aluminium cocatalyst has yet to be decided.

Solvay et Cie have made progress by adopting a slightly different approach.[30] A β-$TiCl_3$, produced by $AlEt_2Cl$ reduction of $TiCl_4$, is treated with a complexing agent, particularly di-isoamyl ether, which selectively removes the cocrystallised $AlCl_3$ as a soluble ether complex. The β-$TiCl_3$, now largely free of aluminium compounds, gives a catalyst which is active but has poor stereoregularity. This can be transformed into an active, highly stereoregular, catalyst by heating with $TiCl_4$ at 60°C. This layer-structured form of α-$TiCl_3$ is highly porous with a high surface area (often $> 150\,m^2/g$), and has a very small crystallite size (ca. 10 nm) containing small amounts of complexed ether. These features presumably aid fragmentation during polymerisation.

For many years it was considered necessary for good polypropylene catalysts to be based upon $TiCl_3$. However, extremely active catalysts have now been developed by Montedison and Mitsui Petrochemical Company[31] based upon titanium complexes supported on solid $MgCl_2$. Although activities for propylene polymerisation are often 100-fold less than ethylene, with these catalysts polypropylene polymers can be obtained which contain as little as 1 ppm Ti. The catalysts are made by milling solid anhydrous $MgCl_2$ with ethyl benzoate (5 : 1 mole ratio), digesting the mixture with $TiCl_4$ then washing with hydrocarbon solvent. The resultant high surface area solid contains $\simeq 1\%$ titanium and ester, the remainder being $MgCl_2$. When treated with a special cocatalyst (e.g. a 3 : 1 mixture of $AlEt_3$ and ethylanisate) a highly active and stereoregular catalyst is obtained. One of the reasons for this may be the layer-structure of the $MgCl_2$ which acts as a template for the titanium centre, although very little mechanistic work has been carried out on these recent catalysts. None of the second generation catalysts can yet meet the criteria of catalyst residues, particularly with respect to corrosive chlorine compounds, and all the polymers require de-ashing. However, with progress being made on several different types of catalysts, these drawbacks will soon be overcome.

10.9 Future trends in polymerisation catalysts

No matter how efficient a catalyst is for a particular polymerisation, the dominant aspects of the polymerisation process will be governed by the nature of the monomer. This includes its chemical, physical and mechanical properties, but also its price. It appears unlikely that new, large-tonnage polymers such as polyethylene, PVC and polystyrene will appear, since most of the cheap monomers are already being used. In the future, however "monomers" such as CO_2, SO_2 and synthesis gas (CO/H_2) could be used, possibly with other conventional monomers. It has recently been found for example that polymethylene can be prepared from synthesis gas using ruthenium catalysts.

Emphasis will undoubtedly be placed on improving the existing range of polymers. Reduction in capital costs and processing costs (energy consumption) will be features of all new processes and this is the area where catalyst research (together with chemical engineering) is of great importance. This is well illustrated by the new gas-phase LDPE and PP processes. Better control of catalysed polymerisation pathways may also lead to improved properties. Thus, most free radical polymerisations yield atactic polymers whereas stereo control is often possible with coordination polymerisation. Although polymer physicists and chemists can usually predict properties like toughness and stiffness, UV and thermal stability are often intimately connected with the microstructure and improvements in these areas may be achievable.

Control of sequencing the monomers in the preparation of copolymers will continue to attract attention. Although physical blends of polymers can be made relatively easily (but not cheaply) the degree of mixing, chain entanglement, size of crystalline domains etc., can be achieved with much more control by copolymerisation. These copolymer materials will certainly become of increasing importance in speciality markets, e.g. controlled drug release in medicine.

In the longer term, both biopolymers and enzyme catalysis of conventional monomers will appear. Micro-organisms are able to synthesise a range of polymers from cheap feedstocks, e.g. plant and agricultural wastes, the properties of many of which have been little studied. The major problems to be tackled include polymer extraction, purification and selection of efficient organisms by selective or hybrid breeding or genetic engineering.

REFERENCES

Further information on most of the topics covered in this chapter can be obtained from M. S. M. Alger, *Classified Index of Reviews in Polymer Science*, published by RAPRA, Shrewsbury SY4 4NR, England. Other useful sources of information are *Polymer Preprints* and *Organic Coatings and Plastics Chemistry* which are preprints of papers presented at National Meetings of the American Chemical Society.

1. D. C. Allport and W. H. Janes, eds, *Block Copolymers*, Applied Science Publishers, Barking, England (1973).
2. R. J. Ceresa, ed., *Block and Graft Copolymerisation* Vols. I and II. Wiley, London (1973, 1976); L. H. Sperling, ed., *Recent Advances in Polymer Blends, Grafts and Blocks*, Plenum Press, New York, 1974.
3. J. A. Brydson, *Plastic Materials*, Butterworths, London (1975).
4. G. Odian, *Principles of Polymerisation*, McGraw-Hill, New York (1970).
5. K. J. Saunders, *Organic Polymer Chemistry*, Chapman and Hall, London (1973).
6. H-G. Elias, *Macromolecules*, Vols. I and II, Plenum Press, New York (1977).
7. H. S. Kaufman and J. J. Falcetta, *Introduction to Polymer Science and Technology*, Wiley-Interscience, New York (1977).
8. C. E. Schildknecht and I. Skeist, eds., *Polymerisation Processes*, Vol. 29 of *High Polymers*, Wiley, New York (1977).
9. J. N. Henderson and T. C. Bouton, eds., *Polymerisation Reactors and Processes*, ACS Symposium Series, No. 104, 1979.
10. T. C. Bouton and D. C. Chappelear, eds., *Continuous Polymerisation Reactors*, AIChE Symposium Series, **72**, No. 160, 1976.
11. U.S. Patents 3,642,749; 3,687,920.
12. U.S. Patents 3,965,083; 3,971,768.
13. U.S. Patent 4,212,847.
14. A. D. Jenkins and A. Ledwith, eds., *Reactivity, Mechanism and Structure in Polymer Chemistry*, Wiley, London (1974).
15. R. N. Haward, ed., *Developments in Polymerisation*, Vols. I and II, Applied Science Publishers, Barking (1979).
16. C. H. Bamford and C. F. H. Tipper, eds., *Comprehensive Chemical Kinetics*, Vols. 14A and 15, Elsevier, Amsterdam (1976).
17. J. P. Kennedy, *Cationic Polymerisation of Olefins, A Critical Inventory*, Wiley-Interscience, New York (1975).
18. J. Boor, *Ziegler–Natta Catalysts and Polymerisations*, Academic Press, New York (1979).
19. H. Sinn and W. Kaminsky, "Ziegler–Natta Catalysis" in *Adv. Organomet. Chem.* **18**, 99 (1980).
20. R. W. Lenz and F. Ciardelli, eds., *Preparation and Properties of Stereoregular Polymers*, Reidel Publishing Company, Dordrecht (1980).
21. P. J. T. Tait, in Ref. 15, Vol. II, p. 81; L. L. Bohm, *Angew. Makromol. Chem.*, **89**, 1, 1980.
22. M. Johnson, *Polyolefins*, Regional Technical Meeting, Houston, Texas, (Feb. 1978), published by Society of Plastic Engineers.
23. *Polyolefins, Recent Improvements, Applications and Comparisons*, 1st Euretec Conference, Ghent, Belgium (June 1979), published by Society of Plastic Engineers.

24. M. Sittig, *Polyolefin Production Processes*, Noyes Data Corporation, New Jersey (1976).
25. R. L. Magovern, *Polymer-Plastics, Technology and Engineering*, **13**, 1, 1979.
26. U.K. Patents 1,209,825; 1,399,294: U.S. Patent 4,076,698: German Patent 2,818,657.
27. U.S. Patents 4,120,820; 4,500,873.
28. U.S. Patent 4,011,382.
29. European Patent 4647.
30. U.K. Patents 1,391,067; 1,391,068.
31. U.K. Patent 1,559,194.
32. Interesting historical viewpoints by two of the original discoverers of PE and PP have recently been provided: H. R. Sailors and J. P. Hogan, *Polymer News*, **7**, 152, 1981; P. Pino and R. Mulhaupt, *Angew. Chem. Int. Ed. Eng.*, **19**, 857, 1980.

For a comprehensive list of commercial polymers, see W. J. Roff and J. R. Scott, *Fibres, Films, Plastics and Rubbers*, Butterworths, London (1971), and H. G. Elias, *New Commercial Polymers, 1969–1975*, Gordon and Breach, New York (1975).

CHAPTER ELEVEN

SELECTIVE HYDROCARBON OXIDATION

W. R. PATTERSON

11.1 Introduction

In this chapter we shall consider processes involving reactions where oxygen and a hydrocarbon (usually an olefin or aromatic) interact in the presence of a catalyst.

Without a catalyst, such reactions can be very complex indeed, leading to the formation of many products, often as a consequence of branched or unbranched chain reactions. The purpose of a catalyst is to "temper" the reactivity of oxygen, to precipitate reaction under mild conditions which can be more easily controlled and, above all, to introduce selectivity, so that only a very small number of products are formed. It is all too easy to lose control of the oxidation process so that carbon dioxide and water are the only significant products formed. The free energy and enthalpy changes are so favourably large that the balance between a selective oxidation process and more complete oxidative degradation is a very fine one indeed. The process-reactions described in this chapter illustrate the many different approaches that must be made to ensure that valuable hydrocarbon feedstock is not (over)oxidised to completion and that the reactants are just sufficiently activated to ensure the selective formation of desired product.

Catalytic oxidation reactions are known in both the homogeneous liquid phase and the heterogeneous gas–solid and liquid–solid states. Four key types of reaction are observed in this type of catalytic action:

(1) The introduction of oxygen as a functional group

$$\text{e.g. } CH_3{-}CH{=}CH_2 \xrightarrow[\text{cat.}]{O_2} CH_2{=}CH{-}CHO + H_2O$$

$$CH_2{=}CH_2 \xrightarrow[\text{cat.}]{\frac{1}{2}O_2} \underset{\diagdown \; O \; \diagup}{CH_2{-}CH_2}$$

(2) The oxidative removal of hydrogen

$$\text{e.g. } CH_2{=}CH{-}CH_2{-}CH_3 \xrightarrow[\text{cat.}]{O_2} CH_2{=}CH{-}CH{=}CH_2 + H_2O$$

Table 11.1 The scale of manufacture of hydrocarbon oxidation products and their principal uses

Product of oxidation	*1980 world capacity 10^3 tpa*	*Approx. plant size 10^3 tpa*	*Principal uses* (m.i. = manufacturing intermediate)
Ethylene oxide	7300	50–250	ethylene glycol → polyester fibres, anti-freeze: di-, tri- and polyols → lubricants, brake and deicing fluids, textile chemicals
Propylene oxide	3100	30–200	
Formaldehyde	2770[1]	10–200	Phenol-, urea-, and other thermoplastic resins
Acetaldehyde	2900	80–200	m.i. for acetic acid and anhydride, crotonaldehyde, ethyl acetate, cellulose acetate
Vinyl acetate	2600	50–200	polyvinyl acetate—emulsion paints, adhesive for wood and packaging
			polyvinyl alcohol—textile, paper sizing, cosmetics, food packaging; also fibres in Japan
Acrolein	—	—	m.i. for DL-methionine for animal foods, allyl alcohol
Acetone from:			m.i. for solvents for paints, resins
isopropanol	940		
cumene	2300		
others	170		
Acetic acid	2500[2]	—	m.i. for vinyl acetate, acetic anhydride, esters
Acetic anhydride	—	—	m.i. for cellulose (tri)acetate for textile fibres, photog. film base, cigarette filters: m.i. for aspirin
Acrylic acid	—	50–100	m.i. for polyacrylate or methacrylate esters (not polymethyl methacrylate)
Methacrylic acid	—	—	
Maleic anhydride	600	10–30	m.i. for resins, agricultural chemicals, fumaric acid
Phthalic anhydride from:			
naphthalene	400	20–80	m.i. for resins (esp. for GRP), phthalate esters (for plasticisers)
o-xylene	2200	20–50	
Terephthalic acid	3500	50–250	polyester fibres
Hydrogen cyanide	600	—	m.i. esp. for polymethyl methacrylate
Acrylonitrile	4000	60–200	acrylic fibres, synthetic rubbers, resins
Vinyl chloride:			
from acetylene	1400	—	polyvinyl chloride,—m.i. for trichloroethylene (solvent)
from ethylene	15900	200–500	

[1] W. Europe + U.S. only.
[2] Figure for 1978.

(3) The formation of nitriles through oxidative coupling with ammonia

$$\text{e.g. } CH_3{-}CH{=}CH_2 \xrightarrow[\text{cat.}]{HN_3,\ 3/2O_2} CH_2{=}CH{-}CN + 3H_2O$$

(4) Complex oxidation to carbon dioxide and water as the major or only products.

Complete combustion reactions (type 4) are not relevant to this chapter and will not be discussed. However, there is a strong interest in total combustion catalysts for pollution-control applications. The well-known example of this is the provision of a catalytic converter in car exhaust systems to completely oxidise unburnt fuel and carbon monoxide.

Catalytic oxidation reactions are employed to provide a large number of products of widely differing properties. More often than not, these products find further widely varying application in the chemical industry as intermediates and as such form a significant bridge between the primary building blocks and the ultimate or penultimate commodity products of the industry. The size of this area of process chemistry and the diversity of its application can be judged from Table 11.1. From this table, it is readily appreciated that the incorporation, of oxygen as a functional group (type (1) reaction) leads to products which are very widely dispersed throughout the industry.

Under this heading, the more important processes are: (a) olefin epoxide synthesis; (b) aldehyde and ketone synthesis; (c) acid and acid anhydride synthesis. Most of the familiar end-products of the chemical industry exist as a result of this type of oxidation. Consequently, a large part of this chapter will be devoted to a description of the chemistry which has been adopted to promote the combination of oxygen with olefins etc. as selectively as possible.

11.2 Olefin epoxidation

The only olefin oxides of significance are ethylene and propylene oxides, both of which are important primary intermediates due to the reactivity of the oxirane ring. Almost all that is manufactured is converted to diols, triols and polyols. The major diol product, ethylene glycol, finds universal use as an automotive anti-freeze agent and in the manufacture of polyesters. Propylene glycol also finds a use as an anti-freeze agent, and may be safely used in the food, pharmaceutical and cosmetic industries in contrast to ethylene glycol which may be adsorbed through the skin and converted to oxalic acid *in vivo*. The major uses of the oxides and diols are given in Table 11.1.

11.2.1 *The manufacture of ethylene and propylene oxide*

Ethylene oxide and propylene oxide were first prepared by Wurtz in 1859–1860 by the base-promoted dehydrochlorination of the appropriate chlorohydrin. This reaction formed the basis of the commercial production of the epoxides, first introduced in Europe by BASF in 1916. In the process chemistry, the

$$2Cl_2 + H_2O \rightleftharpoons HOCl + HCl$$

$$2CHR{=}CH_2 + 2HOCl \rightarrow RHCHOH{-}CH_2Cl\ (90\%) + RCHCl{-}CH_2OH\ (10\%)$$

$$\xrightarrow{Ca(OH)_2} CaCl_2 + 2RCH{-}CH_2 \text{ (epoxide, bridged by O)}$$

$$\text{Overall } Ca(OH)_2 + Cl_2 + CH_2{=}CHR \rightarrow CH_2{-}CHR\ (\text{bridged by O}) + CaCl_2 + H_2O$$

chlorohydrin is prepared *in situ* from aqueous chlorine which generates HOCl to form the chlorohydrin by addition. The major drawback of the process is that elemental chlorine is required for the reaction, only to emerge as low-value calcium chloride. There was thus a strong incentive to find an alternative process which did not involve such a costly step as downgrading the value of elemental chlorine, particularly when the demand for ethylene oxide derivatives leapt after World War II. The first direct synthesis of ethylene oxide from ethylene and oxygen was reported in 1931, when metallic silver was employed as the catalyst. This reaction formed the basis of a process introduced by UCC in 1937. From then on, the utilisation of the chlorohydrin process steadily declined and it is now no longer used for the manufacture of ethylene oxide. However, the silver-catalysed epoxidation is restricted to ethylene—no equivalent process exists for propylene. Earlier, as ethylene chlorohydrin plants became available these were converted without difficulty to manufacture propylene oxide.

11.2.2 *The direct oxidation of ethylene to ethylene oxide*

The process for ethylene epoxidation has hardly changed since its initial commercialisation. A mixture of ethylene (20–30% v/v), oxygen (6–8%) and methane (to 100%) at 10–30 bar is passed through a multitubular fixed-bed reactor maintained at 250–330°C. With a contact time of 1–4 sec, the ethylene conversion is 8–10% at 65–70% selectivity, the only other products being carbon dioxide and water. Trace quantities of a chlorine-containing substance such as ethylene dichloride are added to the reactant feedstream. This has the effect of increasing the selectivity. Recovery and purification of product epoxide is relatively easy and unreacted gas is recirculated. Careful control of the reactor temperature is important as the catalyst is very susceptible to sintering.

Thus, although some sacrifice in selectivity is made in the direct oxidation process compared to the chlorohydrin route, this is worthwhile for several reasons: (1) expensive chlorine is not required for the process; (2) there are no chlorinated by-products to consider; (3) the by-product carbon dioxide and water are easy to separate from the epoxide and are easily disposable.

The catalyst usually consists of 5–30% silver metal supported on a low surface area ($<1\ m^2\ g^{-1}$) refractory material such as alpha alumina (corundum), silica or silicon carbide and containing barium or calcium as an additive. Hucknall[1] has tabulated the composition of several commercial catalysts and has reviewed the more important kinetic and mechanistic investigations.

11.2.3 *The mechanism of the reaction*

The uniqueness of the silver-catalysed epoxidation of ethylene has ensured that it has been extensively studied in the past. The significant developments on the mechanism will be presented here and the reader is directed to recent reviews of the subject[1,2] for a more detailed account.

The basic feature of the reaction is that the reaction on silver can be represented by the scheme:

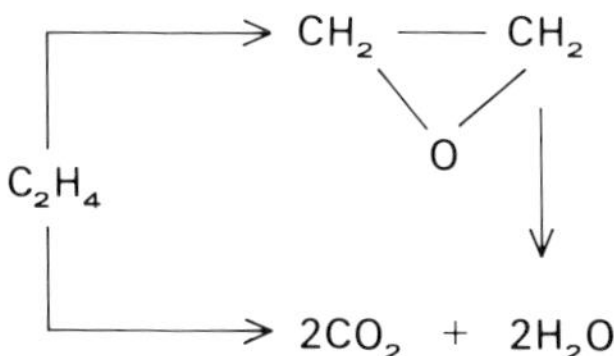

Ethylene oxide is formed by direct oxidation of ethylene in competition with complete oxidation of the latter. It is now generally accepted that the reaction occurs via an Eley-Rideal mechanism. The original view advanced was that dissociative adsorption of oxygen occurred and that the epoxide was formed as a result of the interaction of ethylene with one adsorbed oxygen atom. If reaction with two oxygen atoms occurred, complete combustion resulted.

$$\underset{\substack{|\\ \text{Ag}}}{\overset{CH_2{=}CH_2}{O}} \rightarrow \underset{\substack{\\ \text{Ag}}}{\overset{CH_2{-}CH_2}{\diagdown O \diagup}} \quad \text{but} \quad \underset{\substack{|\quad\ \ |\\ \text{Ag}\quad \text{Ag}}}{\overset{CH_2{=}CH_2}{O\quad\ \ O}} \rightarrow \underset{\substack{|\quad\ \ |\\ \text{Ag}\quad \text{Ag}}}{\overset{CH_2{-}CH_2}{O\quad\ \ O}} \rightarrow CO_2 + H_2O$$

However, it is now believed that diatomic oxygen is involved in the formation of the epoxide and that complete combustion occurs as a result of reaction with atomic oxygen.

$$CH_2{=}CH_2 + \underset{\substack{\|\\ O\\ |\\ \text{Ag}}}{O} \rightarrow \underset{\diagdown O \diagup}{CH_2{-}CH_2} + \underset{\substack{|\\ \text{Ag}}}{O}$$

$$CH_2{=}CH_2 + 6\underset{\substack{|\\ \text{Ag}}}{O} \rightarrow 2CO_2 + 2H_2O$$

This scheme places a value for the maximum possible selectivity at 6/7, i.e. 85.7%. However, in the laboratory, selectivities greater than this are obtained fairly easily. Sachtler[3] believes that recombination of atomic oxygen can occur to reduce the surface concentration of this species. The role of the added halide in enhancing selectivity, never satisfactorily explained, is interpreted in a similar way. Strongly bonded chloride competes with sites which promote the dissociation of molecular oxygen to the undesired atomic species and thus reduces the concentration of this on the surface.

The ability of silver to catalyse the epoxidation of ethylene is quite unique; no other catalyst has yet been discovered which will promote this reaction. Unless some totally new chemistry emerges, it is likely that this process will continue to be used for ethylene oxide manufacture. However, current achievable plant selectivities leave much room for improvement and future progress in this area will be aimed at realising laboratory selectivities, which may only be short-lived, in a catalyst robust enough for plant use.

11.2.4 *The co-product and other routes to propylene oxides*

Up to the present, extensive and vigorous attempts to discover a direct epoxidation route for propylene have not proved successful. However, an alternative approach based on indirect oxidation already accounts for about 40% of the propylene oxide produced in the US. This process, the Oxirane process, has been described in section 7.6 where the economics of the various routes to propylene oxide are compared. The crucial reaction is the catalysed transfer of one oxygen atom from a hydroperoxide to the olefin double bond. This occurs in the liquid phase under mild conditions, usually 120–140°C at 35 bar pressure when a selectivity of 90% at 10% propylene conversion may be obtained. The hydroperoxides, usually *t*-butyl or ethyl benzene, are reduced to the primary co-products, *t*-butanol and 1-phenylethanol respectively, also at a selectivity which exceeds 90%. These in turn may be readily dehydrated by acid catalysts to isobutene and styrene respectively. The hydroperoxides are produced initially by the liquid phase uncatalysed oxidation of isobutane or ethylbenzene, operated continuously.

The overall process is operated on a continuous recycling basis on a scale of 30 000 to 200 000 tonnes p.a.

Catalysts with high activity for the epoxidation reaction are transition metal complexes from Groups IVB, VB and VIB, especially molybdenum, tungsten, vanadium, chromium and titanium. Molybdenum coordination compounds such as molybdenum hexacarbonyl and molybdenum oxyacetylacetonate have been found to be most effective. However, for plant use, catalysts are more likely to be the naphthenate or oxyacetylacetonate.

The general reaction (below) involves heterolysis of the peroxo bond, but it is not clear if the epoxidation involves the direct involvement of the hydroperoxide.

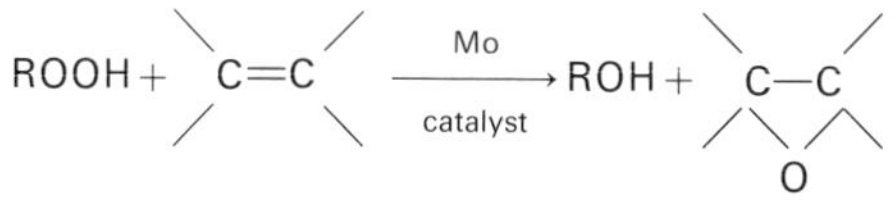

$$\text{ROOH} + \text{C}{=}\text{C} \xrightarrow[\text{catalyst}]{\text{Mo}} \text{ROH} + \text{C}{-}\text{C}\ (\text{epoxide, bridging O})$$

Indeed, despite a lot of work in the last twelve years or so, the nature of the oxygen transfer step and the reactive molybdenum intermediate is not resolved.

Sheldon[4] examined the epoxidation of a variety of olefins with *t*-butyl hydroperoxide using $Mo(CO)_6$ or $Mo(acac)_2$ as catalyst and found that the reaction showed a short induction period with the initial rate depending on the molybdenum species. Final steady-state rates were the same and only depended on the olefin used. From the reaction mixture he isolated molybdenum complexes which did not contain any carbonyl or acetylacetonate ligands; instead, they were Mo^{VI}-1,2 diol complexes whose structure was directly related to the reacting olefin. The complex obtained in the epoxidation of propylene had the structure I.

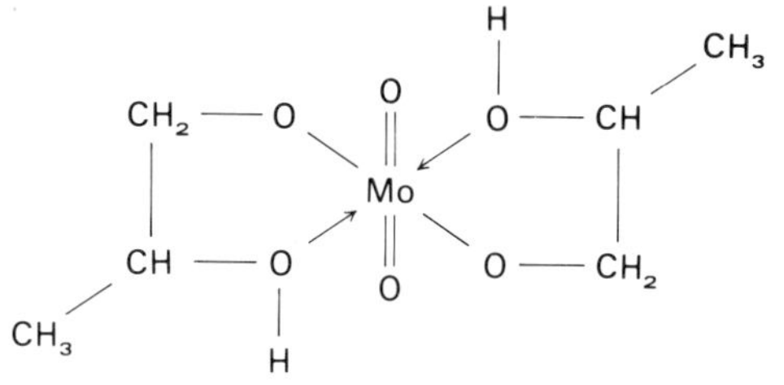

I

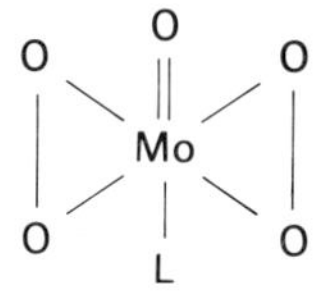

II

Significantly, the steady-state rate of epoxidation when the Mo^{VI}-1,2 diol was used as catalyst was the same as found with $Mo(CO)_6$ or $Mo(acac)_2$. It has since been found that hydroperoxides will decompose $Mo(CO)_6$ at moderate temperatures give molybdic acid, H_2MoO_4, which will react with cyclohexene to give the corresponding Mo^{VI}-1,2 diol.

Sheldon[4] proposed that the Mo^{VI}-1,2 diols were the active species in the epoxidation, reacting with further hydroperoxide to yield epoxide and molybdic acid from which the Mo^{VI}-1,2 diol was regenerated. This proposal for the oxygen transfer step is based on attack on (I) by hydroperoxide, forming a reactive intermediate, which on reaction with olefin yields an epoxide. Species I is regenerated by ejection of an alcohol molecule (page 258).

The alternative structure (II) has been proposed by Mimoun[5,6] for the active catalytic species. This is based on the observation that covalent molybdenum^{VI}-peroxo complexes react with olefins to produce epoxides in high yield. The reactivity of olefins towards these complexes is in the same order as the reactivity for epoxidation using hydroperoxides. The molybdenum peroxo species is easily formed from molybdenum trioxide, hydrogen peroxide and neutral coordinating ligands such as phosphines, aromatic amines, tertiary amides, etc. Mimoun favours the formation of a five-membered ring complex from (II) (page 259).

$\frac{C_3H_6}{R.OOH}$

$-ROH$

O O O Mo O O L $\xrightarrow[-L]{CH_3CH{=}CH_2}$ O O O Mo O O CH=CH$_2$ CH$_3$

O O Mo O O + CH$_3$CH—CH$_2$ (O) ⟵ O O O Mo O O CH—CH$_2$ CH$_3$

ROOH, L ⟶ O O O Mo O O L + ROH

Thus although the detailed chemistry is far from being clearly defined, it is generally accepted that molybdenum is in its highest oxidation state and is believed to be surrounded by oxygen ligands in the active catalytic intermediate.

The reader's attention is drawn to section 7.6.3 where a non-catalytic process for making propylene oxide is discussed. This is the Daicel process in which peracetic acid is used to epoxidise propylene. Unless a captive market exists for the co-product, acetic acid, this process will be very vulnerable to the new Monsanto process for acetic acid from methanol and carbon monoxide (cf. 8.5.1).

A new route which avoids the problems of co-product disposal has recently been announced. This is based on the acetoxylation of propylene and is under development by Chem Systems. First, propylene, oxygen and acetic acid are reacted in the liquid phase in the presence of an undisclosed catalyst to yield about 50% propylene glycol monoacetates and 50% propylene glycol and glycol diacetate. The product stream is then passed into an alkaline medium where partial hydrolysis occurs and the propylene glycol monoacetates are cracked to propylene oxide and acetic acid in a previously unknown reaction. Propionaldehyde and acetone are formed as minor by-products. Acetic acid and unreacted diacetate are recycled. The overall yield of propylene oxide is said to be above 80%.

Finally, direct epoxidation using hydrogen peroxide is being contemplated. Uncatalysed, this reaction is too slow to be of significance but PCUK, world leaders in hydrogen peroxide production, have recently patented the use of arsenic, boron and molybdenum derivatives which are reported to yield propylene oxide at 95 % selectivity in a continuous one-step process, no further details of which are yet available.

Clearly, the potential for a reasonably straightforward process for propylene oxide is great. Problems associated with effluent and co-product disposal are likely to become more severe and, in the absence of a direct epoxidation catalyst equivalent to silver (for ethylene oxide), the developments of Chem. Systems and PCUK are probably indicative of the way ahead.

11.2.5 *The direct synthesis of ethylene and propylene glycol*

Ethylene and propylene glycol are the only industrially significant 1,2-diols. The key oxidation step is in the generation of their precursors, the corresponding epoxides. Clearly, the avoidance of the epoxidation step would be very advantageous, not simply because of process simplification but, in the case of ethylene oxide, the yield of epoxide achievable on plant requires substantial improvement. With propylene oxide the current manufacturing situation is far from ideal with neither the chlorohydrin process or the Oxirane process free from criticism. A significant step forward in process chemistry would be achieved if an epoxide-free route to ethylene and propylene glycol were discovered. The ideal process would simply be

$$C_2H_4 \text{ or } C_3H_6 + \tfrac{1}{2}O_2 + H_2O \xrightarrow{\text{cat.}} CH_2OH.CH_2OH \text{ or } CH_3.CH(OH)CH_2OH$$

A step in this direction is to form the diol diacetate from the olefin directly and hydrolyse this to the diol in a two-stage process according to the reactions

$$R.CH{=}CH_2 + 2AcOH + \tfrac{1}{2}O_2 \xrightarrow{\text{cat.}} CH_2.OAc.CH_2OAc + H_2O$$

$$CH_2.OAc.CH_2OAc \xrightarrow[\text{cat.}]{2H_2O} 2AcOH + CH_2OH.CH_2OH$$

Here, no co-products are produced and hydrolysis of the di-ester and recycling of the acetic acid are the only problems. Oxirane started up a 360 000 tonne per year fibre-grade ethylene glycol plant in 1978 using a homogeneous catalyst system (containing tellurium, lithium and bromide) operating in acetic acid at 170°C. Selectivity of 96 % on ethylene conversion of 98 % was claimed. However, this plant has recently been shut down because of corrosion from acetic acid and bromine, in spite of the use of titanium in the construction of the plant. In addition, it is believed that the catalyst decomposed with deposition of tellurium metal, and doubt has been expressed as to the future of the plant.

Future developments may provide a solution to the problem of ethylene glycol synthesis which frees manufacturers from the difficulties of employing oxidation chemistry. This arises because, in principal, ethylene glycol may be made from synthesis gas. Current developments in this area are discussed in section 8.3.2.

11.3 The manufacture of aldehydes and ketones

The synthesis and manufacture of the lower aliphatic aldehydes and ketones forms an industrially important and scientifically instructive area in the field of oxidation catalysis, both homogeneous and heterogeneous. The introduction of the aldehyde or keto functional group can in most cases be performed directly on olefins using oxygen and a suitable catalyst. It is natural, therefore, that this is an area which has grown in importance as a result of the growth of petrochemistry.

Industrially, the most important aliphatic aldehydes and ketones are formaldehyde, acetaldehyde, acrolein, acetone and methyl ethyl ketone. Their uses are varied, either directly as end-products, or, principally, as commercial intermediates. All are formed from olefin oxidation except, obviously, formaldehyde which is manufactured almost exclusively by methanol oxidation.

11.3.1 *The manufacture of formaldehyde*

A minor proportion (approx. 8 %) of the world's supply of formaldehyde is produced by the non-catalytic partial oxidation of the lower hydrocarbons (mostly methane, propane and butane) but by far the major route lies in the catalytic removal of hydrogen from methanol. In the reaction ($CH_3OH \rightleftharpoons HCHO + H_2$) the free energy change is zero at 400°C. The formation of formaldehyde will be favoured above this temperature; for example, at 400° and 700°C the equilibrium yields of formaldehyde are 50 % and 99 % respectively. If oxygen is added to the mixture, then oxidation ($CH_3OH + \frac{1}{2}O_2 \rightarrow HCHO + H_2O$) occurs, in which hydrogen on the right-hand side of the equation is replaced by water. This has two important consequences.

(1) The endothermic dehydrogenation reaction is replaced by an exothermic oxidation reaction, and heat is generated rather than consumed.
(2) The irreversible removal of hydrogen (as water) from the system effectively removes any equilibrium constraints.

Thus the dehydrogenation of methanol would be complete at, say, 400°C or lower in the presence of oxygen, rather than at 700° in its absence.

In practice, formaldehyde is manufactured according to both basic methods. In one, a mixture of methanol and air in proportions at least exceeding the upper explosive limit (36.5 % v/v) is passed over a silver catalyst, dehydrogenation occurs and most of the hydrogen is subsequently converted to water. In the other (oxidation) method, the catalyst is usually an iron/molybdenum mixed oxide catalyst, and the reacting feedstream of methanol and air lies below the lower explosive range (6.7% v/v). The operating conditions for both processes are given in Table 11.2. The product formaldehyde is stored as a concentrated solution (37% w/w) since the pure material has a tendency to polymerise. Clearly, it is desirable to manufacture formaldehyde close to plants associated with its further utilisation, since it is easier and cheaper to transport methanol than formaldehyde solution. Therefore, large-scale users of formaldehyde are the chief manufacturers.

Table 11.2 Process conditions for the manufacture of formaldehyde

Feedstream	methanol/air	methanol/air
catalyst	silver as gauze, crystals or granules, 0.5–3 mm diam in a bed < 3 cm thick	pellets in salt cooled tubular reactors
catalyst life	2–6 months	1–2 years
pressure	atmospheric	atmospheric
reactor temp	450–650°C; optimum ~635°C	≯430°C
contact time	0.01 sec	0.3–0.5 sec
selectivity	85–95%	91–94%
conversion	60–73%	99%

11.3.2 *The silver-catalysed process*

The original process was developed in Germany about 1910 and employed silver gauze as catalyst. Prior to this, some formaldehyde manufacture had been conducted using copper as catalyst but the increased activity and selectivity of silver quickly displaced this. The original process, or modifications of it, is still in use today, no better catalysts having been discovered.

The methanol/air feedstream is preheated to about 100°C and then fed to the shallow catalyst beds usually arranged in parallel. The heat liberated from the exothermic reaction is sufficient to sustain a reactor temperature of about 600°C (i.e. dull red heat). The supply of air is regulated to control the heat liberated and thus the temperature of the bed may be controlled to within 5°C. After passage through the reactor the gas stream is cooled as rapidly as possible (0.1–0.3 sec) by counter-current passage against a spray of water/methanol. Thereafter, distillation, addition of water etc. is used to recycle the unreacted methanol and convert the gaseous formaldehyde to the 37% solution for transport or, better, piped in gaseous form directly to an adjacent utilisation plant. Hader *et al.*[7] have provided a very readable description of a typical formaldehyde plant.

Reaction occurs largely in the initial part of the layer of silver catalyst. The actual form of silver does not appear to be significant since high selectivities are reported for all catalyst types. The life of the catalyst is usually a few months after which it is recovered with minimal loss of metal.

The process uses a mixture of methanol with insufficient air for complete conversion. The excess methanol acts as a heat sink in the highly exothermic reaction. However, this means that distillation and recycling stages are necessary to handle unreacted methanol. Further, some hydrogen is also formed by simple dehydrogenation and has to be removed. BASF have found that the reaction mixture can be adjusted to ensure almost complete conversion by replacing some of the excess methanol by water.

Despite the extreme reaction conditions very high yields and selectivities based on methanol are obtained, the only by-products being a trivial amount of CO, CO_2 and a trace of formic acid.

11.3. *The iron molybdate catalysed process*

The search for improvements in the basic Karl Fischer process led to the discovery in 1931 of the highly selective iron/molybdenum oxide catalyst system, and processes based on this are in use today as alternatives to the silver-based ones.

The catalyst consists of a mixture of Fe_2O_3 (0.2 moles) and MoO_3 (1.0 mole) calcined at 300–450°C after which it consists of a two-phase product containing MoO_3 and $Fe(MoO_4)_3$, which is believed to be the active phase. Patent variations on this catalyst relate to claimed improvements in preparation and doping of this basic formulation. Although other oxide catalysts are known, particularly vanadium pentoxide-based catalysts, the iron-molybdenum system has not yet been surpassed in performance.

The operating temperature of the catalyst is around 400°C, with a maximum at 430°. At temperatures greater than this, sublimation of MoO_3 occurs and molybdenum blue is deposited downstream in the cooler parts of the reactor. Loss of molybdenum from $Fe(MoO_4)_3$ can also occur with a resultant loss of selectivity in the catalyst. It is essential to prevent reduction of the catalyst so, in contrast to the silver process, the process is run below the lower explosive limit (6.7% v/v). With this air-rich mixture very high conversions are desirable since it would be very expensive to separate unreacted methanol for recycle. In fact, the performance of metal oxide processes are claimed to be higher than the silver ones, conversion of 95–99% at selectivities of 91–94% being quoted.

11.3.4 *Factors affecting the choice of a formaldehyde process*

The potential buyer of a methanol-to-formaldehyde process must weigh up many factors in making a choice between the two systems described above.[8] It is instructive to look at the more important of these for they illustrate some issues which arise when a chemical manufacturer considers the installation of a process.

Will the manufacturer use the formaldehyde he makes or is he a merchant producer who will merely sell it? If he is going to use it directly, he may well desire a steady supply of pure gaseous formaldehyde in which case the oxide process is to be preferred, since the product stream contains virtually no methanol. An example would be in the continuous manufacture of urea-formaldehyde glues (used in the timber composites industry for plywood, chipboard etc.). Batch manufacture (e.g. formaldehyde resins) require stocks of (methanol) stabilised solution so methanol specification of the product stream is not critical.

Apart from the catalyst and operating conditions the most important differences in the two processes are materials of construction and in downstream handling of the product stream. Silver catalysts are very quickly poisoned by iron, sulphur and halogens, so extensive use is made of stainless steel in construction and the process requires methanol of higher specification than the more tolerant oxide catalysts. In the treatment of the product stream the silver process requires highly efficient distillation equipment to separate the formaldehyde from the unreacted methanol which is recycled. Since conversion is complete in the oxide process, this stage is not required but such an advantage is offset by the need for much larger scrubbing and condensing stages due to the larger volume of air used. In general investment costs are lower for silver plants if the capacity is under 10 000 tpa. Although operating costs for silver-based plants are higher for silver plants up to 20–25 000 tpa, there is little to choose above this capacity. Licensing costs for the oxide process are a little higher than for the silver process.

11.3.5 *The mechanism of methanol oxidation*

The conversion of methanol to formaldehyde in the presence of oxygen on silver catalysts is often described as "oxidative dehydrogenation", while the same reaction on iron molybdate is called "direct oxidation". The application of these terms to set apart the two systems is rather indefinite, since a knowledge of mechanistic differences, or assumed differences, is implicit in their use. The meaning of "oxidative dehydrogenation" will be further discussed in section 11.6.1 but, for the present, the term will be avoided and the simple word "oxidation" will be used to describe the reaction on either catalyst, when oxygen is present.

Methanol dehydrogenation will occur on silver at 600–700°C in the absence of air, but the yield gradually declines due to coking of the catalyst. Thus, as well as providing the necessary heat for the process, the addition of oxygen helps to maintain a clean working surface.

No mechanistic investigations have been made on the silver catalyst at anything like plant operating temperatures. Kinetic and mechanistic work has been reported for polycrystalline material at 420°C[9] and on single-crystal silver at 70°C,[10] but whether these investigations have any relevance at 600°C remains conjectural. The key feature which emerges at these temperatures is that methanol will not dehydrogenate on a clean silver surface and that adsorbed atomic oxygen is required to facilitate the dissociation of the hydroxyl group, believed to be the first step in the reaction. Current mechanistic and kinetic work in this area has been recently reviewed.[2]

In contrast to the reaction on silver, the properties of the iron-molybdenum catalyst have been well documented in the working temperature range. As already described, the optimum catalyst is a mixture of ferric molybdate, $Fe_2(MoO_4)_3$, and molybdenum trioxide. This catalyst has a specific activity almost two orders of magnitude greater than the constituent oxides. The activity and selectivity of the catalyst are very similar in the presence and absence of air. Thus any direct interaction between adsorbed oxygen and methanol must be minimal, the role of the former being merely to replenish the supply of lattice oxygen. This important concept of alternate reduction and re-oxidation of the catalyst was first advanced by Mars and van Krevelen in 1954 and has occupied an important position in mechanistic discussions on metal-oxide catalysed (oxidation) reactions. The area is discussed in more detail in the context of olefin oxidation (11.9.3).

Results of more detailed examination of the possible modes of decomposition of methanol have not yet been published and it remains to be seen whether there is any similarity with the proposed mechanisms on silver.

11.3.6 *The manufacture of acetaldehyde*

Apart from speciality uses, acetaldehyde is used primarily for the manufacture of acetic acid and acetic anhydride (11.5.1). It can also provide a useful route to certain C_4 derivatives through aldol dimerisation and related reactions. Before the early 1960's it was made by three methods:

(i) by dehydrogenation of ethanol on silver or, preferably, copper catalysts in the temperature range of 250–300°C or by oxidation on silver at 450–550°C. The conversion in both cases is limited to 30–50 % when selectivities of 85–95 % are achieved.

(ii) by the hydration–dehydrogenation of acetylene over various acid catalysts, particularly phosphates and tungstates containing zinc or cadmium. Using ambient pressures and at 250–350°C overall yields of 90–95 % could be achieved at conversion levels of 30–50 %.

(iii) the non-catalytic oxidation of propane, butane or mixtures of the two at 400–500°C.

During the early sixties, the industrial importance of the above methods fell rapidly due to the successful emergence of new chemical technology which not only made possible the direct oxidation of ethylene to acetaldehyde but which made use of transition-metal olefin chemistry for the first time in catalytic oxidation on an industrial scale.

The chemistry in question is the oxidation of ethylene to acetaldehyde by aqueous palladium chloride during which hydrogen chloride and metallic palladium are formed. This reaction had been reported in 1894, but it was not until 1956 that Wacker was reminded of this during attempts to find alternative catalysts to silver for the production of ethylene oxide.[11] Trace amounts of acetaldehyde were formed when ethylene/oxygen mixtures were passed over a supported palladium catalyst which contained traces of hydrogen chloride from the catalyst preparation.

The reaction as originally reported is stoichiometric,

$$C_2H_4 + PdCl_2 + H_2O \rightarrow CH_3CHO + Pd + 2HCl$$

but the Wacker workers discovered how to make the reaction catalytic by (a) reoxidising the palladium with cupric chloride and (b) reoxidising the reduced cuprous ion with oxygen.

$$Pd + 2CuCl_2 \longrightarrow PdCl_2 + 2CuCl$$

$$2CuCl + \tfrac{1}{2}O_2 + 2HCl \longrightarrow 2CuCl_2 + H_2O$$

overall,

$$C_2H_4 + \tfrac{1}{2}O_2 \xrightarrow[CuCl_2]{PdCl_2} CH_3CHO$$

Such was the significance of this reaction that it was developed to full process status by 1960. (At that time, ethanol and acetylene, the accepted raw materials for ethanol, cost about twice as much as ethylene). A useful review of the patents covering this process is given by Szonyi.[12]

There are two different process options—one-step or two-step. Both are run on a continuous basis and, under their respective operating conditions, both offer the same selectivity in ethylene conversion to acetaldehyde of about 95%, the by-products being acetic acid, crotonaldehyde and carbon chlorides. A little chlorine is lost as methyl, ethyl chloride etc.

In the one-step process a mixture of ethylene and oxygen, devoid of all but a trace of inert gases, are introduced at a pressure of 3 bar to the reactor containing a solution of catalyst maintained at about 100°C. The composition of the gas mixture is kept well above the upper flammability limit so ethylene is present in large excess. The catalytic solution might be expected to contain about 200 g/l of cupric chloride and 4 g/l of palladium chloride and, when hydrogen chloride is present, is highly corrosive. Conditions are adjusted to give about 40% ethylene conversion and total oxygen consumption. The heat produced by the reaction is used to distil off the acetaldehyde. Unreacted ethylene is recycled, hence the necessity for the absence of inert gases. In the two-stage process this is not a constraint, and in this mode of operation air and ethylene are compressed to 10 bar and in proportions so as to ensure almost complete conversion of ethylene when exposed to the catalyst solution. This solution is constantly circulating through another reactor where regeneration by oxidation with air at 10 bar takes place at 100°C. The presence of nitrogen can be tolerated since only the liquid phase is recirculated and any excess inert gas can be vented. In either process, the acetaldehyde is easily distilled off. Periodic recovery and regeneration of the catalyst is necessary due to build up of high-boiling organic chlorides, acetic and oxalic acid by-products. A simplified diagram of both types of process is shown in figure 4.2.

The choice of operating a one- or two-step process depends largely on the quality of the feed gases since it transpires that the overall costs of operating both processes are similar. Thus ready access to high-purity, low-pressure ethylene and pure oxygen will favour a one-step process whereas availability of low-purity, high-pressure ethylene—or cheap power to compress it—will favour the two-step process. In fact, because of the large-scale production of polymer-grade ethylene of high purity, the one-stage process is almost exclusively employed in the Western hemisphere. The usual operating scale is 80–200 000 tonnes per year.

The success of the Wacker process during the 1960's and 70's was due to the availability of cheap ethylene. Although the corrosive nature of the catalyst solution required extensive use of titanium in the reactor zones, it may be surmised that there were few technological difficulties in scaling up the process from laboratory to plant operation. The process remains the most elegant application of homogeneous catalytic oxidation, and any attempts to develop a "heterogeneous" version of the process have yielded little. One reason why the homogeneous process has maintained its commercial competitiveness is that the product is volatile and therefore easily separated from the catalyst.

The major use for acetaldehyde hitherto has been in the manufacture of acetic acid and anhydride. However, the escalating cost of ethylene has now made this route very uncompetitive with the Monsanto process for the carbonylation of methanol (see section 8.5) and it is more likely that the peak manufacturing level for Hoechst–Wacker acetaldehyde has been passed.

11.3.7 *The mechanism of the chemistry in the Wacker process*

In addition to palladium, other Group VIII metals such as platinum (II), ruthenium (III), iridium (III), and rhodium (III) oxidise ethylene but at much reduced rates. The product from the oxidation is always acetaldehyde—further oxidation to acetic acid is not observed to any extent. Most olefins will react in an analogous manner to give ketones but rates and yields can be much lower than for ethylene. Propylene may be oxidised selectively to acetone and all linear butenes to a single main product, methyl ethyl ketone. Higher olefins yield mixed ketones due to isomerisation but by regulating the solvent, excellent yields of methyl ketones from α-olefins have been reported.

The oxidation of ethylene according to the Wacker chemistry has attracted much interest since 1960 and considerable effort has gone into establishing the general scope of the reaction and its mechanism. Many review articles have appeared.[13–15] However, the chemistry is still incompletely described and recent publications by Bäckvall[16] and Stille[17] have renewed interest in some crucial aspects of the mechanism.

The main features of the reaction are as follows:

(a) The rate of ethylene oxidation is proportional to the concentration of $PdCl_4^{2-}$ and ethylene, but is inhibited by chloride ion and the presence of acid.
(b) Ethanol cannot be an intermediate because it is oxidised to acetaldehyde under Wacker conditions much more slowly than is ethylene.
(c) Vinyl alcohol cannot be an intermediate, since acetaldehyde formed from ethylene in $PdCl_2/D_2O$ does not contain any deuterium.

(d) A negligible isotope effect is observed in the relative rates of oxidation of C_2H_4 and C_2D_4.
(e) Free radicals to not appear to be intermediates. Moreover, the rate of oxidation of olefin decreases with increasing substitution at the double bond suggesting that nucleophilic, rather than electrophilic, attack is important.

It is generally accepted that the first step in the reaction is the reversible formation of an ethylene–palladium π-complex.

$$[PdCl_4]^{2-} + C_2H_4 \longrightarrow [PdCl_3(CH_2{=}CH_2)]^- + Cl^-$$

(The formation of such a complex would reduce the electron density at the double bond and thus facilitate nucleophilic attack.)

Water is then coordinated in the next step

$$[PdCl_3(CH_2{=}CH_2)]^- + H_2O \quad [PdCl_2(H_2O)(CH_2{=}CH_2)] \rightleftharpoons [PdCl_2(CH_2{=}CH_2)(OH_2)] + Cl^-$$

(ethylene is known to be a strongly *trans*-directing ligand but Pd(II) complexes are labile, and in the polar aqueous medium considered here isomerisation to the more polar *cis* configuration would be possible). These two steps satisfactorily account for the inhibiting effect of chloride ion in the rate expression.

Inhibition by acid is explained by the next step which is crucial to the mechanism, for it constitutes irreversible nucleophilic attack on the olefin resulting in the transformation of the π-bonded ethylene ligand to a σ-bonded hydroxy ethyl ligand, the decomposition of which produces acetaldehyde.

$$[PdCl_2(CH_2{=}CH_2)(OH_2)] \rightleftharpoons [PdCl_2(CH_2{=}CH_2)(OH)]^- + H^+$$

$$[PdCl_2(CH_2{=}CH_2)(OH)]^- + H_2O \longrightarrow [PdCl_2(CH_2CH_2OH)(OH_2)]^-$$

This nucleophilic attack can be *trans*-directed by an *external* (uncoordinated) OH^- entity, or *cis* attack by the *internal* coordinated OH^- entity. Up until 1977 the latter was held to be more likely. However, in almost simultaneous publications, Bäckvall and Stille have provided fairly conclusive evidence that, in fact, addition of the nucleophile is exclusively *trans*, or, in other words, the π–σ transformation occurs through attack by a free hydroxyl ion. The proof for this has come from an examination of the deuterium configuration after reacting 1,2-dideuteroethylenes to form 1,2-dideuteroethanol and β-propiolactone.

The final decomposition of the β-hydroxy ethyl palladium complex liberates acetaldehyde. The chemistry of this step has still to be fully defined but it has been accepted that the 1,2-hydride shift which occurs in the hydroxy ethyl ligand and precedes the formation of free acetaldehyde is not a rate-determining step. This would be consistent with either a *cis*-insertion mechanism or the Bäckvall–Stille *trans*-attack mechanism for the formation of the σ-complex.

$$\mathrm{Cl_2Pd(OH_2)(CH_2{-}CH(H){-}O{-}H)} \longrightarrow CH_3CHO + H^+ + Pd^0$$

11.3.8 *The manufacture of vinyl acetate*

At the present time, the major route for the industrial production of the important ester, vinyl acetate, is a close derivative of the Wacker acetaldehyde chemistry. Virtually the whole of the world production of vinyl acetate is consumed in the manufacture of polymers and co-polymers. By far the most important is simple polyvinyl acetate and its derivative polyvinyl alcohol.* The major uses of these polymers are given in Table 11.1.

The history of the manufacture of vinyl acetate reflects that of acetaldehyde. Until the arrival of cheap, large-tonnage ethylene, there were two methods generally available; one, the addition of acetic acid to acetylene, catalysed by zinc or cadmium salts at 180° to 210°C; the other, the combination of acetaldehyde with acetic anhydride to form ethylidene diacetate which, on pyrolysis, yields vinyl acetate and acetic acid. In 1960, it was discovered that vinyl acetate could be formed by passing ethylene into palladium chloride contained in a solution of sodium acetate in acetic acid.

* Since vinyl alcohol does not exist in the free state the hydrolysis of vinyl acetate yields acetaldehyde and acetic acid. However, polyvinyl acetate may be hydrolysed to polyvinyl alcohol and acetic acid by, e.g. reacting a methanol solution of the ester with a mineral acid according to the reaction

$$(CH_2{-}\underset{|\atop O.CO.CH_3}{CH})_n \longrightarrow (CH_2{-}\underset{|\atop OH}{CH})_n + n.CH_3COOH$$

This observation aroused industrial interest and within a few years further research developments resulted in the emergence of two quite separate processes. One was a liquid-phase process, developed primarily by ICI in the United Kingdom although Hoechst also held patents in the same area. The chemistry employed was essentially a Wacker reaction performed in non-aqueous media. In ICI's process, ethylene, oxygen and acetic acid were reacted at 100–130°C in the presence of the catalyst solution which consisted of palladium chloride and cupric acetate in acetic acid containing hydrogen chloride. Overall yields were stated to be 90% on ethylene and 95% on acetic acid. The essential chemistry is governed by the reactions

$$CH_2{=}CH_2 + 2CH_3COO.Na + PdCl_2 \rightarrow CH_2{=}CH.O.CO.CH_3 + CH_3CO.OH + 2NaCl + Pd$$

$$Pd + 2CuCl_2 \rightarrow PdCl_2 + 2CuCl$$

$$2CuCl + \tfrac{1}{2}O_2 + 2HCl \rightarrow 2CuCl_2 + H_2O$$

overall

$$CH_2{=}CH_2 + CH_3COOH + \tfrac{1}{2}O_2 \rightarrow CH_2{=}CH.CO.OCH_3 + H_2O$$

Water is produced in the reaction and this leads to the formation of acetaldehyde, which after recovery could readily be oxidised to acetic acid. The addition of water to the reacting mixture can therefore be used to increase the amount of acetaldehyde until the point when, if the latter is converted to acetic acid in an adjoining reactor, the process will be virtually self-sufficient in acetic acid. The ICI process was operated in this manner and was therefore free of a dependence on an external source for acetic acid.

However, after commercialisation of the process, unforeseen corrosion problems were encountered and the severity of these led to the abandoning of the process.

The second process which was being developed proved successful in its operation and is the means whereby, at the present time, vinyl acetate is made from ethylene. This is a gas-phase process employing a supported catalyst and is free from corrosion effects. Several companies independently discovered this reaction. Bayer and Kurashiki developed supported palladium catalysts while Hoechst and USI developed supported palladium acetate catalysts; all required the addition of alkali metal acetates to maximise the selectivity.

A typical process catalyst might contain 0.1 to 2% w/w palladium, 0.5 to 5% w/w added acetate, on a support such as alumina and present as lumps or pellets in tubular reactors 5 to 15 metres long. Since the reaction is conducted in the temperature range of 100–200°C, boiling water may be used to remove heat from the exothermic reaction. A reaction mixture of ethylene, oxygen and acetic acid in proportions above the upper flammability limit is supplied to the reactors at 5–10 bar and conversions of 10–15% ethylene, 15–30% acetic acid and 60–90% oxygen are achieved at selectivities of 91–94% based on ethylene or 98–99%, based on acetic acid. About 1% ethylene is converted to acetaldehyde and carbon dioxide is the other main product. Unreacted ethylene, oxygen and acetic acid are recycled after recovery. In contrast to the liquid phase process, external acetic acid is required.

It is evident that the adoption of this process on such a large scale owes much to its relative simplicity. Comparisons can be drawn with the one-stage homogeneous acetaldehyde process: very selective oxidation with no troublesome by-products, easy recovery of product and, importantly, ease of separation of reacted mixture from the catalyst.

Future developments in vinyl acetate synthesis are rather difficult to foresee. Reference has been made (11.2.6) to the doubts concerning the future of Oxirane's process for the manufacture of ethylene glycol. If the operating problems can be overcome, Halcon, the parent company, plan to exploit the utility of the diacetate intermediate further by the reaction

$$CH_3.CO.O.CH_2{-}CH_2.O.CO.CH_3 \xrightarrow{\text{heat}} CH_2{=}CH.O.CO.CH_3 + CH_3COOH$$

A selectivity of 87 % at 20 % conversion at 535°C has been claimed. Every unit weight of ethylene converted to vinyl acetate requires 2.14 units of acetic acid. In the early seventies, when ethylene was cheap, the price of vinyl acetate was largely determined by the cost of acetic acid. However, at the present time, the cost of ethylene is related to the (escalating) cost of naphtha but cheap acetic acid is available from synthesis gas chemistry—an independent source. Sooner or later, an ethylene-free route will be attractive and there are aspirations towards a non-oxidative route to vinyl acetate based on the synthesis-gas chemistry. Halcon have, in fact, developed a process based on methyl acetate and carbon monoxide and are working on the direct route from synthesis gas (8.5.2).

11.3.9 *The mechanism of vinyl acetate synthesis*

The overall reaction representing the liquid-phase and gas-phase processes is the acetoxylation of ethylene by acetic acid

$$CH_2{=}CH_2 + CH_3COOH + \tfrac{1}{2}O_2 \rightarrow CH_2{=}CH.O.CO.CH_3 + H_2O$$

In the liquid phase it is generally believed that the mechanism is similar to that involving the formation of acetaldehyde, except that instead of forming the incipient, unstable vinyl alcohol by nucleophilic substitution of OH^-, the more stable vinyl acetate is formed by substitution of the nucleophile, CH_3COO^-. Since the essential mechanistic chemistry for the acetaldehyde reaction has been described (11.3.7), no further discussion of the vinyl acetate reaction will be given here, but reviews of the chemistry involved are available.[13–15].

The mechanistic investigation of the heterogeneous, gas-phase reaction has received relatively little attention. Nakamura and Yasui[18] have reported some results from reactions at 150°C on palladium black, palladium acetate and 5 % w/w palladium supported on alumina. From their results, the authors propose a mechanism for the acetoxylation reaction which can be summarised thus:

(1) ethylene is dissociatively adsorbed on palladium (i.e. a σ-bonded vinyl palladium species (I) is formed)
(2) oxygen is dissociatively adsorbed on palladium
(3) in the presence of oxygen, hydrogen is abstracted from associatively adsorbed acetic acid to form a surface acetate species (II)
(4) the surface acetate species combines with the adsorbed ethylene species to form vinyl acetate, according to the mechanistic scheme

H CH2 C Pd + O Pd C CH3 O → O C CH3 O Pd H C CH2 Pd

I II

↓

CH3 C O O H C = CH2 Pd Pd

It will be seen that this mechanism is rather different from those proposed for the liquid phase reaction. One important difference is the suggestion that the vinylpalladium species (I) is the active ethylene intermediate. This is deduced from experiments performed in the absence of oxygen under conditions not experienced by the working catalyst. Even if such a species were present during the acetoxylation reaction, this does not necessarily mean it is the reactive intermediate, nor is the existence of other adsorbed forms excluded.

11.3.10 *Other palladium-catalysed acetoxylation reactions*

The oxidation of higher olefins in acetic acid leads in most cases to acetoxylation. The products formed (see opposite) depend critically on the conditions and whether chloride is present or not. Maitlis[13] has reviewed this and similar reactions of propylene, 1-butene, 2-butene, butadiene as well as benzene, toluene, *p*-xylene and styrene. Although patent property exists for some of the reported reactions, no detailed processes involving these have been described.

$PdCl_2$, $Pd(OAc)_2$ / AcOH

OAc (76%) AcO (24%)

but

$Pd(OAc)_2$ / AcOH

OAc (100%)

Although chemistry in liquid acetic acid can be complicated, some reactions have been found to follow the simplicity of the vinyl acetate reaction in the gas phase, and two of these have aroused industrial interest.

The first is the synthesis of allyl alcohol, through the intermediate allyl acetate which is made in a manner completely analogous to the Bayer/Hoechst process for vinyl acetate.

$$CH_2{=}CH.CH_3 + CH_3COOH + \tfrac{1}{2}O_2 \xrightarrow{Pd} CH_2{=}CH.CH_2COOCH_3 + H_2O$$

Similar catalysts and conditions are employed and the reported overall selectivity of allyl alcohol based on propylene is 90%.

The second reaction concerns the direct synthesis of phenol from benzene. There are at present, numerous ways of manufacturing phenol, but all of them are, at least, two-stage processes (see section 7.7). The only direct oxidation route is the non-catalytic autoxidation of cumene to the hydroperoxide which on decomposition yields phenol and co-product acetone. In the acetoxylation route, phenyl acetate and subsequently phenol can be made from benzene in the same way that allyl alcohol is made from propylene. Alternatively, phenyl acetate may be cracked to phenol and ketene, itself a valuable chemical intermediate.

11.3.11 *The manufacture of acetone and methylethyl ketone*

The only simple saturated ketones of industrial significance are acetone, methyl ethyl ketone (MEK) and methyl isobutyl ketone. The principal use of acetone is in the production of methacrylic acid, methyl methacrylate and higher molecular weight acrylates (11.5.2).

The current major source of acetone is as co-product in the cumene to phenol process (section 7.7.1). This accounted for 66% of production in the United States in 1974, and the capacity in 1976 was 60%. These figures probably reflect average world trends. The only other significant source for acetone, and the major source for MEK, are processes based on the corresponding alcohols,

2-propanol and 2-butanol (most of the 2-butanol which is manufactured is converted to MEK). As with the conversion of methanol to formaldehyde (11.3.1), processes for both dehydrogenation and oxidation have been designed, although the straight dehydrogenation route seems to be the preferred process. Thermodynamically, the dehydrogenation of the lower alcohols to their ketones is quite feasible at ordinary temperatures. Equilibrium conversion to the ketone is 84% at 225°C, 97% at 325°C, and effectively 100% at greater than 500°C. However, side reactions such as the dehydration of 2-propanol occur at these higher temperatures so the optimum temperature is in the range 300 to 400°C. Both dehydrogenation and oxidation reactions are promoted by the same catalysts but slightly lower yields are obtained from the oxidation method (75–85%) than with dehydrogenation (85–90%). The best catalysts for both methods are the oxides of copper or zinc or, in the case of dehydrogenation, copper metal, either alone or alloyed with zinc (i.e. brass). Dehydrogenation will also proceed on a number of metals including palladium and platinum, and the use of Raney nickel has been described in a liquid phase process operating at 100° to 250°C under pressure.

The direct oxidation of propylene to acetone and either 1-butene or 2-butene to methyl ethyl ketone is possible by an extension of the Wacker ethylene chemistry. In an outline of the ketone processes, Smidt[19] claims that yields of 92–94% acetone and 85–88% MEK are attainable. In the case of propylene the by-products consist of 0.5–1.5% propionaldehyde, 2–4% chlorinated compounds, 0.8–1.4% carbon dioxide and 0.5–1.5% other substances; the overall reaction rate is about one-third that of ethylene. From butene oxidation, by-products obtained are 4% *n*-butyraldehyde, 4–6% chlorinated material, 0.5–1% carbon dioxide and 2–2.5% other substances, the overall reaction rate being about one-quarter that of ethylene.

The separation of ketone from the crude product presents more problems with MEK than with acetone, partly because of its lower volatility and partly because it forms an azeotrope with water. Smidt[19] advances the view that a two-stage process is to be preferred because low-grade olefin can be utilised, but in the case of acetone this may not necessarily be an advantage if polymer-grade propylene is available. However, there is an abundance of acetone from the cumene–phenol process in the West—the Wacker route to acetone is not known to be in use in the United States and this probably reflects the European position, but two plants (55 000 and 36 000 tpa) are known to be operational in Japan. There is nothing to suggest that the MEK process has yet been commercialised.

The basic mechanistic features of the Wacker chemistry have been outlined (11.3.7). Broadly, the reaction of olefins under these conditions is general for the lower members at least. In the case of α-olefins, methyl ketones are formed.

$$CH_2{=}CH.R \xrightarrow[\text{chemistry}]{\text{Wacker}} CH_3.CO.R \qquad (R = Me, Et, Pr\ etc.)$$

This is in consequence of nucleophilic attack leading to the simple Markownikoff product. Aldehydes are formed as a result of anti-Markownikoff attack, but to a degree which depends on a number of factors including the structure of the olefin (where steric effects are important), and on the pH and temperature of the reacting system. Clearly electron-withdrawing and -releasing substituent groups present in the olefin will also have an effect on the ketone/aldehyde ratio

11.4 The manufacture of unsaturated aldehydes and ketones

By now it will be evident to the reader that the chemical industry sustained many changes in manufacturing technology at the beginning of the petrochemical era, that is, with the arrival of cheap naphtha. The direct manufacture of acrolein from propylene is another example of such change. It is also important in that its discovery and development also led to the discovery and utilisation of heterogeneous oxidation processes employing "mixed oxide" catalysts which have played an important part in the full exploitation of many petrochemical feedstocks. Briefly, many metal oxides exhibit the ability to catalyse hydrocarbon oxidation, albeit rather poorly. When some of these are intimately mixed, greatly enhanced catalytic behaviour is observed not only in terms of activity but also in selectivity.

As illustrations of this dramatic improvement, two results for the oxidation of propylene to acrolein may be quoted.

(i) *Tin oxide, SnO_2 and antimony oxide, Sb_2O_4*

SnO_2	20% propylene conversion at 11% selectivity at 380°C
Sb_2O_4	20% propylene conversion at 5% selectivity at 520°C
$SnO_2 + Sb_2O_4$ (in 1:1 ratio)	20% propylene conversion at 59% selectivity at 339°C

(ii) *Bismuth oxide, Bi_2O_3 and molybdenum oxide, MoO_3*

propylene oxidation at 450°C

Bi_2O_3	1.0 relative specific rate at 0% selectivity
MoO_3	0.28 relative specific rate at 87% selectivity
$Bi_2O_3 + MoO_3$ (in 1:2 ratio)	13.1 relative specific rate at 94% selectivity

The discovery that simple mixtures such as these could lead to such improvements in catalysis led, inevitably, to multicomponent systems and an enormous patent literature, largely based on claimed improvements in performance, has accumulated over the years. However, such multicomponent oxide catalysts are the basis for the majority of processes described in the remainder of this chapter so it is useful to consider the acrolein process as a detailed example of this chemistry. Moreover the oxidation of propylene to acrolein on metal oxide catalysis is by far the most closely-studied reaction in catalytic heterogeneous oxidation (the corresponding reaction in homogeneous oxidation is the Wacker conversion of ethylene to acetaldehyde).

11.4.1 *The chemistry of acrolein and its manufacture*

Acrolein, the simplest unsaturated aldehyde, has a high capacity to react in many different ways due to the presence of an aldehyde carbonyl group in conjugation with an olefinic double bond. However, commercial utilisation of this reactive intermediate is perhaps surprisingly low—no useful applications of acrolein polymer have been found. This has consequently limited its large-scale manufacture. Nevertheless, in a relatively minor role, its usefulness should not be underestimated (see Table 11.1). In addition, α-methacrolein, $CH_2{=}C(CH_3)CHO$, is very similar to acrolein and may be made from the oxidation of isobutene in an analogous manner to acrolein. No unique applications of methacrolein are known and it is understood that it is not commercially produced at the present time.

Before the advent of the direct, propylene-based route, acrolein was manufactured by promoting the vapour phase aldol condensation reaction between acetaldehyde and formaldehyde with a calcium phosphate catalyst. The direct propylene to acrolein route was first disclosed in a Shell patent in 1948[20] using cuprous oxide supported on silicon carbide as the catalyst. Selectivities of 60–85% to acrolein were reported at 10–20% propylene conversion in the range 300–400°C. Shell developed and commercially operated a process based on this catalyst from 1960, but very little information on this has appeared in the literature. However, to judge from the patent literature, it is reasonable to suppose that the process would have been operated at low propylene conversion and would have required recycle of unreacted olefin.

The disclosure by Shell stimulated much research on the catalytic action of metal oxides for olefin oxidation, and since then the patent literature in this area has grown enormously in the search for more selective catalysts. By far the most important of these are based on the mixed oxides of bismuth and molybdenum. Other notable catalysts discovered during this period are the antimony oxide–tin oxide system and the antimony oxide–uranium oxide system, both of which have been used commercially. As far as is known, most acrolein plants today use a catalyst based on bismuth and molybdenum oxides which are often referred to as "bismuth molybdate" catalysts.

The bismuth molybdate catalyst was patented by Sohio in 1959–1960 for the oxidation of propylene to acrolein, and (in the presence of ammonia) to acrylonitrile. These discoveries are taken by many to mark the beginning of real growth in the petrochemical era. The ammoxidation of propylene to acrylonitrile, which has overshadowed the acrolein process in commercial importance, is discussed in section 11.7.2. An account of the early Sohio work on acrolein and acrylonitrile has been published.[21]

The original Sohio catalyst was a simple silica-supported bismuth phosphomolybdate having a typical empirical formula $Bi_9PMo_{12}O_{52}$. In a fluidised-bed reactor at 425°C and employing a feed-stream of propylene/air/water vapour for a contact time of 3 secs, 57% conversion of propylene was measured at a selectivity to acrolein of 72%. Thus one of the advantages of the new catalyst compared to the Shell catalyst was that the reaction could be run at much higher single pass yields. Another significant advantage was that the catalyst was fairly tolerant to the amount of oxygen in the gas phase and in fact could be operated with almost the stoichiometric requirement. This enabled Sohio to design a high-conversion process in which almost all the propylene was consumed, thus avoiding costly recycle. During the history of the Sohio process, catalyst performance has been developed and improved by Sohio or its licensees with the result that present-day bismuth molybdate catalysts are quite complex systems containing, in addition to phosphorus, elements such as nickel, iron, manganese and alkali and alkaline earth metals. Acrolein catalysts are suitable for fixed-bed or fluid-bed operation, the latter being preferred for large capacity installations. The life of the catalyst can be as long as three years.

In the operation of a typical process, a mixture of propylene, steam and air or oxygen are pre-mixed and passed through a stainless steel reactor at slightly greater than atmospheric pressure. At a reaction temperature of 400°C, olefin conversion is about 90% with a selectivity to acrolein of 78% (i.e. giving an acrolein yield of 70%). By-products are carbon dioxide, acrylic acid, acetaldehyde and acetic acid. The acrolein eventually recovered is 98% pure and if acrylic acid is recovered, it can be processed to a final product of 99.5% purity.

11.4.2 *The mechanism of the acrolein synthesis*

The results of mechanistic studies of acrolein coxidation have many features in common with other so-called "allylic" oxidation reactions forming the basis of industrial processes and which will be described in the following sections. In order to present a coherent account of current mechanistic views for this type of reaction on metal oxide catalysts, a discussion of the relevant acrolein chemistry will be given in a more general account in sections 11.9–11.9.4.

11.5 The manufacture of acids and acid anhydrides

Of the lower aliphatic monobasic acids, acetic acid is the only one produced in large quantities. Formic acid lies a long way behind, with the rest having only speciality interest.

11.5.1 *Formic and acetic acid and acetic anhydride*

It has been said that in the world of chemicals manufacture, acetic acid is to organic chemicals what sulphuric acid is to inorganics—its rate of production is an index of the prosperity of the industry. Some figures reflect its utility. The West European capacity in 1979 was 1.24 million tpa (to be increased to 1.45 million tonnes). Of this, 33% went into the manufacture of polyvinyl acetate, 20% into acetic anhydride, 16% into solvent esters. The other 31% was absorbed in the production of chloroacetic acids and other miscellaneous derivatives. By contrast, very limited application has been found for formic acid (current world production about 100 000 tonnes), its principal use being as a texturing agent in the leather industry.

Formic acid is not, perhaps surprisingly, directly manufactured by the further oxidation of formaldehyde but by the carbonylation of water or methanol under pressure at about 150°C.

$$CO + ROH \xrightarrow{\text{base}} H.COOR \qquad (R = H, CH_3)$$

Suitable base catalysts include sodium or calcium hydroxide, or if methanol is used, sodium methylate. In the latter case, free formic acid may be obtained easily by the hydrolysis of the methyl ester product. However, the major source of formic acid is its by-product formation in various non-catalytic oxidation reactions or oxidative degradations, examples of which will be mentioned in connection with acetic acid manufacture.

The enzymic oxidation of ethanol to acetic acid is probably second only to the production of the alcohol itself in man's development of the use of chemistry and this biological process is still the basis of much of the acid consumed as vinegar and in other feedstuffs applications. However, the acid obtained by this method is dilute and such a method would be clumsy and inefficient applied to the manufacture of glacial acetic acid. This may be synthesised by three major routes (see also section 8.5.1). The relative contribution to overall production in 1980 is given in parentheses:

(a) non-catalytic oxidation of acetaldehyde (70 %);
(b) catalytic and non-catalytic oxidative degradation of alkenes and alkanes (25 %);
(c) the carbonylation of methanol (3 %).

Competitive with petrochemical-based routes even when it was introduced in 1970, the methanol-based process has no equal today and undoubtedly will become the only significant route for the next few years. It is estimated that it will account for 48 % of acetic acid manufacture by 1983. The next logical development will be the direct synthesis of acetic acid from carbon monoxide and hydrogen, and patents have been issued to UCC covering such a process using a supported rhodium catalyst at pressures of about 40 atm.

Although the other methods of manufacturing acetic acid are rapidly becoming obsolete, it is worth mentioning them in view of their current usefulness as a source of other co-products.

Briefly, the current ethylene/acetaldehyde route is based on the oxidation of acetaldehyde by peracetic acid with or without a redox catalyst such as cobalt or manganese acetate. The peracetic acid may be generated *in situ* non-catalytically. By-products of the reaction are carbon dioxide, methanol, formaldehyde, formic acid and methyl formate.

Before 1980, the oxidative degradation of the lower hydrocarbons has been the other major route to acetic acid and the process is conducted in such a way that the major by-products are also recovered. For example a Celanese process which has been operated on a large scale involves the liquid phase oxidation of *n*-butene with oxygen at 170–180°C at 50 bar pressure with cobalt acetate as catalyst. As well as acetic acid (the major product), methyl ethyl ketone, ethyl acetate and acetone are recovered, minor by-products being recycled for further oxidation to acetic acid or carbon dioxide. BP operate a non-catalytic process whereby light naphtha containing C_5 to C_7 hydrocarbons is oxidised at about 180°C at 50 bar. One advantage of the process is the use of a feedstock taken from the oil refinery without any separation to individual components. Formic and propionic acids and acetone are recovered as by-products, in total between 0.35 and 0.75 tonnes per tonne of acetic acid, depending on the operating conditions.

Acetic anhydride is currently manufactured by one of two methods; one involves a modified acetaldehyde oxidation process and the other starts with acetic acid. It would be fair to expect the latter route to replace the former owing to the availability of acetic acid from synthesis gas. However, the acid-based process depends on the addition of ketene to acetic acid. Ketene is made by dehydrating acetic acid at 700°C, that is, in a very energy-intensive step.

$$CH_3.COOH \xrightarrow[(EtO)_3PO]{700^\circ C} \underset{\text{ketene}}{CH_2{=}C{=}O} \xrightarrow{CH_3COOH} (CH_3CO)_2O$$

A recent announcement by Halcon and Eastman Kodak has indicated that the intermediate ketene step may be bypassed and acetic anhydride made directly from methyl acetate and carbon monoxide.

$$CH_3.CO.OCH_3 \xrightarrow{CO} (CH_3CO)_2O$$

No details have been disclosed beyond the statement that there are no recoverable by-products, the selectivity being "well over" 90%. Eastman are commercialising this process and the first plant is to come on-stream in 1983 (cf. 8.5.2).

11.5.2 *The manufacture of acrylic and methacrylic acid*

Acrylic and methacrylic acids are produced almost exclusively for their conversion via their esters to the polyacrylate or polymethacrylate resins. Acrylates, the important ones being methyl, ethyl and butyl, are made directly by esterification with the appropriate alcohol. The most important methacrylate ester is methyl methacrylate which is made by a non-oxidative method from acetone. Higher methacrylates are made either by transesterification with the appropriate alcohol or by direct esterification using the free acid. Examples of these are the stearyl and lauryl esters, made by both methods, which are used to control the viscosity of lubricating oils. Polymethyl methacrylate is by far the most important acrylic resin and the sheet form is known by various trade names such as "Perspex" (ICI) and "Lucite" (du Pont). It is also used as a moulding resin for the manufacture of a wide variety of utility objects.

Various non-oxidative processes are in current use for the manufacture of acrylic acid, which is captively processed to the polyester. These are multistage processes based on ethylene oxide, acetylene or acrylonitrile. New plant installations, however, increasingly rely on the direct oxidation of propylene. This is achieved with optimum selectivity by using a tandem two-stage reactor where propylene is first converted to acrolein using catalysts and conditions described in section 11.4.1. The product stream, containing mostly acrolein and a little acrylic acid, is passed to a second reactor where conversion of the aldehyde to acrylic acid occurs. Here the catalyst is usually based on molybdenum-vanadium mixed oxides containing one or more of elements such as phosphorus, tin, antimony, tellurium and cobalt. Operation is at a significantly lower temperature (250–300°C) than the acrolein first stage. The overall conversion of propylene to acrylic acid is greater than 95 % with a selectivity of 85–90 % (note: to obtain this overall selectivity a minimum selectivity of about 92–95 % is required in each stage).

The slow step in the reaction is the primary oxidation of propylene, the oxidation of acrolein being at least an order of magnitude faster. Because a rise in temperature nearly always results in reduced selectivity due to combustion, it is clear that a direct one-stage process would at best be a compromise of the optimum conditions for each step.

The oxidation of isobutene to methacrylic acid can be carried out in a similar fashion. Indeed, there are many incentives for replacing the acetone-based route for methyl methacrylate which has been the sole method for forty-five years.

$$CH_3COCH_3 \xrightarrow{HCN} (CH_3)_2C(CN)(OH) \xrightarrow{H_2SO_4} CH_2{=}C(CH_3){-}CO{-}NH_3.HSO_4$$

$$CH_2{=}C(CH_3){-}CO{-}NH_3.HSO_4 \xrightarrow{CH_3OH} NH_4HSO_4 + CH_2{=}C(CH_3){-}COOCH_3$$

Handling hydrogen cyanide in quantity is hazardous. Furthermore, its current major source is as byproduct in the manufacture of acrylonitrile (11.7.2) and hence its availability is dependent on the demand for the latter. Finally sulphuric acid is consumed and a low-value by-product, ammonium hydrogen sulphate, produced. Consequently, there has been considerable development work toward alternative routes to methacrylic acid. However, it has been difficult to achieve economic yields involving two-stage isobutene oxidation, mainly because of poor selectivity in the second, methacrolein oxidation, stage. To be economic, first-stage performance should exhibit about 87% selectivity at 96% conversion, with second-stage figures of 75–85% selectivity at 85% conversion. The first successful process of this type has been announced by Mitsubishi Rayon who hope to have a plant on-stream in 1982, following successful pilot-plant operation where yields of methacrylic acid greater than 60% (based on isobutene) have been obtained. The process does not require pure isobutene, but utilises the C_4 fraction from a naphtha cracker after butadiene extraction.* Isobutene is selectively extracted by liquid-phase hydration to *t*-butanol which is then oxidised at 360° to methacrolein in a fixed-bed reactor containing a multi-component mixed oxide catalyst. How complex the catalyst is can be judged from an example from Mitsubishi patents; excluding oxygen, the empirical formula is $Mo_{12}Ni_6Bi_2Fe_2Sb_2Co_2Cs_{0.5}Zn_{0.5}$. The second stage catalyst is of similar complexity, containing predominantly molybdenum, arsenic and phosphorus. Further developments in isobutene oxidation are to be expected, as Sumitomo, Oxirane and Sohio, among others, are known to be active in process development.

11.5.3 *The oxidation of aromatics*

In the area of large tonnage chemicals, the oxidation of aromatics is limited to but a few. These are benzene (to maleic anhydride) toluene (to benzoic acid), *o*-xylene and naphthalene (to phthalic anhydride) and *p*-xylene (to terephthalic acid).

Commercially very important resins† are manufactured from the branched polyesters of polyhydric alcohols and polybasic acids modified by a variety of additional materials such as vinyl acetate, styrene and fatty acids. These give the resins a very wide range of applications. The most widely-used resins contain phthalic acid and (rather less) maleic acid.

Maleic acid is manufactured as the anhydride largely by the oxidation of benzene. Although accounting for only 4% of benzene consumption, this route is being challenged by processes that utilise *n*-butenes or, more attractively *n*-butane. As well as theoretically being more efficient in the use of carbon, the C_4 route also makes further use of the C_4 raffinate (the linear butenes in this application). The move away from benzene as feedstock began about 1960 and currently about 17% of the world's maleic acid is made from a C_4 source. Originally phthalic anhydride was made by the oxidation of coal-tar naphthalene, but once cheap *o*-xylene became available, the inevitable change-over in feedstock occurred. Today, only about 16% of the world's capacity for phthalic anhydride is based on naphthalene, and it is located mostly in the United States and Japan.

* This is often referred to as "C_4 raffinate"—raffinate means "spent liquor".

† The word resin is used to describe polymer formulations that harden through cross-linking on exposure to air, light, heat or a catalyst.

Commercial processes for maleic and phthalic anhydrides are quite similar. Usually oxidation of gaseous feedstock mixed with air (in proportions below the lower explosive limit) is performed in a fixed bed multi-tubular reactor to high conversion. The reactor is operated isothermally and heat from the considerably exothermic reaction is removed by a circulating molten salt bath. The alternative to the multi-tubular salt-cooled reactor, a fluidised bed reactor is used to a much lesser extent.

In recovery of the product, crude phthalic anhydride is condensed (m.p. 130°C) and purified by vacuum distillation. A preliminary stage in the case of maleic anhydride consists of scrubbing the product stream with water, yielding maleic acid solution, from which maleic anhydride is extracted before final distillation.

Oxidation reactions forming maleic and phthalic anhydrides are characterised primarily by oxide catalysts containing vanadium. Indeed, vanadium pentoxide based catalysts have been used since the earliest commercial exploitation of these reactions and no better catalysts have yet been discovered. One of the most commonly used catalysts, both commercially and for scientific study, is composed of a mixture of 70 mole % vanadium pentoxide and 30 mole % molybdenum trioxide. It is believed that molybdenum (VI) forms a solid solution in vanadium pentoxide with a maximum solubility of 15 atom %. Charge balancing is maintained with the corresponding reduction of vanadium (V) to vanadium (IV). The unique position of the vanadium-molybdenum series is demonstrated by comparing the relative specific activities of simple molybdenum ternary systems for the oxidation of benzene at 400°C.

Catalyst	V/Mo	Ti/Mo	Sn/Mo	Sb/Mo	U/Mo	Fe/Mo	Bi/Mo
Relative spec. act.	100	11.6	35.1	3.6	6.4	2.1	0.17
Initial selectivity (maleic anhydride %)	55	47	32	51	36	30	12
Max. yield (maleic anhydride %)	30	30	23.5	20	6.2	0.12	0.13

It is noteworthy that bismuth molybdate and the vanadium/molybdenum catalyst lie at opposite ends of the series, reflecting important differences in the basic chemistry of oxidising an aromatic structure to the anhydride, or for that matter, oxidising linear butenes to maleic anhydride rather than butadiene.

The chief process licensor and manufacturer of phthalic anhydride using fixed-bed reactors is von Heyden who have a process which can utilise either naphthalene or *o*-xylene and accounts for 45 % of the world's output of phthalic anhydride. BASF are the other major producer, using *o*-xylene as feedstock. UCB and Bayer are important manufacturers of maleic anhydride from benzene, and Bayer also operate a process based on butene oxidation. Table 11.3 summarises general features of the various processes and illustrates the similarities between them.

Table 11.3 Principal features in the manufacture of phthalic and maleic anhydride

Process			*Temp.* (°C)	*Catalyst*	*Contact time* (sec)	% *Molar selectivity*	% *Molar yield*
o-xylene	⟶	PA	375–410	V_2O_5/TiO_2 on quartz or SiC spheres	0.1–1	78	70–75
naphthalene	⟶	PA	350–400[1]	V_2O_5 on Al_2O_3	0.1–1	86–90	80
			400–550[2]	V_2O_5/K_2SO_4 on Al_2O_3		60–74	
benzene	⟶	MA	400–450	V_2O_5/MoO_3 on Al_2O_3	0.1	50–60	43–57
			355–375	V_2O_5/MoO_3	—	95	—
C_4's	⟶	MA	350–450	V_2O_5/H_3PO_4	—	50–60	—

[1] Using pure naphthalene
[2] Using impure naphthalene feedstock, when 5–10% maleic anhydride by-product is formed.

11.5.4 *Terephthalic and benzoic acid*

These acids are manufactured by the oxidation of *p*-xylene and toluene respectively and differ from the aromatic oxidations described above in that the conversions are performed in the liquid phase with homogeneous catalysts. Terephthalic acid, or its dimethyl ester, is produced almost exclusively for the production of polyethylene terephthalate fibres and to a much lesser extent, films. Benzoic acid finds many uses as an intermediate. To a very minor extent it is used in phenol manufacture through oxidative decarboxylation.

The methyl groups in both toluene and *p*-xylene can be converted to carboxylic groups by a free radical reaction, catalysed by transition metal salts such as cobalt and manganese acetate. In the case of *p*-xylene the oxidation stops at the formation of *p*-toluic acid due to the retarding effect of the existing carboxylic group at the *p*-position. The oxidation of the second methyl group is achieved in one of two ways. In one the catalysed oxidation is supplemented by the addition of bromine compounds or by hydroperoxides generated *in situ* by co-oxidising acetaldehyde or methyl ethyl ketone. In the second method, the electron-withdrawing effect of the *p*-toluic acid group is reduced by esterification with methanol, either in a step subsequent to the initial oxidation, or simultaneously if methanol is used as a solvent in this stage. In this case dimethyl terephthalate is produced directly in one step.

Each of these methods is used commercially to produce terephthalic acid or dimethyl terephthalate. Although either product can be used in the manufacture of polyethylene terephthalate either by esterification or transesterification, the commercial production of each has advantages and disadvantages. The crude terephthalic acid product produced by any of the methods described above is recovered from the product stream in crystalline form and is about 99 % pure, containing trace amounts of *p*-toluic acid and 4-carboxy benzaldehyde. The removal of these is essential for polymer grade material. However, purification by repeated recrystallisation is costly as terephthalic aciid is relatively insoluble in water and would require solvents such as dimethyl sulphoxide. Dimethyl terephthalate, on the other hand, has a much greater solubility in a range of solvents. The two moles of methanol added during manufacture are recovered during polymerisation and, although this makes purification simpler, it constitutes a basic economic drawback; each tonne of dimethyl terephthalate contains only 0.86 te of useful monomer. This has resulted in improvements and amendments being made to the above processes for the production of polymer grade acid. One of these, the Amoco process, accounts for 80 % of the world output of fibre-grade terephthalic acid.

The Amoco route is a continuous process in which the catalyst is a mixture of cobalt and manganese acetate together with ammonium bromide and/or tetrabromomethane as co-catalysts and dissolved in acetic acid. In the first stage of the process the oxidation of *p*-xylene takes place at about 200°C at 15–30 bar using air. During reaction, terephthalic acid crystallises and is separated from the product-slurry. In the second (purification) stage the crude acid is dissolved in water and passed through a bed of a hydrogenation catalyst such as palladium on charcoal and super-heated to about 250°C (10 te of water are required to dissolve each tonne of crude acid at this temperature). On addition of hydrogen any aldehyde groups present are converted to methyl groups. Cooling and washing results in terephthalic acid of 99.99 % purity, suitable for polymerisation. Process selectivity of 90 % is achieved at 95 % *p*-xylene conversion.

CHO / CH_3 CHO / COOH $\xrightarrow{Pd/H_2}$ CH_3 / CH_3 CH_3 / COOH

Typical scale of operation is about 100–300 000 tpa.

Benzoic acid is manufactured from toluene by similar methods. The Amoco process is operated for toluene conversion where 96 % selectivity is achieved at virtually complete conversion. In the similar Snia Viscosa process, toluene containing 0.002 % w/w cobalt acetate is oxidised by air at 165°C and 9 bar, when the yield of benzoic acid is about 94 %.

11.6 The oxidative removal of hydrogen

In the next few sections we shall consider processes which involve the combination of a hydrocarbon and oxygen and possibly ammonia, but where the organic product of the reaction does not contain oxygen. The essential role of oxygen is to "strip" hydrogen from the reacting species, so that co-product water is formed. The free energy of formation of water is large enough to displace any equilibria involving loss of hydrogen in favour of products and the heat of formation is sufficient to convert an otherwise endothermic process into an exothermic one.

11.6.1 *The manufacture of butadiene by oxidative dehydrogenation*

Oxidative dehydrogenation is the term used to describe any reaction of the type where a dehydrogenation reaction is performed in the presence of oxygen. Water, not hydrogen, is formed as a co-product. There are two important consequences of this. The first, and most important, is that for a given temperature the free energy change accompanying simple dehydrogenation may be so unfavourable that negligible equilibrium amounts of product are formed. However, if oxygen is added and water is formed as product, the free energy change is now so favourable that the reaction will proceed to completion, provided a suitable catalyst is present. This is seen in the dehydrogenation of 1-butene to butadiene at 500°C and atmospheric pressure.

	$\Delta G_{500°C}$ (kJ/mole)	*equil. % conv.*	$\Delta H_{500°C}$ (kJ/mole)
$C_4H_8 \rightleftharpoons C_4H_6 + H_2$	28	11.5	117
$C_4H_8 + \frac{1}{2}O_2 \rightarrow C_4H_6 + H_2O$	−303	100*	−240
	(*$K = 3 \times 10^{20}$)		

The second consequence of forming water rather than hydrogen as co-product has been illustrated for methanol dehydrogenation (11.3.1). That is, even if the equilibrium is favourable with respect to product at a given temperature, dehydrogenation reactions are endothermic. Valuable savings in heating costs can therefore be made if the reaction can be made exothermic by the introduction of oxygen to augment the system with the heat of combustion of hydrogen.

The problems of large scale production of butadiene by catalytic dehydrogenation are described in section 12.10.1. Apart from the high temperatures required (600–700°C), cracking, catalyst deactivation by coking and relatively poor yields (about 50%) have led to the replacement of such processes by ones based on oxidative dehydrogenation. In the United States more than 75% of butene-derived butadiene is now manufactured by oxidative dehydrogenation, where better yields at lower temperatures, and robust long-life catalysts have made this process very attractive. The most widely adopted process is that due to Petrotex—the so-called "Oxo-D" process. In this, a fixed-bed reactor is operated at about atmospheric pressure. Air or oxygen or a combination of the two is mixed with butene and reacted so that all the oxygen is consumed. A mixture of linear butenes may be used, but isobutene must be extracted beforehand as it oxidises largely to carbon dioxide and water. Not much detailed information is available on the Petrotex process but judging from published pilot plant figures the adiabatic reactor temperature rises from about 340° to 580°C. Selectivities of nearly 80% are obtained at 75% conversion with about 10% loss of carbon as carbon oxides. Trace quantities of vinyl acetylene, furan and formaldehyde are formed. Unreacted butenes may be flared or recycled. The catalyst used is very probably an iron spinel. Petrotex have a wide patent coverage in these materials, which are very selective for olefin dehydrogenation, even at high conversion. Such catalysts are based on the basic spinel structure $A^{III}Fe^{III}O_4$ where A could be magnesium and/or zinc etc. In practice the catalysts are likely to contain other bivalent and trivalent cations such as cobalt, chromium and copper. Hucknall[1] has tabulated the more important oxidative dehydrogenation catalysts. In the very similar, Phillips O—X—D process, the catalysts are believed to be similar to those which are active in propylene and isobutene oxidation to acrolein and methacrolein.

11.7 The manufacture of nitriles by ammoxidation

Ammoxidation or, less commonly, oxidative ammonolysis, is the oxidation of hydrocarbons, especially olefins, in the presence of ammonia according to the reaction

$$R.CH_3 + NH_3 + \tfrac{3}{2}O_2 \xrightarrow[\text{cat.}]{} R.C{\equiv}N + 3H_2O$$

where R=H, alkyl, alkenyl.

With olefins, the nitrile functional group is introduced at the carbon atom adjacent to the double bond. In commercial practice, the only olefin reaction of significance is propylene ammoxidation to acrylonitrile.

11.7.1 *The manufacture of hydrogen cyanide*

Hydrogen cyanide is required as an intermediate in many applications, some of which are mentioned in this chapter and in other parts of the book. Although the simplest example of this class of reaction, the ammoxidation of methane to hydrogen cyanide is quite different in its characteristics from the corresponding olefin chemistry, and forms the basis of the long-established Andrussow process. The application of this process has declined somewhat due to the availability of by-product hydrogen cyanide from propylene ammoxidation (11.7.2) but it is nevertheless the most important direct method of production at the present time. Furthermore, for captive use, it may be beneficial to have a reliable supply based on an independent route, rather than rely on availability from acrylonitrile plants whose output is tied to the notoriously cyclic fibre-production sector.

A mixture, consisting approximately of 12% ammonia, 13% methane and 75% air at a pressure of about 2 bar is passed at high velocity through a catalyst in the form of a wire-gauze made from platinum, or better, a (10%) rhodium–(90%) platinum alloy; the latter has greater thermal and mechanical strength. The reactor is operated adiabatically at 1100°C with a catalyst contact time of about 10^{-4} sec. The resulting reaction is complex and, although little studied, is believed to proceed through mixed surface-gas phase stages in which 60–65% of the ammonia is converted to cyanide and 10% to nitrogen. A minor amount of carbon monoxide is also formed. After reaction, the product stream is rapidly quenched to below 400°C and unreacted ammonia is either removed as ammonium sulphate or recovered for recycle. The world capacity for direct synthesis of hydrogen cyanide is about 600 000 tonnes.

11.7.2 *The manufacture of acrylonitrile*

The importance of acrylonitrile as a raw material in the polymer field is immense. This can be judged from the numerous applications described in chapter 10 which embrace a whole range of products from elastomers and fibres to thermoplastics. The first commercial application in the 1930's was in the manufacture of synthetic rubber (nitrile-rubber or "buna-N") from the co-polymerisation of acrylonitrile and butadiene. This was manufactured by the addition of hydrogen cyanide to ethylene oxide in a two-stage process.

$$\underset{\diagdown O \diagup}{CH_2{-}CH_2} + HCN \xrightarrow[60°]{\text{aq. base cat.}} \underset{\text{ethylene cyanohydrin}}{HOCH_2.CH_2CN} \xrightarrow[200°C]{} CH_2{=}CHCN + H_2O$$

This gave way to a route based on the direct addition of hydrogen cyanide to acetylene in dilute hydrochloric acid at about 90°C. This process had two major disadvantages, namely, the use of high-cost raw materials and the tendency of the reactor system toward deactivation through polymerisation reactions. Nevertheless, this was the major route to acrylonitrile until the 1960's when it was quickly rendered obsolete by the direct ammoxidation of propylene to acrylonitrile in a relatively simple, single-pass, one-stage process, a development which transformed this area in less than a decade. The economic effect of this on the business is illustrated in section 4.5.

The reaction,

$$CH_2{=}CH.CH_3 + NH_3 + \tfrac{3}{2}O_2 \xrightarrow{\text{cat.}} CH_2{=}CN.CN + 3H_2O$$

was first reported in a U.S. patent to Allied Corp. in 1949, but it was not until 1957 that Sohio found that the recently discovered acrolein catalysts such as bismuth molybdate could also catalyse the ammoxidation reaction in high yield. The first patent was granted in 1960 and the first plant came on stream in the same year—an example of remarkably rapid development. The advent of the Sohio acrylonitrile process is one of the milestones in the story of catalytic petrochemistry. Virtually all of the world's acrylonitrile (nearly four million tpa) is now made by propylene ammoxidation, mostly by Sohio technology. Excellent accounts of the process and its chemistry exist,[1,22] so an abbreviated treatment will be given here, since in many respects the reaction and technology are similar to other olefin oxidation processes which have already been described in more detail in this chapter.

In the operation of the process approximately equimolar quantities of ammonia and propylene are mixed with air and exposed to the catalyst at 400–500°C at about atmospheric pressure and at a contact time of between 1–15 sec. At total conversion of propylene the selectivity is 65–70 %. The by-products are carbon dioxide, acetonitrile and hydrogen cyanide. Additionally, trace amounts of acrolein and nitrogen (from ammonia combustion) may be formed. The acetonitrile and hydrogen cyanide by-products, although small in percentage terms, represent considerable quantities of these substances. Thus for every 1000 tonnes of acrylonitrile produced, about 40 tonnes of acetonitrile and 180 tonnes of hydrogen cyanide are formed. Plant sizes range from 60 000 to 200 000 tpa so the acrylonitrile process provides a useful supplementary source for hydrogen cyanide. There is, however, little demand for acetonitrile, beyond a minor use as a solvent, so this is usually disposed of by burning. Acrylonitrile is recovered from the product stream with other organic products by scrubbing with water and after several distillation stages is obtained at greater than 99 % purity.

Both multi-tubular and fluidised bed reactors are in use, although the latter are favoured by Sohio. The original catalyst employed by them was bismuth phosphomolybdate. Since then a number of catalysts have been discovered and over the years, improvements in selectivity have been obtained by continued catalyst research and development. The catalyst currently employed in a major way is believed to be multicomponent containing, in addition to bismuth and molybdenum, alkali metal, alkaline earth metal, nickel or cobalt and phosphorus and/or arsenic and antimony. Previously, antimony-based oxide catalysts have been used commercially. An antimony–uranium catalyst was discovered by Sohio and employed for a short time. Antimony–iron catalysts developed in Japan by Sohio licensees have been used in a number of plants there. In addition, an antimony–tin catalyst was used commercially by Distillers in early competition with Sohio. Sohio pursue a policy of encouraging licensees to develop their own catalysts and, while this has generated many variants it has, no doubt, played a part in the continued improvements observed during the history of the process so far. Hucknall[1] has tabulated some of the more important commercially-used catalysts.

The ammoxidation of isobutene yields α-methacrylonitrile and although polymers based on this nitrile have some potential, and methacrylonitrile is a possible source for methacrylic acid, it is not certain whether isobutene ammoxidation has ever been commercialised.

11.7.3 *Aromatic ammoxidation reactions*

By and large, selective olefin oxidation catalysts are also good ammoxidation catalysts, albeit at higher temperatures. The same parallel exists with aromatics, where catalysts active enough to promote oxidation of alkyl benzenes will form the corresponding nitriles in the presence of ammonia. Vanadium and molybdenum are usually present in the more active catalysts. Thus, toluene and *p*-xylene may be ammoxidised to benzonitrile and *p*-tolunitrile. However, processes based on aromatic ammoxidation are not of any commercial significance, due either to alternative and cheaper routes or low demand for the products. One reaction which could have large industrial potential is the ammoxidation of benzene to mucononitrile (NC.CH=CH.CH=CH.CN). This could then be hydrogenated directly to hexamethylenediamine for nylon manufacture and replace the multi-stage chemistry which is now employed by a relatively simple alternative. However, the catalyst which will selectively attack the benzene ring in this manner still awaits discovery.

11.8 The manufacture of vinyl chloride by oxychlorination

Vinyl chloride is the fourth largest tonnage chemical in the world and, excluding petroleum refining, is the largest organic product of a catalytic reaction. It is, of course, produced for the manufacture of polyvinyl chloride, PVC, that most ubiquitous of polymers. In the past it was manufactured by the addition of hydrogen chloride to acetylene, but today is manufactured from chlorine and ethylene in a three-stage process, again reflecting the change to oil-derived feedstocks in the late 1950's. In principal, ethylene can be chlorinated directly to vinyl chloride in a catalyst-free reaction at 300–500°C. In practice there are a multitude of products including soot and this route does not merit attention for commercial operation. The chlorination is therefore accomplished indirectly by making the 1:2 dichloride followed by the elimination of hydrogen chloride.

$$CH_2{=}CH_2 + Cl_2 \rightarrow CH_2Cl\,.\,CH_2Cl \rightarrow CH_2{=}CHCl + HCl$$

Chlorine for the process is obtained by the electrolysis of brine.

Although the reaction can be conducted in the gas phase, most processes employ liquid phase conditions where EDC, containing ferric chloride as catalyst, is used as the reaction medium. Chlorination is effected at 40–100°C at 1–10 bar. The efficiency of this process is very high and can result in crude EDC of 99.8 % purity.

The second stage (dehydrochlorination) consists of a simple thermal process carried out at 450–600°C at 10–35 bar. The reaction occurs by a free-radical chain mechanism, so careful control of the reactor residence time and temperature is necessary. The conversion is usually limited to about 50% at which a selectivity to vinyl chloride of greater than 98% may be realised. The product stream is quenched and unreacted EDC recycled after separation. Purification of vinyl chloride is accomplished by distillation.

What makes the total process economically attractive is the fact that in the third stage, use is made of the hydrogen chloride product, representing 50% of the chlorine consumption, which would otherwise be of low value. In this stage, oxychlorination of ethylene to EDC is accomplished according to the reaction

$$CH_2{=}CH_2 + 2HCl + \tfrac{1}{2}O_2 \xrightarrow[CuCl_2]{250^\circ} CH_2ClCH_2Cl + H_2O$$

In this, the third stage, ethylene, anhydrous (i.e. gaseous) hydrogen chloride and air or oxygen are reacted in the presence of a catalyst at 200–300°C and 2–10 bar. The basic catalyst is cupric chloride supported on, for example, alumina and contains additives such as alkali metal chlorides and, occasionally, rare earth oxides or chlorides. The alkali chlorides reduce the melting point and vapour pressure of the cupric chloride—the catalyst is more stable and active in this form. The rare earth additives are believed to enhance these qualities further. The reactor may be a fluidised-bed or multi-tubular, fixed-bed type. The reaction is highly exothermic and careful control of temperature is required to prevent over-chlorination or combustion of ethylene. However, overall conversions and selectivities in the mid-90's are attained (based on ethylene), the by-products being carbon oxides and C_1 and C_2 chlorocarbons. EDC is purified by distillation. Clearly, the overall process can be operated in balance with respect to hydrogen chloride or to product or consume it by simply altering the ratio of oxychlorination to direct chlorination.

Not much has been written on the mechanism of the ethylene oxychlorination reaction.[23] It is accepted that the role of the catalyst in this reaction is to supply chlorine according to the following cycle

$$2CuCl_2 + C_2H_4 \rightarrow 2CuCl + CH_2ClCH_2Cl \qquad \text{(A)}$$

$$2CuCl + 2HCl + \tfrac{1}{2}O_2 \rightarrow 2CuCl_2 + H_2O \qquad \text{(B)}$$

Reaction A is the rate-determining step below 300°C while above this temperature, reaction B is slower than reaction A. The regeneration of the cupric chloride is believed to occur via a bridged oxychloride species

$$2CuCl + \tfrac{1}{2}O_2 \rightleftharpoons ClCuOCuCl \overset{2HCl}{\rightleftharpoons} 2CuCl_2 + H_2O$$

The exact nature of the chlorine to ethylene transfer step is not known, nor indeed that of the steady-state composition of the catalyst. One view is that the former involves the formation of a bridged chloronium ion which is subsequently *trans*-attacked by a chloride ion.

$$CH_2{=}CH_2 + CuCl_2 \rightarrow \underset{\diagdown \oplus \diagup \atop Cl}{CH_2{-}CH_2} + Cu^+Cl \xrightarrow{Cl^-} ClCH_2{-}CH_2Cl$$

In the general current move toward cheaper processes, possible future developments in vinyl chloride manufacture might include simplification of the existing oxychlorination process

$$C_2H_4 + HCl + \tfrac{1}{2}O_2 \rightarrow CH_2{=}CHCl + H_2O$$

if sufficiently effective catalysts are discovered. In another approach, the oxychlorination of ethane is potentially attractive if it is available in sufficient quantities and at an attractive price. Direct conversion of ethane to vinyl chloride is the subject of a process offered by Lummus. Known as the Transcat process, the oxygen required for the reaction

$$2C_2H_6 + Cl_2 + \tfrac{3}{2}O_2 \rightarrow 2CH_2{=}CHCl + 3H_2O$$

is supplied solely by the catalyst which is a molten mixture of $CuO.CuCl_2$ at low concentration in $CuCl_2/KCl$. This melt is maintained at 500–600°C and is transported between the oxychlorination reactor and an oxidiser unit. The yield of vinyl chloride is said to be 90%. The major drawback is that each tonne of ethane converted requires the circulation of several hundred tonnes of molten catalyst between the bottom and top of the reactor. The process has not to date been commercialised. Other fixed-bed and fluidised bed processes have been suggested where iron seems to be a common catalyst ingredient. It is not clear if dehydrogenation of ethane to ethylene occurs as a preliminary reaction.

11.9 The mechanism of olefin oxidation and ammoxidation on mixed oxide catalysts

In this section we focus attention on the appropriate reactions and processes described in sections 11.4 to 11.7 and present the established beliefs as to how these reactions occur on mixed oxide catalysts. These reactions are, as a class, one of the most complex known in catalysis. In order to treat the subject fully, many aspects of physical, inorganic and organic chemistry must be brought to bear. The catalysts are fairly complex materials and often the identity or structure of the active phase is not clearly defined. Adequate descriptions of the surface layers of these phases are even less likely to be available. The organic chemistry which occurs on the surface is difficult to examine in detail owing to the absence of suitable probe techniques—reactive intermediates are not easily detectable in these reactions. Furthermore, there is a tendency for given reactions to occur on certain groups of catalysts and it is not easy to see why this should be so. Despite these difficulties, the study of this area of catalysis has been well sustained since its appearance on the industrial scene and there are encouraging signs that perhaps some general rationalisation of the various, sometimes disparate, observations and speculations may be possible in the not-too-distant future. There are many thorough reviews of the subject currently worth further study.[22,24,25]

A fully comprehensive account of such a complex subject is however beyond the scope of this book and the treatment here will be restricted to the principal mechanistic features of the reactions which are widely accepted.

11.9.1 *The general reaction scheme*

The bringing together of combustible material (such as an olefin) and oxygen in close proximity on a catalyst surface requires some very careful thermodynamic and kinetic balancing. The formation of carbon dioxide and water is so favourable that arresting the oxidation process short of these ultimate products demands highly restrained catalyst reactivity, and obtaining a partial oxidation product in high yield can require some very fine "tuning" of catalyst selectivity indeed. This is the basic reason why industrial oxidation catalysts seem to be such complex mixtures of major and minor components.

By and large, the catalytic oxidation of hydrocarbons proceeds via series and consecutive reactions which may be represented by the overall scheme

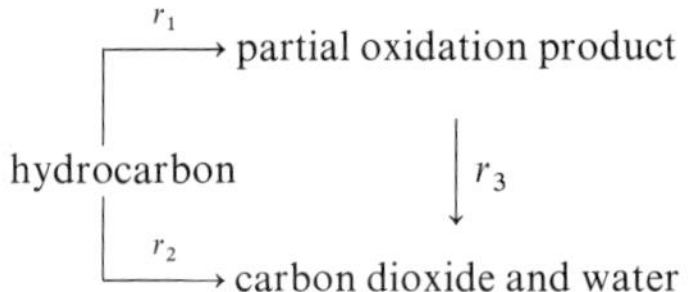

Thus at high conversion it is to be expected that carbon dioxide and water may be formed from the further oxidation of desired product by route r_3, as well as from direct combustion, route r_2. At low conversion, r_2 will predominate. Clearly, by extrapolating to zero conversion the ratio of the rate constants of r_1 and r_2 reflects the *intrinsic* or *primary* selectivity of the catalyst. Kinetic control of the reaction can be used to minimise r_3. Operating at low conversion with subsequent recycle, and using low porosity catalysts thus prevents over-oxidation of product trapped within the pores. The many illustrations in the above sections show how much value is placed on operating at high conversion, so these methods have a limited utility and the exploration for high activity, high selectivity catalysts is vigorously pursued. In other words much catalyst development is taken up with attempts to minimise the value of r_2 and r_3.

As will be seen from the above sections, the oxidation of the lower olefins is of particular interest to the chemical industry. Apart from ethylene, which tends to combust on mixed oxide catalysts, oxidation usually entails attack on the carbon atom adjacent to the double bond. Such reactions have become classed as "allylic" oxidations to distinguish them for example, from epoxide or double-bond fission reactions. Further, because of their industrial prominence and, possibly, because of their relative simplicity these reactions have been extensively studied. Unfortunately, the aromatic oxidation reactions have not been subjected to the same endeavour. Thus in the following sections much of the mechanistic information described has resulted from a study of allylic oxidation reactions and in particular, the oxidation of propylene. This is perhaps an unfortunate state of affairs and it is to be hoped that, now that some major tenets have been established in this area, more broadly based chemistry will be explored in an effort to instil greater generality into this important field.

11.9.2 *The adsorption of olefins and the allyl intermediate*

It has been established beyond doubt that the oxidation and ammoxidation of propylene to acrolein and acrylonitrile proceeds through a symmetric intermediate. The proof of this has come from the use of carbon-13, carbon-14 and deuterium labelling experiments on cuprous oxide, bismuth molybdate and phosphomolybdate, tin oxide–antimony oxide and uranium oxide–antimony oxide catalysts.

Results obtained for the reaction of ^{13}C-labelled propylene on cuprous oxide at 300°C illustrate this quite clearly.

C_3H_6 before reaction:	$CH_2{=}CH{-}^{13}CH_3$	0.96 (fraction of total ^{13}C)
	$^{13}CH_2{=}CH{-}CH_3$	0.02
acrolein product:	$CH_2{=}CH{-}^{13}CHO$	0.50
C_3H_6 unreacted:	$^{13}CH_2{=}CH{-}CH_3$	0.20

The equal distribution of the carbon-13 atom between the two end carbon atoms in the product acrolein clearly implies that oxygen attack at either end of the adsorbed propylene molecule is equally probable.

A similar effect is observed with the reaction of 3-d_1-propylene on a bismuth molybdate catalyst, $Bi_2O_3 . MoO_3$, at 460°C. Further, a primary isotope effect is observed showing that the removal of hydrogen is a slow step in the reaction. It is generally accepted that the common symmetrical species involved in these reactions is an adsorbed allyl entity. The other possible symmetrical species, an isopropyl entity, has been eliminated by oxidising propylene on a deuterated bismuth phosphomolybdate catalyst. Negligible deuterium was found in the acrolein subsequently recovered, compared with 50% deuterated acrolein expected from the isopropyl intermediate.

$$CH_2{=}CH{-}CH_2 \xrightarrow{-H} CH_2{-}CH{-}CH_2 \longrightarrow CH_2{=}CH{=}CHO$$

$$CH_2{=}CH{-}CH_2 \xrightarrow{+H} CH_3{-}CH{-}CH_3 \longrightarrow CH_2{=}CH{=}CHO$$

Exactly the same results are obtained under ammoxidation conditions. Interestingly, in carbon-13 work the distribution in the acetonitrile product also confirmed that this product was formed by carbon–carbon bond fission *after* the formation of the allyl species.

It is believed that the initial removal of the allylic hydrogen is the rate-determining step. Measurement of the rates of oxidation of alkyl-substituted propylenes has revealed that the order of reactivity for bond breakage is

$$\text{C—H}_{tert\ \text{allyl}} > \text{C—H}_{sec\ \text{allyl}} > \text{C—H}_{prim\ \text{allyl}}$$

This has been taken to indicate the absence of any carbanion character in the allyl intermediate.

There is little that can be said with certainty about the subsequent steps in the oxidation. Early experiments with deuterium-labelled olefins on bismuth molybdate catalysts suggested that a second hydrogen abstraction step was necessary before the addition of oxygen to form acrolein and a kinetic isotope effect of 2.5 was observed for the removal of this second hydrogen. However, as a result of further work, proposals have ascribed cationic character to the allyl species, which is then open to attack by nucleophilic lattice oxygen before further hydrogen abstraction. Rationalisation of this apparent conflict has been offered in the recent finding that in the oxidation of deuterated allyl alcohol, the removal of the allylic hydrogen is the rate-determining step. This hydrogen corresponds to the second hydrogen in the oxidation of propylene and exactly the same kinetic isotope effect has been observed. If these recent results are not refuted by further investigations it seems that a possible reaction scheme along the following lines may be proposed.

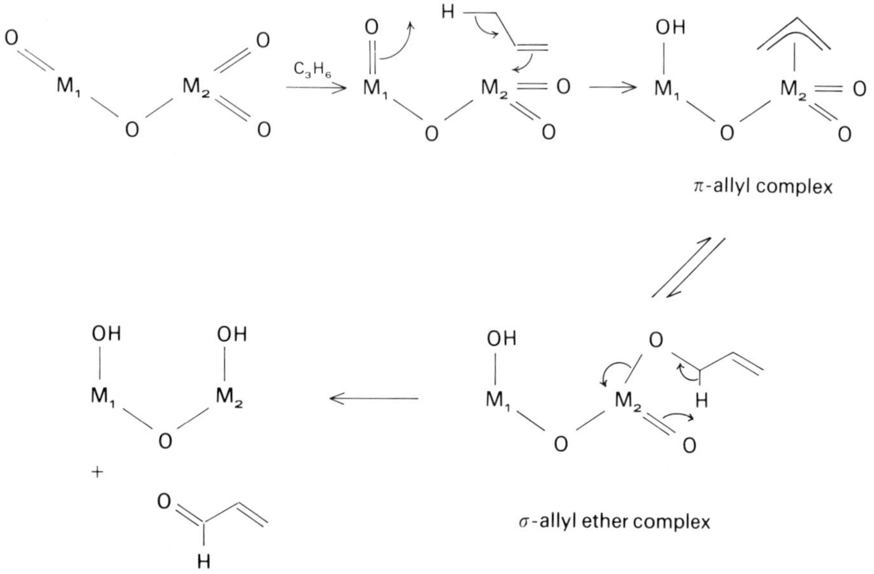

No mechanism has been firmly established for the oxidation of olefins to the unsaturated acids or anhydrides. It is highly probable that aldehydes are formed in the process but the further insertion of the hydroxyl group to yield the acid product, or the manner in which the acid anhydrides are formed are processes that as yet have not been generally explained. For example, there is no clear explanation why, on "good" acrolein catalysts, the oxidation should in fact stop at acrolein. As has been seen in section 11.5.2, the conversion of propylene to acrylic acid requires a two-step process only because different catalysts are required for each stage.

11.9.3 *The role of oxygen and the Mars–van Krevelen Cycle*

It is now fairly well established that the oxygen contained in the solid catalyst—so-called "lattice" oxygen—has an active role in selective hydrocarbon oxidation. It has been shown for a variety of reactions and catalysts that, having brought the catalyst to a steady-state performance level with a stream of oxygen and hydrocarbon, the catalyst will continue to operate with the same selectivity if the oxygen supply is discontinued. Eventually the activity will decline as the catalyst is reduced, but the original steady-state operation can be regained by restoring the oxygen supply. It is generally believed that under steady-state conditions the catalyst may be partially reduced to a degree which has an optimum value depending mainly on the catalyst.

The use of isotopic oxygen is the most direct approach and the experimental results tend to confirm that lattice oxygen is the primary oxidant, at least in allylic oxidation.

The oxidation of propylene to acrolein on bismuth molybdate and antimony–tin oxide catalysts has been investigated using gaseous oxygen-18. Both the carbon dioxide and the acrolein products initially contained only oxygen-16 (present only in the catalyst) but, as reaction continued, oxygen-18 appeared in the products due to replacement in the lattice (figure 11.1). The amount of oxygen-16 taken up in the products showed that, in the case of bismuth molybdate, essentially the whole of the lattice oxygen was mobile enough to be available for reaction at the surface. In the antimony–tin catalyst, only a few surface layers participate in the reaction.

Thus, it is widely accepted that the formation of acrolein is dependent on available lattice oxygen and, indeed, more generally that selective oxidation requires the latter. In the case of complete oxidation to carbon dioxide and water, there is as yet no firmly established view. There is some confusion over the interpretation of the experiments with oxygen-18 since both carbon dioxide and water are known to exchange rapidly with the lattice oxygen in bismuth molybdate. Hence the preponderance of oxygen-16 in carbon dioxide could be explained by reaction involving adsorbed oxygen-18, followed by rapid isotopic equilibrium with a large pool of oxygen-16 atoms. From non-isotopic experiments, the occurrence of carbon dioxide as a product in reaction where there is

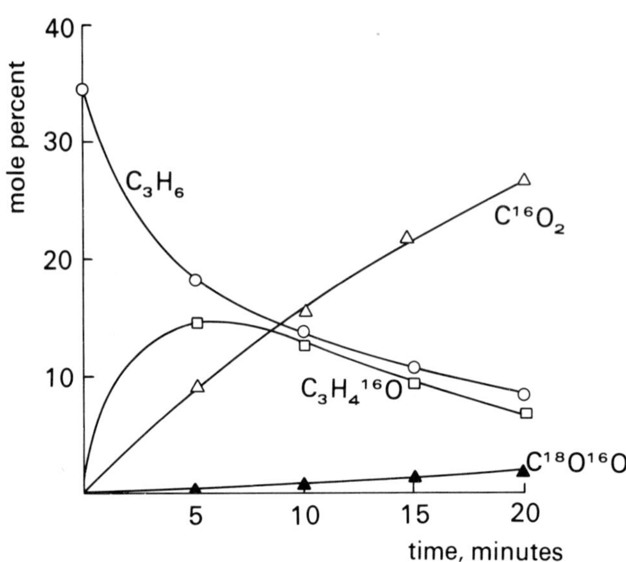

Figure 11.1 The oxidation of propylene with gaseous oxygen-18 on a bismuth molybdate catalyst at 425°C (from G. W. Keulks, *J. Catalysis*, **19**, 232, 1970).

no gas-phase oxygen clearly indicates that total oxidation involving lattice oxygen is possible. The consensus viewpoint at present is that lattice oxygen certainly leads to selective oxidation but that both adsorbed or lattice oxygen can precipitate complete oxidation.

It will be realised that the removal of oxygen from the lattice and its subsequent replenishment constitutes a cyclic redox system involving metal ions in the lattice. As suggested in section 11.3.5, this concept was first proposed in 1954 by Mars and van Krevelen in a study of the vanadium pentoxide-catalysed oxidation of benzene to maleic anhydride. Consequently, mechanistic proposals involving lattice oxygen often bear the names of these authors. In the case of a single metal oxide the situation is relatively straightforward. There is only one metal cation species present and it can therefore be identified with the redox cycle. Where there is more than one cation present, each of which can be reduced, the possibilities become more complicated. At present it is becoming recognised that one metal may be responsible for activating and oxidising the hydrocarbon. Reoxidation of this cation takes place by migration of oxygen anions through the lattice leaving the other cation in a reduced state. This in turn is reoxidised by the uptake of gaseous oxygen. This dual-redox system is completely analogous to the homogeneous Wacker oxidations in which palladium is reoxidised indirectly through the agency of cupric chloride. Unfortunately experimental work to date, again concerning bismuth molybdate, has produced somewhat contradictory views on the roles of bismuth and molybdenum.

11.9.4 *Active phases and active sites*

Mixed metal oxide catalysts pose enormous problems in relation to the identity of the essential components. Firstly, the optimum catalyst may be a multiphase system, so there immediately arises the question of identifying the active phase, if there is indeed only one. Secondly, catalysis is a surface phenomenon, and a knowledge of the bulk structure may or may not be helpful in elucidating the structure of the active surface as it is possible that the composition of the surface will be different from that of the bulk. Although modern techniques such as X-ray and UV photoelectron spectroscopy (XPS and UPS respectively) and secondary ion mass spectrometry (SIMS) can distinguish between surface and bulk composition, surface-structural tools such as low energy electron diffraction (LEED) are not as yet capable of analysing such complex materials. Thirdly, a further major problem lies in the identification of those sites on the surface which are responsible for the adsorption and oxidation of the hydrocarbon. Experimental evidence points to the involvement of lattice oxygen in the reaction (section 11.9.3) but does not reveal the nature of the sites responsible. The possibility therefore arises that the adsorption and oxidation of hydrocarbon by lattice oxygen occur at one type of site and the reoxidation of the catalyst at another. Fourthly, the addition of modifiers and promoters adds a new dimension in complexity. It is clear that for a complex catalyst such as the promoted bismuth phosphomolybdate (section 11.7.2) the task of defining the active surface is truly formidable.

Clearly, the approach would be to take the "basic" catalyst—that is before it is loaded with promoters etc. and derive whatever is possible about its composition and structure. In fact, relatively few catalysts have been studied in any real depth in this way—the simple bismuth molybdate system being by far the most widely investigated catalyst. From those that have been examined, it is clear that, even in the case of a simple ternary system of two metals and oxygen, no common bulk structural feature has as yet emerged which can be associated uniquely with catalytic activity.

Many mixed oxide catalysts and their uses for particular reactions have been described in this chapter and there is not space enough to review what is known about them in terms of structure and reactivity. Moreover, it has been stressed that the common chemistry which links them in this field has yet to be realised, so there is a limited amount which can be learned from the study of one catalyst in isolation. However it is possible to illustrate the more important aspects of the approach to the problem of catalyst structure and reactivity by a few examples.

First of all it is very important to know whether the catalyst is a multiphase system, as the question of developing maximum selectivity may become a problem in maximising or minimising the presence of one or more phases.

One of the earlier Sohio ammoxidation catalysts was a simple uranium oxide–antimony oxide system.[26] In this system, the optimum composition for the conversion of propylene to acrylonitrile corresponds to an antimony–uranium ratio of about 3:1. The overall activity of the catalyst declines as the ratio is raised beyond this value but, below this value, increasing amounts of carbon dioxide are formed at the expense of acrylonitrile. Pure antimony tetroxide is virtually inactive, but uranium trioxide is a complete combustion catalyst. When a range of catalysts of varying composition were subjected to

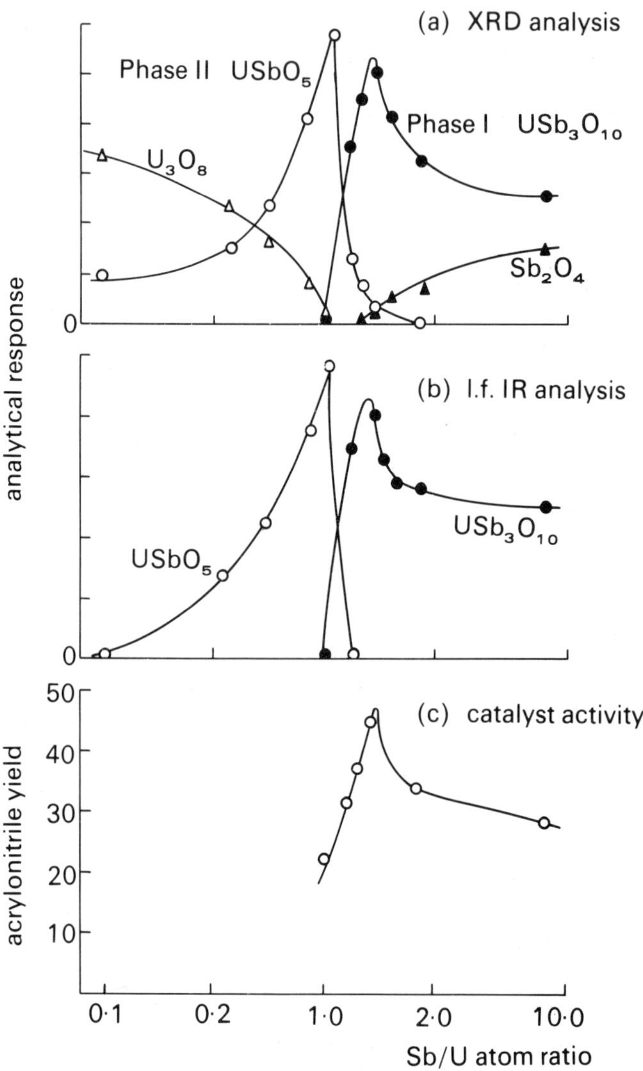

Figure 11.2 The composition of a series of antimony oxide–uranium oxide catalysts as revealed by X-ray diffraction (a) and low-frequency infra-red analysis (b). Catalytic activity for the ammoxidation of propylene to acrylonitrile (c) is clearly coincident with the presence of USb_3O_{10} (from R. K. Grasselli and J. L. Callahan, *J. Catalysis*, **14**, 93, 1969).

X-ray diffraction and low-frequency infra-red analyses it was discovered that the oxides combined to form two phases corresponding to USb_3O_{10} (phase I) and $USbO_5$ (phase II). When each of these phases was made in a pure form, phase I proved to be a very active and selective catalyst having a selectivity to acrylonitrile of 77% at 95% conversion, measured at 460°C. Phase II on the other hand yielded much less acrylonitrile, selectivity to acrylonitrile being 40% at 97% conversion at 460°C. The close connection between catalyst activity and phase composition is shown in figure 11.2. As might be expected from such a result, surface and bulk compositions in each phase were identical as determined by XPS.

This example shows the importance of a knowledge of the phase composition. In some cases, like the example above, it is possible to relate the activity directly to the phase composition and structure in a direct manner. In others, reorganisation of the surface layer(s) due to a change in surface composition may make it impossible to relate bulk properties to catalyst performance. This has been well illustrated by Grasselli[26] who covered the surface of a sample of pure $USbO_5$ (unselective phase II) with antimony tetroxide, in just sufficient quantity so that, after brief exposure to heat, a layer of about three monolayers would have the composition USb_3O_{10} (selective phase I). The overall stoichiometry of the solid corresponded to $USb_{1.036}O_{5.09}$. This material was found to have exactly the catalytic performance expected of bulk phase I, with a selectivity to acrylonitrile of 78% at 97.5% propylene conversion measured at 460°C. It should be emphasised that bulk analysis would have revealed only the presence of the catalytically inferior phase II. Both $USbO_5$ and USb_3O_{10} have similar structures consisting of layers containing cations and oxygen linked by oxygen bridges. In USb_3O_{10} each uranium cation has, as nearest neighbours, only antimony cations, whereas in $USbO_5$ it has both cations as nearest neighbours. There are five cation-containing layers in the unit cell, and the cation positions of some of these layers are shown for both phases in figure 11.3.

Grasselli speculated that the differences in catalytic behaviour in Phase I and Phase II might be accounted for in the following way. Clearly, the existence of U-Sb cation pairs are associated with high, selective, activity, being the only possible configuration in Phase I. In Phase II, lower selectivity might arise as a result of the U-U cation pairs. The latter are also present in UO_3—a total combustion catalyst.

The antimony–uranium catalyst is perhaps the simplest example where the structure of the solid is apparently related to catalytic activity. In other systems, the picture is not at all clear. Structurally, there may be gross differences between catalysts of comparable activity and explanations for reactivity on other grounds must be sought. For example, it has been accepted for some time that the active phase in the tin–antimony oxide catalyst is a substitutional solid solution of antimony (V) ions in the tin oxide lattice. The latter has the hexagonal-close-packed arrangement of oxygen anions as in the rutile form of titanium dioxide with cations present in the anion interstices. Conductivity

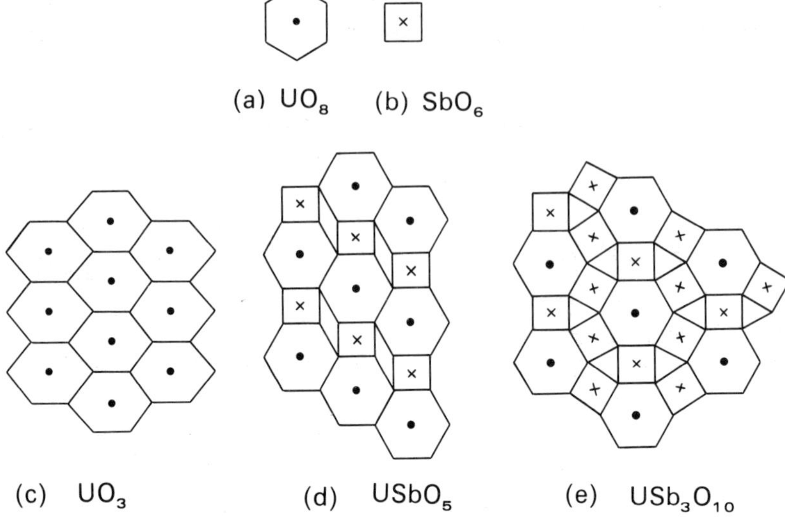

Figure 11.3 An approximate representation of the structures of UO_3, $USbO_5$ and USb_3O_{10} to demonstrate the juxtaposition of the cations. (a) and (b) are the basal planes of a UO_8 dodecahedron and a SbO_6 octahedron respectively. (Redrawn from Gates *et al.*, 1979).

measurements suggest a solubility limit of about 4 atom % antimony. Therefore, each antimony ion will be surrounded by tin ions, with both cations octahedrally co-ordinated to oxygen. It has recently been suggested[27] that the presence of the solid-solution phase may not be essential for selective activity. All that may be required is a high concentration of adjacent antimony- and tin-containing octahedra. In this respect the proposals concerning the nature of the antimony–tin and antimony–uranium catalysts are quite similar, despite the differences in their structures. The antimony–tin oxide system has been the subject of a recent review.[28]

The concept that cation pairs are required as a basis for a selective oxidation catalyst has been examined in more detail with bismuth molybdate catalysts. Good catalytic activity is observed when the bismuth to molybdenum ratio is greater than $\frac{2}{3}$ and less than 2. Three distinct phases exist within this range having the general formula $Bi_2O_3{:}xMoO_3$; the α-phase where $x = 3$, the β-phase where $x = 2$ and the γ-phase where $x = 1$. Although there are differences in structure and reactivity, each phase has significant activity and selectivity. For a full description of the structural features of this well-explored system the reader is directed to any of the listed reviews, but as an example, figure 11.4 shows a representation of the structure of the γ-phase. This is a substance having layers of molybdenum and oxygen separated from layers of bismuth and oxygen by oxygen bridges. Based on experiments involving the oxidation of allyl radicals (obtained by decomposition *in situ* of allyl halides) it has been concluded that bismuth ions have no ability to transfer oxygen to the allyl radicals whereas molybdenum oxide can do so with the selective formation of acrolein. On the other hand molybdenum trioxide has no activity for propylene oxidation to acrolein. Thus, in a bismuth molybdate catalyst it is believed that the bismuth ions accelerate the formation of allyl radicals which are then oxidised by molybdenum ions. It would seem, therefore, that co-operating cation pairs are a feature of active and selective catalysts but

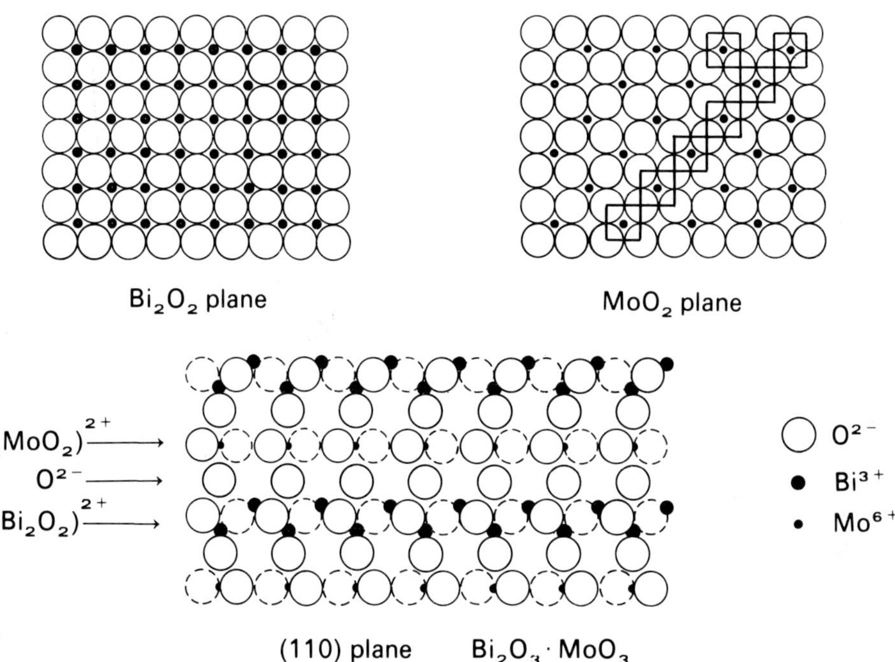

Figure 11.4 The layer structure of Bi_2MoO_6 showing how it is built up from corner-shared MoO_6 octahedra, joined through their apical oxygens to a Bi_2O_2 layer.

the mechanism of oxygen transfer to the organic species remains a matter for speculation.

Finally, it is now well established that the surfaces of mixed oxide catalysts have acidic and basic properties that are reflected in selectivity patterns. Thus in reactions where acidic products are formed, the desorption of such products from basic surfaces will be difficult and further oxidation may occur with a consequent reduction in selectivity.

In summary, the major features which have been established as important aspects of complex mixed-oxide catalysts for allylic oxidation reactions are as follows:

(1) The oxidation step is performed by a metal ion in a redox mechanism within the catalyst.

(2) After initial adsorption, the fission of the first hydrogen–carbon bond is rate determining and an allylic species is formed.
(3) More than one cation may be required for the reaction to proceed and it is likely that different cations are responsible for oxidising the organic substrate and for replenishing the catalyst with oxygen.
(4) The bulk crystallographic structure of the catalyst plays no part in determining catalytic activity but it may have importance in establishing the required surface structure and composition.
(5) Surface basicity and acidity are important in determining selectivity on oxide catalysts.

It will be some time before the complexities of oxidation reactions on metal oxide catalysts are unravelled but meanwhile it is becoming more important to maximise the selectivity as fully as possible. Although the search for a new catalyst continues to be based on empirical exploration, it is therefore necessary today to have as full a description as possible of these complex materials in order to formulate a catalyst composition having the highest intrinsic selectivity.

11.10 Conclusions

The important link between the basic building blocks and many important large-tonnage products lies in the sphere of oxidation chemistry. The ability to practice this chemistry economically depends critically on having access to efficient catalysts. This chapter has shown how selective oxidation has been achieved on an industrial scale. The process chemistry is sometimes elegant in its chemical simplicity, for example, as found in the Wacker process for acetaldehyde production, and sometimes very obscure and complex as in the ammoxidation of propylene. However, given the many and varied reactions possible when hydrocarbons and oxygen are mixed at high temperatures it is perhaps understandable why it is necessary to employ so many different approaches.

The general movement of the industry away from a petrochemical base has already become a reality in this sector, as demonstrated by the demise of oxidation routes to acetic acid. Other replacements will doubtless follow during the next few years. Existing oxidation technology is meeting, and will continue to meet, two major challenges. One is to strive for even more efficient catalysts to maximise feedstock utilisation (selectivity) and to minimise energy requirements (activity). The second is to be able to deal with changing feedstocks. This has happened before when coal-based feedstocks gave way to petrochemical based ones (for example, naphthalene to *o*-xylene for phthalic anhydride). Then, however, the new era provided raw material in apparently limitless abundance. This time the alternative feedstocks are likely to be limited in number, so future oxidation processes are likely to be more demanding of the controlling chemistry than hitherto.

REFERENCES

1. D. J. Hucknall, *Selective Oxidation of Hydrocarbons*, Academic Press, London, 1974.
2. R. W. Clayton and S. V. Norval, *Spec. Per. Rep.* (*Catalysis*), **3**, 70, 1980; X. E. Verykios, F. P. Stein and R. W. Coughlin, *Catal. Rev. Sci. Eng.*, **22**, 197, 1980.
3. P. A. Kilty, N. C. Rol and W. M. H. Sachtler, *Proc. Int. Cong. Catalysis 1972*, **2**, 929, 1973.
4. R. Sheldon, *Rec. Trav. Chim. Pays-Bas*, **92**, 367, 1973.
5. H. Mimoun, I. De Roch and L. Sajus, *Tetrahedron*, **26**, 37, 1970.
6. H. Mimoun, R. Charpentier, A. Mitschler, J. Fischer and R. Weiss, *J. Amer. Chem. Soc.*, **102**, 1047, 1980.
7. R. N. Hader, R. D. Wallace and R. W. McKinney, *Ind. Eng. Chem.*, **44** (7), 1508, 1952.
8. A. R. Chauvel, P. R. Courty, R. Maux and C. Petitpas, *Hydrocarbon Processing*, 179, (September) 1973.
9. D. A. Robb and P. Harriott, *J. Catalysis*, **35**, 176, 1974.
10. I. E. Wachs and R. J. Madix, *Surface Science*, **76**, 531, 1978.
11. J. Smidt, *Chem. and Ind.*, 54, (Jan. 13 issue) 1962.
12. G. Szonyi, *Adv. Chem. Ser.*, **70**, 53, 1968.
13. P. M. Maitlis, *The Organic Chemistry of Palladium, Vol. II: Catalytic Reactions*, Academic Press, New York, 1971.
14. R. A. Sheldon and J. K. Kochi, *Advances in Catalysis*, **25**, 272, 1976.
15. J. K. Kochi, *Organometallic Mechanisms and Catalysis*, Academic Press, New York, 1978.
16. J. E. Bäckvall, B. Åkermark and S. O. Ljunggren, *J. Chem. Soc., Chem. Comm.*, 264, 1978.
17. J. K. Stille and R. Divakaruni, *J. Amer. Chem. Soc.*, **100**, 1303, 1978: *J. Organomet. Chem.*, **169**, 239, 1979.
18. S. Nakamura and T. Yasui, *J. Catalysis*, **17**, 366, 1970.
19. *Chem. Eng. News*, 50, July 8 1963.
20. U.S. Patent 2,451,485 (1948).
21. J. L. Callahan, R. K. Grasselli, E. C. Milberger and H. A. Strecker, *Ind. Eng. Chem. Prod. Res. Develop.*, **9** (2), 134, 1970.
22. B. C. Gates, J. R. Katzer and G. C. A. Schuit, Chapter 4 in *Chemistry of Catalytic Processes*, McGraw-Hill, New York, 1979.
23. J. A. Allen and A. J. Clark, *Rev. Pure and Applied Chem.*, **21**, 145, 1971.
24. K. van der Wiele and P. J. van den Berg, Chapter 2 in *Complex Catalytic Processes* (**20**) *Chemical Kinetics*, ed. C. H. Bamford and C. F. H. Tipper, Elsevier, Amsterdam, 1978.
25. R. Higgins and P. Hayden, *Spec. Per. Rep.* (*Catalysis*), **1**, 168, The Chemical Society, London, 1977.
26. R. K. Grasselli *et al.*, *J. Catalysis*, **14**, 93, 1969; *ibid*, **18**, 356, 1970; *ibid*, **25**, 273, 1972.
27. D. R. Pyke, R. Reid and R. J. D. Tilley, *J. C. S. Faraday I*, **76**, 1174, 1980.
28. F. J. Berry, *Advances in Catalysis*, **30**, 1981, in the press.

CHAPTER TWELVE

CATALYTIC HYDROGENATION AND DEHYDROGENATION

J. M. WINTERBOTTOM

12.1 Introduction

The use of catalytic hydrogenation as a tool in organic synthesis was developed and reported at the turn of the century by Sabatier and his co-workers. Shortly afterwards, the important Haber process for ammonia synthesis was also developed, along with the production of margarine by fat hardening. During and following World War II, the development of hydrogenation/dehydrogenation processes in chemical manufacture was given fresh impetus by the rapid growth of oil processing. This led to the availability of relatively cheap hydrogen from the steam reforming of hydrocarbons. The modern chemical industry is very dependent upon hydrogen transfer processes and Table 12.1[1,2] summarises the most important of these.

In the ensuing sections, some of the more important processes will be discussed in more detail; others, such as platinum reforming, hydrocracking and hydrotreating, are discussed elsewhere in this book.

12.2 Catalytic hydrogenation of unsaturated carbon–carbon bonds

Much of the classical research in catalysis has been carried out in the field of catalytic hydrogenation and a number of texts[3–6] have covered the main findings.

From an industrial standpoint, the hydrogenation of olefins is not of great significance. Olefins are valuable synthetic intermediates and are not generally hydrogenated. An exception to this principle occurs in the hydrogenation of oils and fats (section 12.5) which involves, in essence, the selective hydrogenation of isolated olefinic bonds. For this reason, certain essential features of olefin hydrogenation will be referred to in section 12.5 but for a detailed account of this topic, the reader is referred to more specialized texts.[3–6]

Table 12.1 Major hydrogenation/dehydrogenation processes

Functional group (reaction)	*Typical catalyst*	*Comments*
acetylene → ethylene 1,3-butadiene → butene	Pd/Al_2O_3	removal of acetylene and butadiene from olefin streams to ppm level; very selective catalyst required
triglyceride (isolated polyolefin)	Ni	Conversion of liquid oils to margarine and frying oils
benzene → cyclohexane	Ni/Al_2O_3 $Pt—Li/Al_2O_3$	Production of high purity cyclohexane for manufacture of nylon
$—CHO \rightarrow CH_2H$	Ni Pt, Ru	(i) conversion of aldehydes from olefin hydroformylation to alcohols for detergents (ii) conversion of sugars to sugar alcohols, i.e., D-glucose to sorbitol in Vitamin C manufacture
$—NO_2 \rightarrow —NH_2$ (nitrobenzene → aniline)	Copper	aniline is a multipurpose intermediate e.g., for polyamide resins
$—C{\equiv}N \rightarrow —CH_2NH_2$ (adiponitrile → hexamethylene diamine)	Cu—Co or Ni	hexamethylene diamine for nylon manufacture
$\equiv C—C\equiv \rightarrow \equiv CH + HC\equiv$ hydrocracking	Pt-, Pd-zeolite	occurs with chain branching and converts high m.w. hydrocarbons to more valuable lower m.w. hydrocarbons
hydrodesulphurisation	cobalt molybdate	conversion of sulphur compounds e.g., thiophen in petroleum streams, to H_2S
n-paraffins, naphthenes → aromatics (platinum reforming)	Pt or Pt—Re on acid Al_2O_3	aromatics production from naphtha
n-butenes → butadiene ethylbenzene → styrene	$Fe_2O_3/K_2O/Cr_2O_3$	production of polymer building blocks
hydrodealkylation toluene → benzene	chromia molybdena or cobalt oxide on alumina	production of high purity benzene from toluene

12.3 Acetylene hydrogenation

Acetylene hydrogenation has been studied less extensively than that of lower olefins, although it has attracted the attention of many workers interested in the mechanism of catalytic hydrogenation. The reaction is of industrial importance because acetylene is frequently present in process streams of olefins (ethylene and propylene) and diolefin (1,3-butadiene) which are to be used for polymerisation.

Acetylene is a strong poison of polymerisation catalysts, is usually present in ethylene streams to the extent of about 1% v/v, and must be reduced to concentrations of <10 ppm. The most effective means of achieving this is by selective hydrogenation to ethylene. Quite severe demands are made upon the catalyst, since in reducing the acetylene content to <10 ppm, production of the undesirable (but inert) ethane must be minimised. Most industrial processes for selective acetylene removal employ supported palladium catalysts. Base metals and base-metal sulphides can be employed but are less active. Consequently, larger quantities of catalyst, larger reactors and usually higher reaction temperatures are required.

Catalysts, based upon low surface area γ- or α-aluminas containing typically 0.05% w/w palladium, are very selective. These are employed in two-bed adiabatic reactors, with intercooling, designed to minimise temperature increase in each bed; the number of beds can be increased to accommodate higher concentrations of acetylene. In streams containing acetylene, conjugated diolefins and olefins, the ease of hydrogenation is acetylene > conjugated diolefin > olefin.

12.3.1 *Acetylene chemisorption*

The chemisorbed state of acetylene is not completely understood, compared with that of ethylene. It has been suggested that the following species are formed on metal surfaces (* = active site):

(i) Di-σ-adsorbed

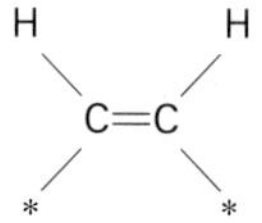

$$\begin{array}{ccc} H & & H \\ & \diagdown \; C{=}C \; \diagup & \\ & \diagup \quad\;\; \diagdown & \\ * & & * \end{array} \qquad (I)$$

(ii) Di-π-absorbed

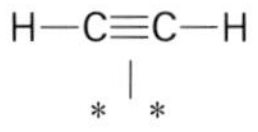

$$\underset{*\;\;|\;\;*}{H{-}C{\equiv}C{-}H} \qquad (II)$$

(iii) Dissociatively adsorbed

$$3C_2H_2{=}4C+C_2H_6$$

(this equation is one example among many possibilities). Spectroscopic evidence has been reported for the ethylene analogues of (I) and (II) and well characterised organometallic complexes of alkynes involving both π- and σ-bonding are cited in the literature.[7]

12.3.2 *Acetylene hydrogenation*

The hydrogenation of acetylene can be represented by the series/parallel reaction scheme:

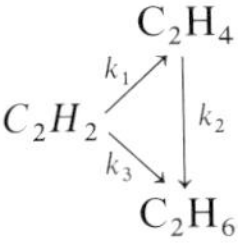

The selectivity for ethylene production depends upon (i) the magnitudes of k_1 and k_2 relative to k_3—this is termed the *mechanistic* factor; and (ii) the relative magnitudes of k_1 and k_2, which is termed the *thermodynamic* factor.[3] The mechanistic factor will be determined by the amount of acetylene which is hydrogenated directly to ethane. The thermodynamic factor is dependent upon the ease of desorption of ethylene and its ability, once desorbed, to compete with acetylene for active sites. If adsorption of acetylene and ethylene is assumed to occur on the same sites then it can be shown that the ratio r_f of the fractional surface coverage of acetylene to that of ethylene is given by

$$r_f = \frac{P_{C_2H_2}}{P_{C_2H_4}} \exp(-\delta\Delta H_{ads}/RT) \exp(\delta\Delta S_{ads}/R)$$

where $\delta\Delta H_{ads}$ and $\delta\Delta S_{ads}$ are the differences in the heat and entropy of adsorption, respectively, of acetylene and ethylene. Thus, for low partial pressures of acetylene, r_f will be large if the difference in heats of adsorption between acetylene and ethylene is significant e.g., for nickel the difference is about 34 kJ/mole.[3] The entropy contribution is more difficult to assess and will depend on the relative entropy losses for each species on adsorption. However, ethylene is reported[3] to be more mobile than acetylene and presumably its entropy loss on adsorption is smaller. Consequently this will offset to some extent the advantage gained from the heat of adsorption term.

Selectivity for the reaction is defined as $S = \%$ ethylene/($\%$ ethane + $\%$ ethylene). Palladium has been found to exhibit superior performance over a wide range of temperatures and pressures compared to other catalysts giving values of $S = 0.95$;[5] other metals exhibit a wide variation in selectivity with temperature and hydrogen partial pressure. Hence palladium is employed industrially as indicated in section 12.2.1.

12.3.3 *The kinetics and mechanism of acetylene hydrogenation*

The kinetics of acetylene hydrogenation have been studied in some detail for a number of base and noble metals and the results have been reviewed.[3–5] The rate of acetylene hydrogenation at a given temperature for the metals studied can be expressed as:

$$\text{rate} = kP^x_{H_2} \cdot P^y_{C_2H_2} \quad (x = 1 \text{ to } 1.5, y = -0.7 \text{ to } 0)$$

This expression indicates that acetylene is much more strongly adsorbed than hydrogen and occupies much of the catalyst surface. The results have been explained in terms of the following mechanism[4,5]

$$H_2(g) + 2* \underset{(2)}{\overset{(1)}{\rightleftharpoons}} 2\underset{*}{H}$$

$$C_2H_2(g) + 2* \xrightarrow{(3)} \underset{*\;\;*}{HC{\equiv}CH}$$

$$\underset{*\;\;*}{HC{\equiv}CH} + \underset{*}{H} \underset{(5)}{\overset{(4)}{\rightleftharpoons}} \underset{*}{HC{=}CH_2} + 2*$$

$$\underset{*}{HC{=}CH_2} + \underset{*}{H} \xrightarrow{(6)} \underset{*}{H_2C{=}CH_2} + *$$

$$2\underset{*}{HC{=}CH_2} + * \xrightarrow{(7)} \underset{*\;\;*}{HC{\equiv}CH} + \underset{*}{H_2C{=}CH_2}$$

$$\underset{*}{HC{=}CH_2} + \underset{*}{H_2} \xrightarrow{(8)} \underset{*}{H_2C{=}CH_2} + \underset{*}{H}$$

$$\underset{*}{H_2C{=}CH_2} \xrightarrow{(9)} C_2H_4(g) + *$$

Most interpretations until relatively recently were based upon the assumption that acetylene and ethylene are hydrogenated on the same type of site. However, it has been shown by use of ^{14}C-tracer studies[6] that there may be as many as three types of site and on this basis it is suggested that the mechanism is as shown opposite.

12.3.4 *Polymer formation*

It is well-established that acetylene will undergo self-hydrogenation and polymerisation. The main polymers are C_4-olefins which can even exceed the yield of C_2-hydrogenated product.[3] There is strong and mounting evidence[6] that hydrogenation occurs on the carbon overlayer rather than on the metal; nevertheless, "coke" formation is always a real problem with industrial catalysts since activity is usually lost through physical blocking of the catalyst surface. In practice, the "carbon" so formed must be removed by steam/air treatment to regenerate the catalyst and this is a well-known practice with platforming and catcracking catalysts (chapter 5).

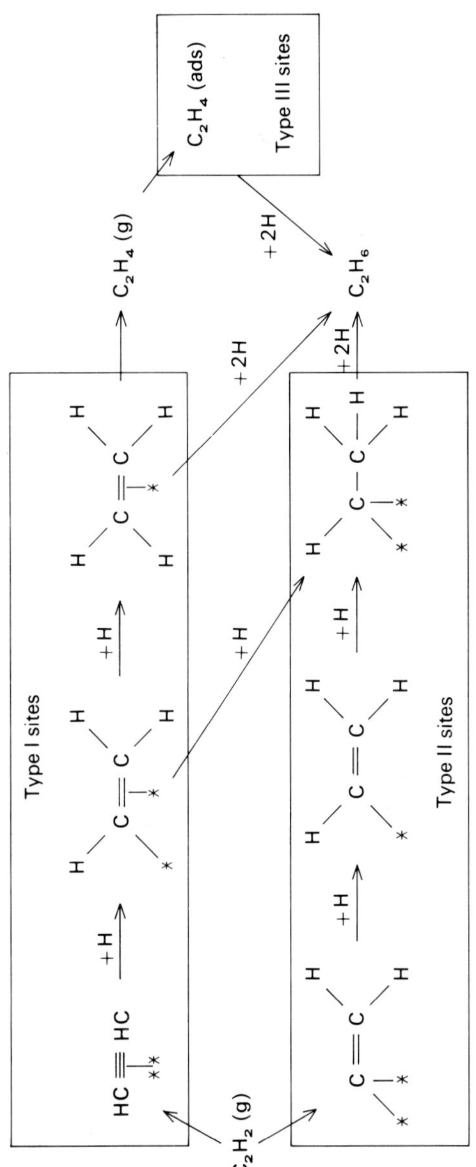
Type I sites
Type II sites
Type III sites
C_2H_2 (g)
C_2H_4 (g)
C_2H_4 (ads)
C_2H_6
+H
+2H

The mechanism of polymer formation from acetylene is not well understood but has been suggested[5] to be as follows:

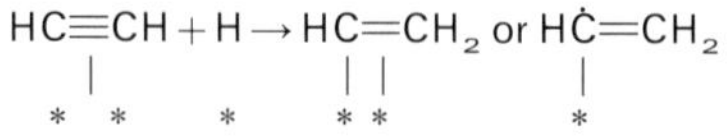

$$\underset{*\quad *}{\mathrm{HC{\equiv}CH}} + \underset{*}{\mathrm{H}} \rightarrow \underset{*\,*}{\mathrm{HC{=}CH_2}} \text{ or } \underset{*}{\mathrm{H\dot{C}}}\mathrm{{=}CH_2} \tag{10}$$

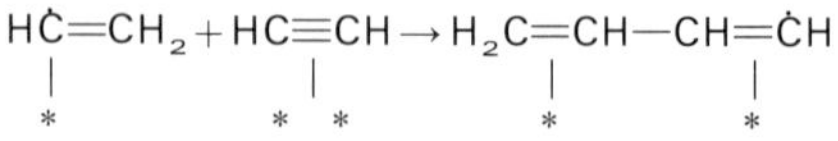

$$\underset{*}{\mathrm{H\dot{C}}}\mathrm{{=}CH_2} + \underset{*\quad *}{\mathrm{HC{\equiv}CH}} \rightarrow \mathrm{H_2C{=}}\underset{*}{\mathrm{CH}}\mathrm{{-}CH{=}}\underset{*}{\mathrm{\dot{C}H}} \tag{11}$$

$$\underset{*}{\mathrm{H\dot{C}}}\mathrm{{=}CH_2} + \underset{*\,*}{\mathrm{HC{=}CH_2}} \rightarrow \mathrm{H_2C{=}}\underset{*}{\mathrm{CH}}\mathrm{{-}\dot{C}H{=}}\underset{*}{\mathrm{CH_2}} \tag{12}$$

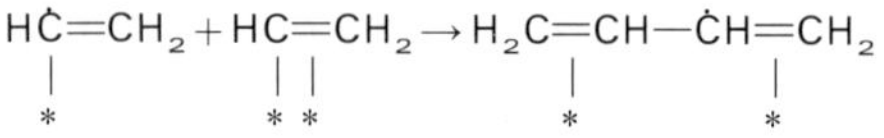

An alternative to the adsorbed free radicals proposed in the above scheme is:

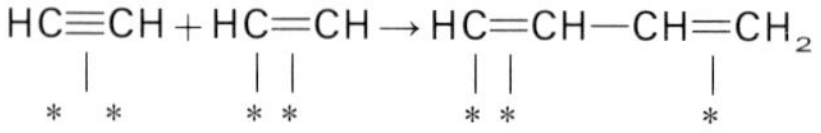

$$\underset{*\quad *}{\mathrm{HC{\equiv}CH}} + \underset{*\,*}{\mathrm{HC{=}CH}} \rightarrow \underset{*\,*}{\mathrm{HC{=}CH}}\mathrm{{-}CH{=}}\underset{*}{\mathrm{CH_2}} \tag{13}$$

Undoubtedly such species would lead to both C_4-olefins and "coke" formation.

12.4 Diene hydrogenation

Conjugated dienes, especially 1,3-butadiene, must be removed from olefin streams which are required for hydroformylation, where they cause catalyst deactivation, and also from gasoline streams derived from steam cracking operations, to prevent polymer formation. Removal is again accomplished by selective hydrogenation.

As is the case for acetylene, low-concentration palladium-on-alumina catalysts are very selective ($S \rightarrow 1.0$) for butene formation and will operate under relatively mild conditions. However, other metals, such as copper and nickel, are also very selective and require only slightly higher temperatures to achieve selectivities approaching or equal to unity.

12.4.1 *The chemisorption of 1,3-butadiene*

Little is known concerning the adsorbed state of 1,3-butadiene in terms of direct studies. Fortunately reasonable deductions can be made from the nature of reaction products, following reaction of the diene with hydrogen or deuterium. 1,3-butadiene exists in two conformations in the gas phase as follows, species (IV) predominating:

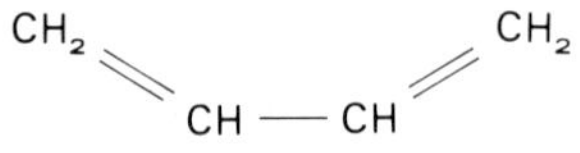

$$\mathrm{CH_2{=}CH{-}CH{=}CH_2}\ \text{(both } \mathrm{CH_2}\text{ on the same side)} \qquad \text{syn (III)}$$

$$\mathrm{CH_2{=}CH{-}CH{=}CH_2}\ \text{(}\mathrm{CH_2}\text{ groups on opposite sides)} \qquad \text{anti (IV)}$$

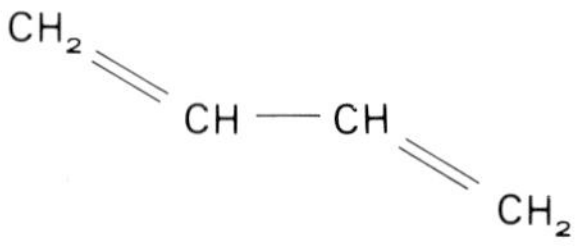

As a result of product distribution studies during hydrogenation over a number of metal catalysts it has been proposed[5] that 1,3-butadiene can form the following adsorbed species:

(i) *π-adsorbed*

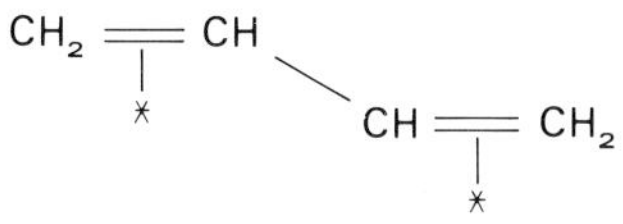

$CH_2{=}CH$ / $CH{=}CH_2$ (each double bond bonded to *)

(V)

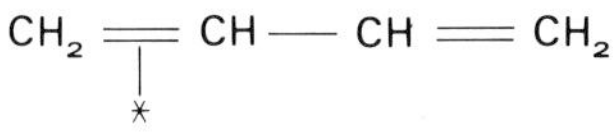

$CH_2{=}CH{-}CH{=}CH_2$ (first double bond bonded to *)

(VI)

$CH_2{=}CH{-}CH{=}CH_2$ (both double bonds bonded to *)

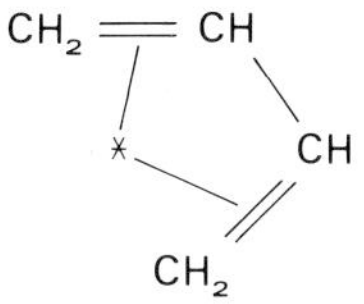

(VII)

The di-$\bar{\pi}$- anti (V) and -syn (VII) forms are interconvertible via the π-adsorbed (VI) form.

(ii) *π-allylic*

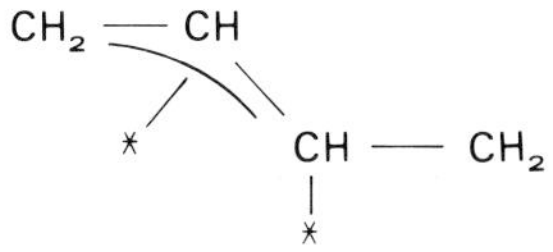

$CH_2{-}CH{-}CH{-}CH_2$ (allylic, with CH bonded to *)

(VIII)

$CH_2{-}CH{-}CH{-}CH_2$ (allylic, ring form bonded to *)

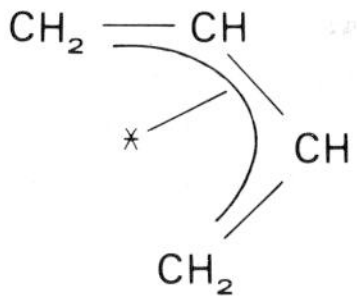

(IX)

The species (VIII) and (IX) are not interconvertible.

12.4.2 *The mechanism of butadiene hydrogenation*

The hydrogenation of 1,3-butadiene gives rise to all three *n*-butenes in the initial products. Two mechanisms have been proposed to account for the behaviour of metal catalysts.[5] The first is termed Type A and occurs with the majority of Group VIII metals (Ni, Co, Cu, Ru, Rh Pt, Ir). 1-Butene is the major product, with *trans*-2-butene/*cis*-2-butene ratios of approximately 1. For this mechanism it is proposed that the diene is adsorbed in the di-π-forms (species V, VI and VII), which are interconvertible. Addition of adsorbed hydrogen gives π-σ adsorbed species, which are also interconvertible:

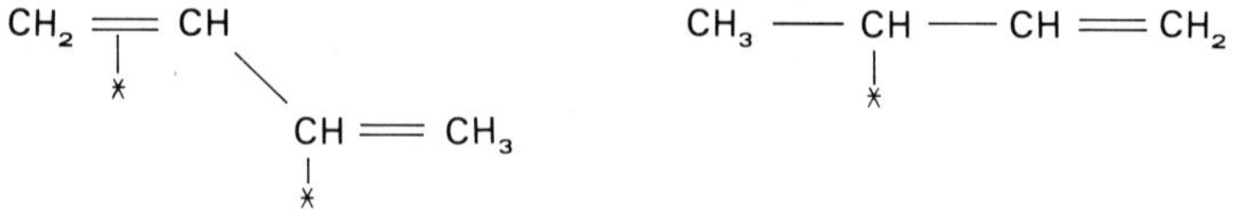

$CH_2{=}CH{-}CH{=}CH_3$ (with * below CH_2 and below the second CH)

(X)

$CH_3{-}CH{-}CH{=}CH_2$ (with * below the first CH)

(XI)

$CH_2{=}CH{-}CH{-}CH_3$ (with * below CH_2 and below the second CH)

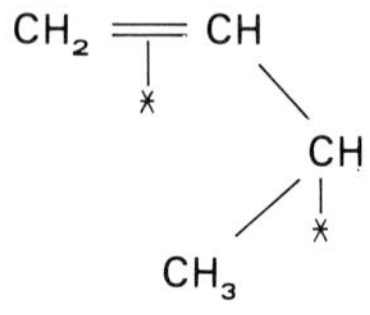

(XII)

Further addition of adsorbed hydrogen will give adsorbed 1-butene. The 2-butenes arise by formation of π-allylic species from X and XII.

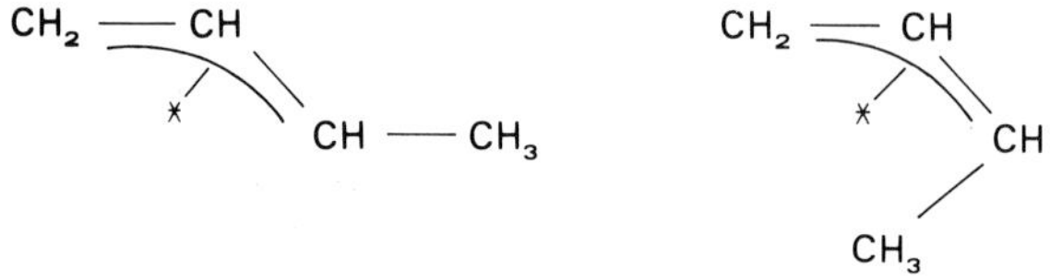

$CH_2{-}CH{-}CH{-}CH_3$ (π-allyl, *)

(XIII)

$CH_2{-}CH{-}CH{-}CH_3$ (π-allyl, *)

(XIV)

Addition of adsorbed hydrogen to XIII and XIV gives adsorbed *trans*- and *cis*-2-butene respectively, but the π-allylic species are not interconvertible. Mechanism B was proposed particularly for palladium and occurs via the π-allylic chemisorption of butadiene (species VIII and IX). These forms and their derivatives do not readily interconvert and the butene distribution is determined by the relative proportion of *syn*- and *anti*-conformers. This usually leads to high *trans*/*cis* ratios (4–14). Mixtures of mechanisms A and B have been observed with cobalt and nickel. The occurrence of this behaviour appears to be associated with sulphur contamination of the metal surface.

12.5 Triglyceride hydrogenation (fat hardening)

12.5.1 *General background*

Fat hardening, one of the oldest commercial catalytic processes, was developed to convert liquid oils into butter substitutes (i.e. margarine). This involves the selective hydrogenation of naturally-occurring triglycerides. These are found in the plant world (oils from soya beans, groundnuts, sunflower seeds, rape seeds) and also in the animal world (e.g. whale, fish oils) as complex mixtures of fatty-acid esters of glycerol: $R^1COOCH_2.CH(OCOR^2).CH_2OCOR^3$. The R groups are straight-chain hydrocarbons containing 8 to 24 (even numbers only) carbon atoms and usually up to three olefinic bonds per chain. Chains of 16 and 18 atoms predominate, the olefinic bonds being non-conjugated and existing almost entirely in the *cis*-configuration. Soya-bean oil, the most abundant of the various sources, has the typical composition (C no./% w/w/no. of double bonds): $C_{16}/11/0$;$C_{18}/4/0$;/$C_{18}/22/1$;$C_{18}/55/2$;$C_{18}/8/3$.

The stability of a given oil is related to the content of chains containing more than two double bonds. The more unsaturated material reduces both the thermal/oxidative stability (important in frying oils) and the melting point. Some oils, such as sunflower-seed oil, have very low levels of triunsaturation and require no further processing, but are available only in restricted quantities. The major sources of supply e.g. soya-bean or rape-seed oil are unsuitable and require selective hydrogenation.

12.5.2 *Selective hydrogenation of soya-bean oil*

The stepwise hydrogenation of C_{18} unsaturated acids is shown in (14)

$$\underset{\text{(3 double bonds)}}{\text{Linolenic }(L)_3} \xrightarrow{H_2} \underset{\text{(2 double bonds)}}{\text{Linoleic }(L_2)} \xrightarrow{H_2} \underset{\text{(1 double bond)}}{\text{Oleic }(O)} \xrightarrow{H_2} \text{Stearic }(S) \tag{14}$$

The major problems in catalyst and process design are those of selectivity. High selectivity is required for the following purposes.

(a) Selective removal of L_3, which is largely responsible for high temperature in stability and must be reduced from an initial 8% to under 2% w/w in the final product. At the same time L_2, which is an essential part of a well-balanced diet, should be retained at as high a level as possible. This is demanding on catalyst selectivity.

(b) The production of S should be avoided as an increase from 4% to beyond 4.5% can give a product which is too hard. Fortunately the selective hydrogenation of L_3 and L_2 to O is easy to accomplish.

(c) Limitation of double bond migration and geometrical isomerisation. Positional isomerisation can lead to conjugation and such species are more prone to oxidation. Geometrical isomerisation leads to a change in physical properties and a good reason for retaining *cis*-isomers is that they pack less regularly than *trans*-isomers in the solid fat/liquid oil matrix and give a desirable wide softening range in the product. In addition, *trans*-isomers may give rise to dietary problems, since they are not present in the natural fats.

Selectivity is all the more difficult to achieve, firstly because the catalyst has to discriminate between molecules with double bonds of similar character (i.e. non-conjugated with *cis*-conformation). Secondly, even on a statistical basis, the double bonds and chains behave as if they are separated, even though there are three chains per triglyceride. However, once adsorbed, there is a strong tendency for multiple hydrogenation to occur since active sites are in close proximity and rates of desorption are similar to rates of hydrogenation. Homogeneous catalysts, with their isolated, uniform active centres may offer some advantages in this respect (see section 12.12).

Selectivity cannot, therefore, be defined as a single, simple parameter as for simple alkynes and dienes and the following empirical parameters are used:

(i) S_I = ratio of pseudo-first order rate constants for polyolefin/mono-olefin hydrogenation; if S_I is high S will not be produced until L_2 and L_3 are removed.
(ii) S_{II} = selectivity for L_3 removal.
(iii) S_i = isomerization selectivity for *trans* isomers vs. hydrogenated bond.
(iv) S_T = triglyceride selectivity—is high if chains behave independently.

Since this is a gas/liquid/solid process, it is subject to diffusional problems, generally caused by hydrogen deficiency at the catalyst surface. These can be caused by inadequate mass transfer of hydrogen in the bulk and/or pore diffusion problems. In general, good selectivity is promoted by low hydrogen concentrations, but this also enhances *cis-trans* isomerisation, which is not generally desirable. These conditions result from low hydrogen partial pressures, high temperature and inefficient agitation in the reactor. Catalyst design can produce similar effects because a wide pore support will allow good access of reactants to the surface, with negligible pore diffusion and little multiple hydrogenation. A narrower pore catalyst support will have the opposite effect.

Thus, by variation of the operating conditions, a higher melting point product (slightly lower S_I and higher S_i) or lower and wider melting range product (high S_I and lower S_i) can be obtained. The higher melting point product is required for production of harder margarines or shortening used in confectionery; the latter for table margarines. The physical properties of the product can therefore be determined to some extent by choice of operating conditions and by catalyst design. In practice, however, there must always be a compromise between attainment of good selectivity on the one hand and low levels of *trans* acids on the other.

12.5.3 *The mechanism of triglyceride hydrogenation*

The main factors contributing to a reaction pathway, namely the mechanistic and thermodynamic considerations, have been outlined for alkynes in section 12.3.2, and are also pertinent to triglyceride hydrogenation. The interesting additional feature is that of the isomerisation of the olefinic bonds, which can significantly affect the chemical and physical properties of the product. This can be illustrated via interconversion of π- and σ-adsorbed intermediates, using *cis*-9,10-monoene as the example

(XV)

(XVI) (XVII)

(i)

(ii)

(XIX)

(XVIII)

(i) rotation around 8, 9 bond ,—H from 8

(ii) rotation around 9, 10 bond,—H from 10

It should be noted that π-allylic species will give an identical result. The sequence above illustrates (i) positional isomerisation (XV, XVI, XVII), (ii) geometrical isomerisation (XV, XVI, XVIII); and (iii) positional/geometrical isomerisation (XV, XVI, XIX).

12.5.4 *Fat hardening catalysts*

Nickel represents the best compromise of economic and selective properties and is used commercially, in either supported or Raney forms. In the latter, it is protected by encapsulation in solid fat (tristearin) since this form of nickel is pyrophoric.

Both palladium and copper possess some advantages. Palladium is much more active and more selective over a wider range of conditions but is expensive.

Copper is much less active but possesses a high S_{II} selectivity. However, expensive post-hardening treatment is required to remove it from the product, otherwise catalytic product oxidation may occur.

12.6 The hydrogenation of aromatics

The general term "aromatic hydrogenation" embraces three classes of reaction in which hydrogen is added to (a) the aromatic nucleus; (b) unsaturated hydrocarbon side chains and (c) other functional groups. Though many catalysed examples of all of these exist in the scientific literature only two reactions have any significance from the industrial viewpoint. One reaction is the reduction of benzene to cyclohexane (12.6.1); the other is the reduction of nitrobenzene to aniline (section 12.7.1).

12.6.1 *The hydrogenation of benzene*

The hydrogenation of benzene is operated on a very large scale to give high purity cyclohexane, which is an important intermediate in the manufacture of nylon 6 and nylon 6,6 (see section 7.3). Cyclohexane is present in petroleum from the naphtha fraction, but its separation by distillation in sufficient purity is too difficult and costly. A better route to cyclohexane involves platinum reforming of naphtha to high purity benzene, which is then hydrogenated to cyclohexane.

The hydrogenation of benzene over heterogeneous catalysts is formally simple in that the only product obtained is cyclohexane. The possible intermediates (1,3- and 1,4-cyclohexadiene and cyclohexene) are not detected. However, it is noteworthy that the latter is observed as a product with some homogeneous systems. The free energy changes at 25°C for benzene conversion to 1,3-cyclohexadiene, cyclohexene and cyclohexane are respectively +55.4, −19.7 and −98.3 kJ/mole and ΔG for the complete reaction is zero at 287°C. This means that the reverse reaction becomes significant at temperatures around 200°C. Thus 1,3-cyclohexadiene is always unstable with respect to benzene and the other products. As temperature increases cyclohexene will also become so. The formation of cyclohexene is theoretically possible but, in fact, does not occur. It is either too reactive toward further hydrogenation or, as a product, is not mechanistically significant.

Several types of process are available for benzene hydrogenation,[8] but the newest processes are designed for benzene with a low sulphur content and employ nickel or platinum catalysts under mild conditions (20–40 bar; 170–230°C). Both gas-phase (short residence time) and liquid-phase processes (with solvent recirculation) are used, usually with multiple reactors to limit conversion and therefore aid temperature control for a very exothermic process.

Studies of benzene chemisorption have shown that it is not as strongly adsorbed as olefins and alkynes.[3] Its heat of chemisorption is similar to that of hydrogen and is low because its resonance energy of stabilisation is lost on chemisorption. Three distinct modes of benzene adsorption have been identified:

(a) associative (π) C_6H_6 + * ⇌

(b) dissociative via associative

C_6H_6 + * ⇌ ⇌ + H *

(c) associative di-σ

C_6H_6 + 2* ⇌

The mechanism of benzene hydrogenation is believed to consist of a stepwise addition of hydrogen to π-adsorbed benzene, in which the π-bond between the catalyst and the C_6 nucleus is progressively weakened. Interconversion of the π-bonded form of cyclohexene with the di-σ species probably occurs to facilitate the addition of the final two hydrogen atoms.

C_6H_{12} ⇌

12.7 The hydrogenation of aromatic nitro compounds[1,2,8]

The hydrogenation of aromatic nitro compounds to the corresponding amines is of considerable importance, as the latter are used in the manufacture of synthetic dyes, polyamide resins, polyurethanes and colour photographic emulsions.

12.7.1 *Nitrobenzene—general considerations*

Nitrobenzene hydrogenation is facile and many metals efficiently catalyse its reduction to aniline, both in the gas and liquid phase.

The nitro group is strongly chemisorbed and this is confirmed by the fact that rhodium, which will catalyse aromatic ring hydrogenation under mild conditions,[3] will preferentially catalyse hydrogenation of the aromatic nitrogroup. The reaction is highly exothermic ($\Delta H = -493$ kJ/mole) and industrial processes have been operated in both gas and liquid phases. The liquid phase operation has some merit as it offers good heat transfer properties. However, the mass transport problems mentioned in section 12 for fat hardening apply in this case also.

The most modern processes employ gas phase operation with either fixed or fluidised catalyst beds. In the former case sulphided nickel or copper catalysts are used at temperatures of 300–475°C. Presumably, these are less active and this aids temperature control. The fluidised-bed system employs a copper catalyst at 270–290°C, 1–5 bar with a large excess of hydrogen (H_2/nitrobenzene = 9). Despite the excellent heat-transfer characteristics of such a reactor it is necessary to have additional cooling built into the bed. Catalysts used in this case are highly selective (99.5 % selectivity to aniline), robust and capable of operating at a high temperature, a reactor feature which facilitates heat removal.

12.7.2 *Mechanism of nitrobenzene hydrogenation*

The mechanism of nitrobenzene hydrogenation is not well understood, although it has been established that hydroxylamines occur as intermediates and therefore nitroso-intermediates must also occur. Kinetic studies have shown that the rate expression is of the form

$$\text{rate} \propto P^1_{H_2} \times P^0_{\text{nitro}}.$$

This indicates that nitrobenzene is strongly adsorbed while hydrogen is very weakly chemisorbed. A possible mechanism is as follows:

$$H_2(g) + * \rightleftharpoons \underset{*}{H_2} + * \rightleftharpoons 2H \quad (15)$$

$$RNO_2 + * \rightleftharpoons \underset{*}{RNO_2} \quad (16)$$

$$\underset{*}{RNO_2} + \underset{*}{H} \rightleftharpoons \underset{*}{RN}\begin{matrix} OH \\ \diagup \\ \diagdown\!\!\diagdown \\ O \end{matrix} \quad (17)$$

$$\underset{*}{RN}\begin{matrix} OH \\ \diagup \\ \diagdown\!\!\diagdown \\ O \end{matrix} + \underset{*}{H} \rightleftharpoons \underset{*}{RNO} + H_2O \quad (18)$$

$$\underset{*}{RNO} + \underset{*}{2H_2} \rightleftharpoons \underset{*}{RNH_2} + \underset{*}{H_2O} \quad (19)$$

$$\underset{*}{RNH_2} \rightleftharpoons RNH_2 + * \quad (20)$$

$$\underset{*}{H_2O} \rightleftharpoons H_2O + * \quad (21)$$

where (17) is suggested as the rate-determining step. The formation of the nitroso intermediate explains (i) the appearance of hydroxylamines and (ii) the tar formation which can sometimes occur. The statement of (19) as such is rather improbable and is probably a combination of steps which includes hydroxylamine formation by

$$\underset{*}{RN{=}O} \xrightarrow{\overset{H}{*}} \underset{*}{RN{-}O}{\diagup}H \xrightarrow{} \overset{\overset{H}{*}}{R{-}N}\begin{matrix}{\diagup}H \\ {\diagdown}O{\diagup}\end{matrix}H \xrightarrow{\overset{+2H}{*}} \underset{*}{RNH_2} + \underset{*}{H_2O} \quad (22)$$

12.8 Hydrogenation of nitriles

12.8.1 *General considerations*

The C≡N group is strongly adsorbed (more so than C≡C) and the catalytic reduction of nitriles can be quite complex (figure 12.1).[1,2,8] Thus secondary and tertiary amines can be produced. Hydrogenation, which can occur under quite mild conditions, (25–50°C, 3–10 bar) is catalysed by most Group VIII metals. However, rhodium, which gives secondary amines, and platinum and palladium, giving tertiary amines, are unsuitable. Cobalt catalysts are most selective for primary amine formation , presumably because the imine is not desorbed and cannot react with product amine. With nickel catalysts, the use of excess ammonia has been found to prevent secondary and tertiary amine formation.

12.8.2 *Adiponitrile hydrogenation*

Adiponitrile ($N{\equiv}C(CH_2)_4C{\equiv}N$) is used in nylon 6,6 manufacture where it is hydrogenated to hexamethylene diamine, $H_2N(CH_2)_6NH_2$, which is condensed with adipic acid $HOOC(CH_2)_4COOH$ (see section 7.3). A range of catalysts is used industrially—copper–cobalt or iron are used at high pressures (300–650 bar,

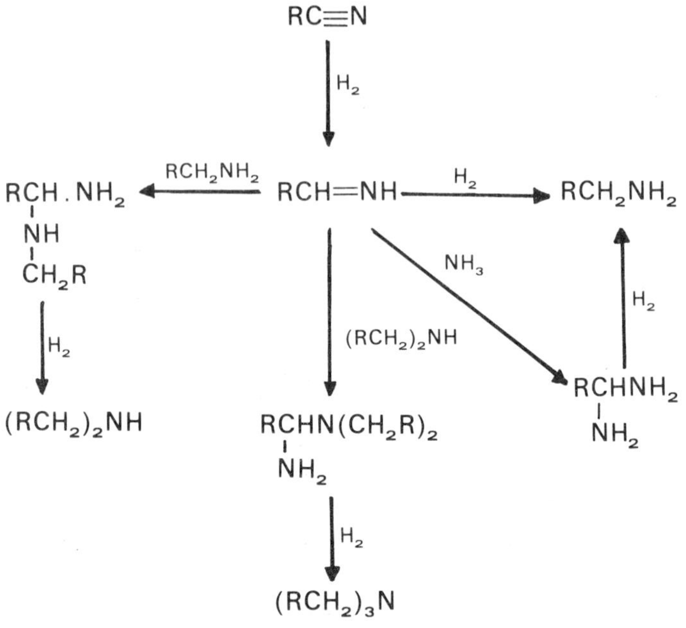

Figure 12.1 Possible products from nitrile hydrogenation.

100–180°C). Nickel-based catalysts (nickel, nickel-iron or chromium-nickel) can be used at lower pressures (30 bar) and temperatures (75°C) but excess ammonia is essential to suppress secondary and tertiary amine formation. Presumably the reaction proceeds through a di-imine intermediate.

12.9 The hydrogenation of sugars

Most of the reactions described in the preceding sections relate to the manufacture of chemical intermediates—often on a very large scale. The hydrogenation of aldo- and keto-sugars to the corresponding alcohols represents a sizeable business area where the product is utilised directly by the consumer.

Sugar alcohols find many uses, particularly in foodstuffs, pharmaceuticals and cosmetics. For example, sorbitol, a hexitol derived from D-glucose, is used to impart body, texture and some sweetness in certain foodstuffs. It is also an important intermediate in the manufacture of L-ascorbic acid (Vitamin C). In fact, both sorbitol and mannitol can be used as sweeteners, when dietary problems dictate that sugars should be absent from certain foodstuffs. For similar reasons, xylitol (a penta-alcohol) has been produced by xylose hydrogenation.[10] Xylitol has a sweetening capacity greater than that of sugar but possesses no insulin requirements and can be used by diabetics.

Manufacture of mannitol and sorbitol is accomplished using an aqueous solution of the chosen sugar. Raney nickel is used extensively as the hydrogenation catalyst, although other Group VIII metals such as ruthenium, rhodium or palladium are active catalysts. The reaction is completed in the temperature range 120–160°C at 70–140 bar pressure. Whereas hydrogenation of glucose yields exclusively sorbitol, fructose yields sorbitol and mannitol in equimolar amounts.

CHO		CH_2OH	CH_2OH		CH_2OH		CH_2OH
H–OH		H–OH	=O		H–OH		HO–H
HO–H	→	HO–H	HO–H	→	HO–H	+	HO–H
H–OH		H–OH	H–OH		H–OH		H–OH
H–OH		H–OH	H–OH		H–OH		H–OH
CH_2OH		CH_2OH	CH_2OH		CH_2OH		CH_2OH
D-Glucose		Sorbitol	D-Fructose				Mannitol

Further, sucrose may be used as the starting material. Inversion of this yields invert sugar (an equimolar mixture of glucose and fructose). Hydrogenation of a neutral aqueous solution of invert sugar leads to the production of sorbitol and mannitol in the molar ratio of 3 : 1. If the solution is made slightly basic this ratio falls to 1.8 : 1.[9]

12.10 Catalytic dehydrogenation

Catalytic dehydrogenation[8] is an important means of manufacturing unsaturated hydrocarbons which are not available in sufficient quantity from various petroleum streams. It is also used in the production of aromatic hydrocarbons from *n*-paraffins and naphthenes via platinum reforming.

Of particular interest are the production of 1,3-butadiene and styrene, both of which are important building blocks in the manufacture of various polymers and resins. In Western Europe most 1,3-butadiene is obtained from C_4 cracker streams by solvent extraction, but in the U.S. there is a shortfall and this is made up by dehydrogenation of butane and butene. Styrene is manufactured by the dehydrogenation of ethylbenzene which is produced by Friedel–Craft alkylation of benzene with ethylene. Dehydrogenation is an endothermic process, favoured by high temperatures and low pressure.

12.10.1 *n-Butene dehydrogenation*

Butadiene can be produced from *n*-butane or *n*-butenes and the respective endothermic heats of reaction are 109 kJ/mole and 126 kJ/mole. High temperatures (600–700°C) are required for economic conversions as can be judged from the equilibrium data in Table 12.2.

Table 12.2 Equilibrium conversions for butane/butenes to 1,3-butadiene

Reaction		% Equilibrium conversion 500°C	700°C	900°C
n-C_4H_8/1,3-C_4H_6	1 bar	12	60	93
	0.1 bar	35	92	98
n-C_4H_{10}/1,3-C_4H_6	1 bar	5	~100	~100
	0.1 bar	—	~100	~100

Low partial pressures of hydrocarbons are desirable for sensible yields, but plant operation at sub-atmospheric pressure is impractical if not hazardous. In modern processes, it is customary to use steam as diluent.

The early chromia-alumina catalysts, developed for butane dehydrogenation, are water-sensitive and unsuitable for n-butenes. For this reason, iron-based catalysts were developed, such as Shell 205: 70% Fe_2O_3, 27% K_2O (promoter), and 3% Cr_2O_3 (stabilizer), which are inactive for n-butane dehydrogenation. This type of catalyst is favoured because:

(i) steam can be added, decreasing the n-butene partial pressure and thus allowing operation at atmospheric pressure;
(ii) steam reacts with coke and extends catalyst life via the reaction

$$C + H_2O = CO_2 + H_2 \tag{23}$$

(iii) steam can be used as a heat source by preheating it to about 725°C, mixing with butenes at 600°C to give adiabatic reaction.

In recent years direct dehydrogenation of n-butene and n-butane to 1,3-butadiene has been rendered obsolete by the advent of oxidative dehydrogenation (section 11.6.1). The advantages of the latter are firstly that reaction equilibrium is forced towards butadiene by continuous removal of hydrogen as water, and secondly, the reaction becomes exothermic due to hydrogen oxidation and is thus more economical in terms of energy usage.

12.10.2 *Ethylbenzene dehydrogenation*

Most styrene is produced via direct dehydrogenation of ethylbenzene, although oxidative dehydrogenation is feasible and may well in future render the direct dehydrogenation obsolete.

The equilibria for ethylbenzene/styrene are such that comparable conversions can be obtained at about 100°C lower than for n-butene/1,3-butadiene. Nevertheless, it is again advantageous to use steam addition to reduce the ethylbenzene partial pressure and again, water-insensitive iron oxide based catalysts are used. The process can be performed either adiabatically or isothermically. The latter is more expensive in terms of capital costs, but gives slightly better yields and uses less steam.

12.11 Hydrodealkylation of aromatic hydrocarbons

12.11.1 *General considerations*

The main use of this process is to manufacture benzene of high purity from toluene. The benzene can be used for cyclohexane (section 12.6.1) or high quality phenol production (section 7.7). Toluene, from platinum reforming processes, is produced in quantities in excess of that required for chemical synthesis.

The main reaction occurring is

$$C_6H_5 . CH_3 + H_2 \rightarrow C_6H_6 + CH_4 \quad (24)$$

but a number of other reactions, which are hydrocracking in type, can occur giving paraffins ranging from methane to *n*-hexane; biphenyl and polycyclics can also be formed.

Fairly high temperatures of operation are employed (550–650°C) with pressures in the range 25–50 atmospheres. Excess hydrogen is used ($H_2/C_6H_5CH_3 = 4$ to 8/1) and this helps to remove heat from the exothermic reaction ($\Delta H = -126$ kJ/mole). Thermal hydrodealkylation is also carried out but requires higher temperatures and is less selective than the catalytic process.

12.11.2 *Hydrodealkylation catalysts*

Catalysts in general use are chromium, molybdenum or cobalt oxides supported on alumina or silica-alumina. In a recent novel approach, toluene and steam were reacted over a rhodium/alumina catalyst at 400–480°C, the reaction being

$$C_6H_5CH_3 + H_2O = C_6H_6 + 2H_2 + CO \quad (25)$$

This reaction has the advantage of being a hydrogen producer, which may be desirable as hydrogen costs increase in the face of diminishing oil stocks.

12.12 Homogeneous hydrogenation

Hydrogenation of olefins using soluble metal complexes has been extensively studied, perhaps more so than any other homogeneously catalysed reaction.[11,12] At first sight, this may seem unusual since this type of catalyst is rarely used either in laboratory organic synthesis or in industrial hydrogenation. Heterogeneous catalysts are usually more active and are preferred for practical reasons, but homogeneous catalysts have one major advantage: selectivity. This manifests itself in several ways, but is perhaps best demonstrated in the (at present) sole commercial application, the production of L-dopa via asymmetric hydrogenation of an unsaturated amino acid precursor (see 13.5).

In contrast to heterogeneous catalysts, these metal complex catalysts usually have well-characterised active centres and well-understood catalytic cycles. For this reason, soluble metal complexes have an important role in the modelling of heterogeneous reaction pathways. Indeed many of the hypotheses concerning the nature of the bonding between olefins and metal surfaces have received strong support from the better-understood bonding in organometallic complexes.

Among the numerous metal complexes that act as homogeneous hydrogenation catalysts four groups stand out (Table 12.3). Differences between these groups are centred on activity and selectivity, and hence practical application, and on the mechanism of hydrogen activation. Some aspects of selectivity are indicated in the table. A particular feature of the well-defined metal centre is that both activity and selectivity can be controlled by changes in the ligand environment. Thus, the cationic $[Rh(diene)(PR_3)_2]^+$ is more active than its neutral counterpart, $RhCl(PR_3)_3$, and is able to hydrogenate more highly hindered olefinic groups and carbonyl groups in aldehydes and ketones. Subtle changes in the nature of the ligand, e.g. incorporation of a centre of chirality, gives asymmetric products (e.g. amino acids) in very high optical purity, often in excess of 90%. This selectivity is subtle indeed: differences in activation energy in the diastereomeric transition states leading to the two possible optical isomers can be as little as 5 kJ/mol.

Table 12.3 Catalysts for homogeneous hydrogenation

Catalyst	*Typical applications*
Metal phosphine complexes e.g. $RhCl(PPh_3)_3$ and related $[Rh(diene)(PR_3)_2]^+$ catalysts	Selective hydrogenation of terminal double bonds in the presence of more highly substituted double bonds and other groups such as C=O and NO_2; asymmetric hydrogenation using chiral phosphine ligands
Platinum chloride/tin chloride e.g. $[Pt(SnCl_3)_5]^{3-}$	Selective hydrogenation of C=C in the presence of C=O; selective hydrogenation of unsaturated triglycerides
$[Co(CN)_5]^{3-}$	Selective hydrogenation of C=C conjugated with C=C, C=O, C=N; catalyst system relatively inactive
Ziegler–Natta systems	Hydrogenation of unsaturation in polymers
Metal carbonyls e,g, $[Co(CO)_3PBu_3]_2$, $Cr(CO)_6$	Selective hydrogenation of polyenes to monoenes

The four groups of catalysts in Table 12.3 are mechanistically distinct. The best studied is the so-called Wilkinson's catalyst, $RhCl(PPh_3)_3$. The kinetically dominant catalytic cycle is shown in figure 12.2. The Rh^I species in this cycle is a solvated species formed via ligand dissociation:

$$RhCl(PPh_3)_3 \rightleftharpoons PPh_3 + [RhCl(PPh_3)_2] \overset{S}{\rightleftharpoons} RhCl(PPh_3)_2S \qquad (26)$$

(S = solvent)

The sequence of events in figure 12.2 shows that some of the most widely studied and fundamental reactions in organometallic chemistry are essential features of the catalytic cycle. These are: (i) formation of the active Rh^I via *ligand dissociation*; (ii) activation of molecular hydrogen by *oxidative addition*;

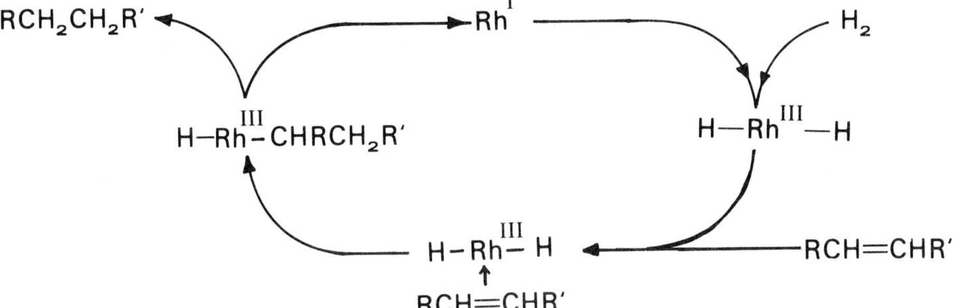

Figure 12.2 Catalytic cycle in olefin hydrogenation by Wilkinson's catalyst.

(iii) olefin activation by *formation of a metal π-complex*; (iv) *insertion* of the co-ordinated olefin into the metal-hydrogen bond to give a metal-hydride; and (v) *reductive elimination* to give alkane and regenerate the active Rh^I species. Different pathways for hydrogen activation are found in the platinum and cobalt systems. In the former hydrogen is split heterolytically and in the latter homolytically:

$$H_2 + [Pt(SnCl_3)_5]^{3-} \rightleftharpoons H^+[HPt(SnCl_3)_4]^{3-} + SnCl_3^- \tag{27}$$

$$H_2 + Co_2(CO)_8 \rightleftharpoons 2HCo(CO)_4 \tag{28}$$

The advantage of improved selectivity with homogeneous catalysts is offset by the difficulties of catalyst-product separation. For example, homogeneous catalysts would be preferred for the selective hydrogenation of triglycerides were it not for the problems of separation of the soluble metal species from the high boiling product. One solution is to anchor the catalyst species by chemically bonding it to an inorganic solid or to a cross-linked organic polymer, a process known as heterogenisation.[13,14] This is usually achieved by chemical modification of a neutral ligand, e.g. a tertiary phosphine. Examples of the most commonly adopted methods are in equation (29), which shows modification of a polystyrene backbone, and equation (30) which shows attachment via reaction of a silylated phosphine with the hydroxyl on the surface of silica.

$$\sim CH_2{-}CH(C_6H_5) \sim \xrightarrow{Br_2} \sim CH_2{-}CH(C_6H_4Br) \sim \xrightarrow{LiPPh_2} \sim CH_2{-}CH(C_6H_4PPh_2) \sim$$

$$\xrightarrow[-PPh_3]{RhCl(PPh_3)_3} \sim CH_2{-}CH(C_6H_4Ph_2P \rightarrow RhCl(PPh_3)_2) \sim \tag{29}$$

$$\begin{array}{l} \equiv Si-OH \\ O \\ \equiv Si-OH \end{array} + (EtO)_3Si(CH_2)_xPPh_2 \rightarrow \begin{array}{l} \equiv Si-O \quad OEt \\ O \qquad\qquad Si(CH_2)_xPPh_2 \\ \equiv Si-O \end{array}$$

$$\xrightarrow[-PPh_3]{RhCl(PPh_3)_3} \begin{array}{l} \equiv Si-O \quad OEt \\ O \qquad\qquad Si(CH_2)_xPPh_2RhCl(PPh_3)_2 \\ \equiv Si-O \end{array} \qquad (30)$$

In this way the desirable features of the homogeneous catalyst can be retained with the additional benefit of ease of separation in a heterogeneous form. In practice, perhaps not unexpectedly, there are some drawbacks, but there are also some additional benefits. Drawbacks include the need to tailor the support material so that substrate species have free access to the catalyst sites. With cross-linked organic resins the matrix must be swelled by the reaction medium. Leaching of the metal from the support into the reaction medium can be a most serious problem resulting in a steady long-term loss of precious metal, and is one of the major factors that have prevented commercialisation of heterogenised catalysts. The addition of the somewhat esoteric support/modified ligand can represent a considerable expense and the long-term thermal, mechanical, or oxidative stability of the support matrix is poor, particularly with the cross-linked resins.

Some benefits include improved selectivity with some catalysts on cross-linked resins—the micro-pores in the resin give preferential access to the more linear substrate molecules. Sulphur resistance is improved, and also activity, with in some cases a complete change in the chemistry. Thus, a palladium-anthranilic acid catalyst when attached to a cross-linked resin is active in the hydrogenation of both aromatic hydrocarbons and nitriles, whereas the equivalent free complex is completely ineffective.

$$\text{(resin)}-C_6H_4-CH_2-NH(-Pd)-C_6H_4-C(=O)-O-Pd$$

The heterogenisation of homogeneous catalysts is being vigorously pursued in both industrial and academic laboratories. Most activity centres on hydrogenation catalysts because they are the easiest to study experimentally. Once the problems can be ironed out, the benefits are most likely to be commercially realised in the synthesis of speciality chemicals (cf. chapter 13).

REFERENCES

1. R. L. Augustine, *Catal. Rev.*, **13**, 285, 1976.
2. P. N. Rylander, *Catalytic Hydrogenation over Platinum Metals*, Academic Press, New York, 1967.
3. G. C. Bond, *Catalysis by Metals*, Academic Press, New York, 1962.
4. G. C. Bond and P. B. Wells, *Adv. Catal.*, **15**, 92, 1964.
5. G. Webb, *Chemical Kinetics*, **20**, 1, Ed. C. H. Bamford and C. F. H. Tipper, Elsevier, Amsterdam, 1978.
6. G. Webb, *Chem. Soc. Spec. Per. Rep.* (*Catalysis*), 145, 1978.
7. J. B. B. Haynes and F. G. A. Stone, *J. Organometallic Chem.*, **160**, 337, 1978.
8. K. Weissermel and H. J. Arpe, *Industrial Organic Chemistry*, Verlag Chemie, 1978.
9. L. W. Wright, *Chemtech*, **4**, 42, 1974.
10. J. Wisniak and R. Simon, *Ind. Eng. Chem. Prod. Res. Dev.*, **18**, 50, 1979.
11. B. R. James, *Homogeneous Hydrogenation*, Wiley-Interscience, New York, 1973.
12. F. J. McQuillin, *Homogeneous Hydrogenation in Organic Chemistry*, Vol. 1, in *Homogeneous Catalysis in Organic and Inorganic Chemistry*, Ed. R. Ugo, Reidel, Dordrecht, 1976.
13. L. L. Murrel, Chap. 8 in *Advanced materials in Catalysis*, Ed. J. J. Burton and R. L. Garten, Academic Press, New York, 1977.
14. R. H. Grubbs, *Chemtech*, **7**, 512, 1977; D. D. Whitehouse, *ibid.*, **10**, 44, 1980; N. L. Holy, *ibid.*, **10**, 366, 1980.

CHAPTER THIRTEEN

SPECIALITY CHEMICALS

D. J. THOMPSON

13.1 Introduction

Speciality chemicals differ from commodity chemicals in one important respect: they are sold on the basis of their ability to produce a desired effect, whereas commodity chemicals are sold primarily as chemical intermediates and have more closely defined chemical compositions. Generally, speciality chemicals are produced in small volume, some important areas being pharmaceuticals, flavours and fragrances, fine chemicals and pesticides.

In the synthesis, of speciality chemicals, a catalytic step is usually only one in a multistep reaction scheme, and hence the catalyst may contribute only a very small part of the cost of the final product. Often, indeed, there is a choice between a catalytic and a reagent process, for example reduction of a ketone using either catalytic hydrogenation or sodium borohydride. The superiority of the catalytic process may not always be obvious; firstly, a catalytic process may involve installing speciality equipment e.g. high pressure autoclaves, whereas use of a reagent can be undertaken in conventional apparatus. This is particularly important if the apparatus is used for a variety of preparations. Secondly, with homogeneous catalysts it may be difficult to separate the product from the catalyst, particularly when the product is non-volatile. Use of a metal catalyst may require an extra purification step, not only to recover the precious metal but to free the final product from traces of harmful metal impurities. This is especially important in the preparation of pharmaceuticals. Work on immobilisation of homogeneous catalysts is nevertheless progressing and is discussed in section 12.12.

However, as the demand for more sophisticated chemicals increases, metal catalysts, and in particular homogeneous catalysts, can offer advantages.

Reaction can proceed under milder conditions and in higher yields, higher selectivity (both regio- and stereo-) can be achieved, and novel synthetic routes can be used.

The types of reaction where metal catalysis, both homogeneous and heterogeneous, has had or may have some impact in the synthesis of speciality chemicals are listed in Table 13.1.

Table 13.1 Metal-catalysed reactions applicable to speciality-chemical synthesis

Type of reaction	*Typical metal involved*	*Application*
Isomerisation	V, Ni, Rh	Synthesis of terpenes and other natural products.
Hydrogenation	Rh, Ru, Pt, Pd	Regio- and stereo-selective reduction of olefins, carbonyls, nitro and nitrile groups. Synthesis of deutero- and tritio-labelled compounds.
Oxidation	Mo, V, Cr	Selective oxidation, regiospecific epoxidation.
Carbene reactions	Cu, Rh	Synthesis of cyclopropanes e.g. chrysanthemate insecticides
Dimerisation, telomerisation, oligomerisation	Co, Ni, Pd	Synthesis of natural products,[1] heterocycles[2]

13.2 Isomerisation

In the synthesis of terpenes (used particularly for flavours and perfumery) isomerisation is often an important step in a synthetic sequence. For example, Rhone–Poulenc Industries have developed a vanadium-catalysed system for the isomerisation of dehydrolinalol (1) to citral (2), the main component of the commercially important lemon grass oil[3]. This process was commercialised in 1972.

OH $\xrightarrow{VO(OR)_3}$ CHO

(1) (2)

The reaction consists of heating the dehydrolinalol at 160° in an inert atmosphere in the presence of 0.1–1 % of a vanadate catalyst, $VO(OR)_3$, for a short time (0.5–2 h), followed by distillation of the product. Fresh dehydrolinalol is charged to the residue and the reaction repeated. The catalyst can be reused several times without serious loss of activity. At 18 % conversion of dehydrolinalol, citral is obtained in 77 % selectivity.

The Kuraray Company of Japan have developed an efficient process for the synthesis of pseudoionone (3), but unfortunately the main product of the reaction is the isomer (4). Pseudoionone can be readily cyclised in good yield to the required ionones, e.g. α-ionone (5), which are commercial odorants smelling strongly of violets. The isomer (4) is not so readily cyclised and it was therefore necessary to find an efficient catalyst for the isomerisation of (4) to (3). A suitable catalyst is the readily available complex nickel acetylacetonate,

$Ni(acac)_2$.[4] Typically the 2,5-dienone (4) is heated at 165° for 2 hours in an inert atmosphere in the presence of 1% $Ni(acac)_2$. Distillation gives in 95% yield a 97/3 mixture of (3) and (4). Moreover the catalyst can be recycled without any decrease in activity.

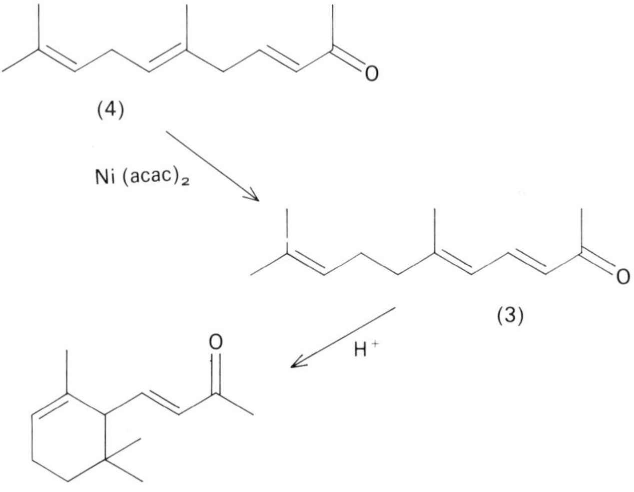

(4)

Ni (acac)$_2$

(3)

H$^+$

(5)

13.3 Epoxidation

Selective epoxidation is often a key step in the synthesis of functionalised natural products. In the presence of vanadium or molybdenum catalysts, e.g. $VO(acac)_2$ or $Mo(CO)_6$, the terpene geraniol (6) undergoes a regiospecific epoxidation with $t\text{-BuO}_2H$, at the olefinic site closest to the hydroxyl group to give the product (7) in good yield.[5] This is in contrast to the uncatalysed process where the more electron rich 6,7-double bond is epoxidised.

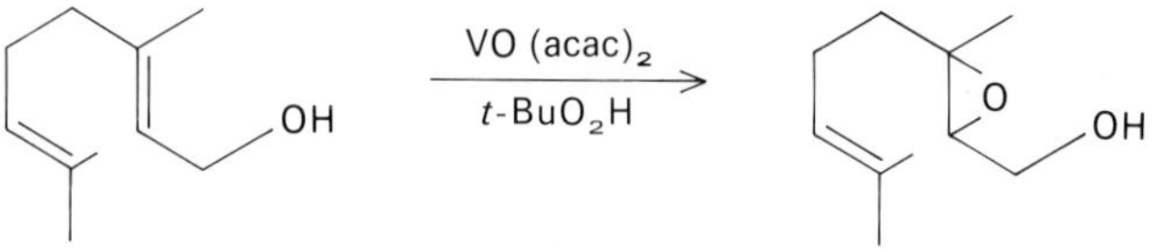

VO (acac)$_2$

t-BuO$_2$H

(6) (7)

13.4 Hydrogenation

There are numerous examples of selective hydrogenations in the literature and the topic is well covered in a variety of reviews.[6] Two examples will illustrate the high level of selectivity that can be achieved with homogeneous catalysts.

Up to 98% cyclododecene is produced[7] when 1,5,9-cyclodecatriene is selectively hydrogenated in the presence of catalytic amounts of $Ru(CO)_2(Ph_3P)_2$. This process has potential industrial application since cyclododecene is a precursor for the polyamide monomers 1,12-dodecanedioic acid, 1,12-diamino-dodecane and 12-aminododecanoic acid lactam. High selectivity is important in the reaction since cyclododecene cannot be separated from the other possible products (cyclododecadiene or cyclododecane) by distillation.

In the synthesis of pharmaceuticals there is plenty of scope for selective hydrogenation but it is difficult to find an actual industrial application. However the type of selectivity that can be achieved using homogeneous catalysts is illustrated by the catalytic hydrogenation of the steroid (8) using $RuCl_2(Ph_3P)_3$ and triethylamine to give the product (9) with less than 4% of the fully saturated product.[8] This selective reduction of one of four possible reducible bonds would be very difficult to achieve using a heterogeneous catalyst or a reducing agent.

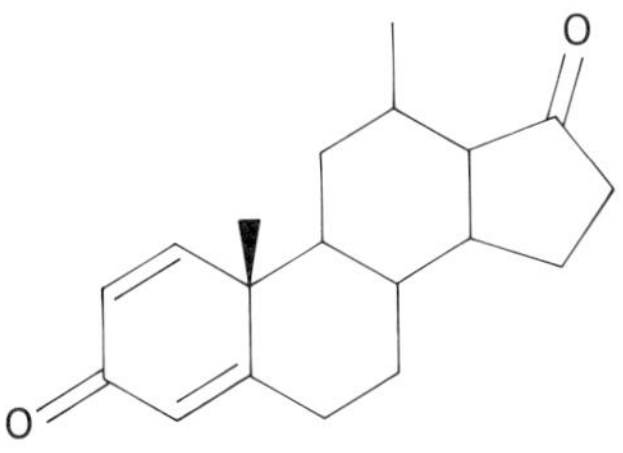

(8) 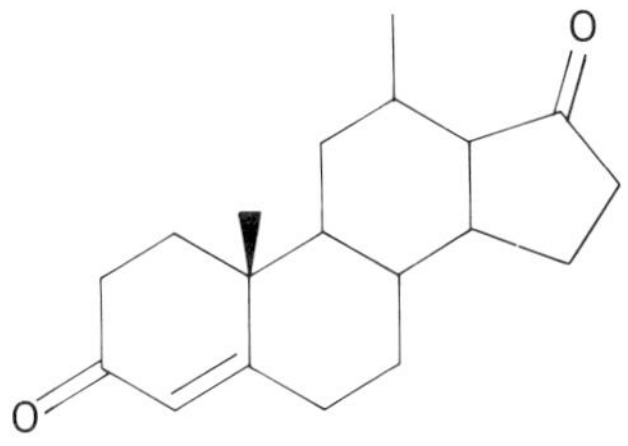 (9)

13.5 Asymmetric hydrogenation

Probably the most exciting development in selective catalysis in recent years has been the discovery of asymmetric catalysis. In many useful chiral compounds which find application as pharmaceuticals, insecticides or perfumes, only one enantiomer is active. Often the other isomer is inactive, but in some cases it can be toxic. Moreover there is a strong possibility that future legislation will demand that drug companies sell only pure enantiomers. Of the three potential routes to optically active compounds, resolution, asymmetric synthesis and asymmetric catalysis, the classical route is to separate the isomers by resolution, i.e. reaction of the mixture with a chiral reagent followed by physical separation of the two diastereoisomers. This is not only tedious and expensive but is limited to the theoretical yield of 50%. In asymmetric synthesis a new chiral centre is created in a prochiral substrate by use of a chiral *reagent*. Although this is a well-developed area,[9] the main drawback is that stoichiometric quantities of expensive chiral reagent must be used and only rarely can these compounds be recycled. An asymmetric reaction involving a *catalyst* that is chiral is a much superior way to accomplish chiral synthesis because with only a small amount of chiral material large quantities of optically active material can, in theory, be produced.

Most progress in asymmetric catalysis has been made in hydrogenation, particularly with systems based on the so-called Wilkinson's catalyst (see 12.12).

By using optically active phosphines co-ordinated to the rhodium catalyst, unsymmetrically substituted olefins can be reduced to optically active products.[10] Although asymmetric hydrogenation is still in its infancy, the Monsanto Company have developed an industrial process for the efficient synthesis of L-amino acids using a rhodium catalyst containing chiral phosphines such as *o*-anisyl-cyclohexylmethylphosphine (ACMP) or the diphosphine DIPAMP(10).[11]

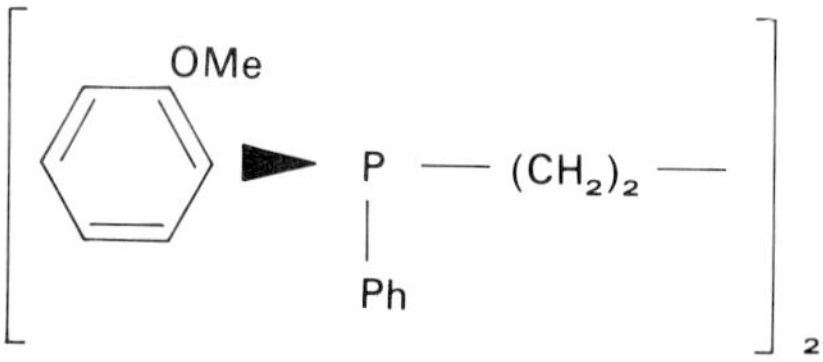

$$\left[(o\text{-MeOC}_6\text{H}_4)\text{P(Ph)}-(\text{CH}_2)_2- \right]_2$$

(10)

Using catalysts of this type the readily available α-acyl amino acrylic acids (11) are reduced to the acyl amino acids (12) in excellent yield and with high stereoselectivity, up to 96% enantiomeric excess, i.e. a 98:2 mixture of the two optical isomers.

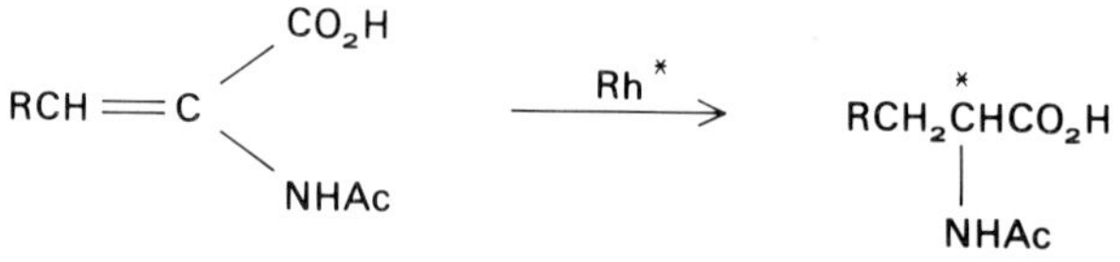

$$\text{RCH}{=}\text{C}(\text{CO}_2\text{H})(\text{NHAc}) \xrightarrow{\text{Rh}^*} \text{RCH}_2\overset{*}{\text{C}}\text{H}(\text{NHAc})\text{CO}_2\text{H}$$

(11) (12)

The reaction is run in methanol at 50°C and 3 bar of H_2, and Monsanto operate a process of this type for the synthesis of L-Dopa (13), which is used to reduce the symptoms of Parkinson's disease. The L-Dopa (13), is isolated by crystallisation from methanol, the catalyst being left in the mother liquors from which the rhodium can be extracted and isolated for re-use if necessary. It is reported that 1 pound of catalyst yields 1 ton of L-Dopa. In systems of this type it can be envisaged that the catalysts are very expensive, not only because they contain rhodium (which can be recovered) but more importantly they contain chiral phosphines which are not only difficult and expensive to synthesise but probably are difficult to recover and re-use. In spite of this the Monsanto process is an enormous improvement over a conventional reduction followed by a resolution of the D- and L-isomers.

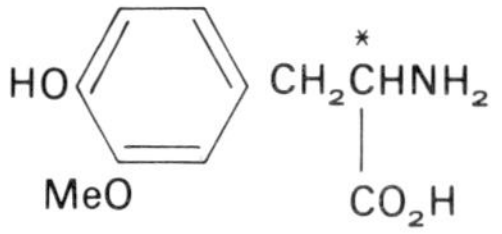

HO, MeO, $CH_2\overset{*}{C}HNH_2$, CO_2H

(13)

13.6 Cyclopropanation

Insecticides based on the chrysanthemate esters (14) are amongst the most potent pesticides known. Like many other speciality chemicals they were initially made by a batchwise process by reaction of an alkyldiazoacetate (15) with 2,5-dimethyl-2,4-hexadiene (16) in the presence of a copper catalyst. This process is hazardous, however, because of the repeated handling of the diazoacetate (15), a reagent which is possibly carcinogenic, and prone to explode. Stauffer Chemicals now operate a continuous process in the presence of a copper bronze catalyst where reaction times are relatively short, thus minimising explosive hazards, and reducing the risk of exposure of operators to the intermediate diazoester (15).[12] High yields (80%) of the chrysanthemate ester (14) are obtained.

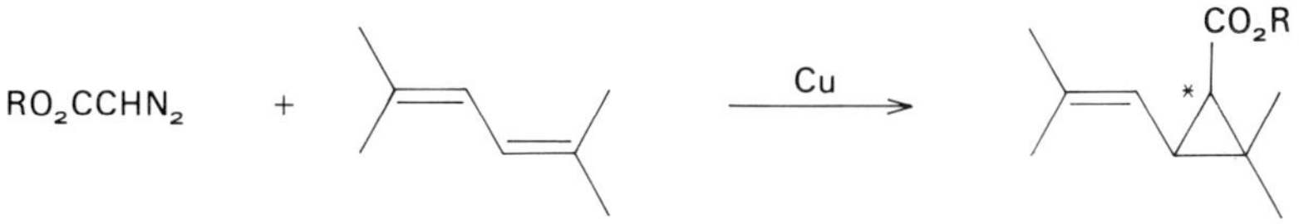

RO_2CCHN_2 + (16) $\xrightarrow{Cu}$ (14) [CO_2R]

(15) (16) (14)

Four possible isomers of the chrysanthemate ester (14) exist, *cis-* and *trans-*geometrical isomers about the cyclopropyl ring and the R and S form of each of these geometrical isomers. Of these four isomers two are generally biologically inactive, one is moderately active and the fourth isomer very active. Work has therefore been directed towards finding catalysts which will selectively give the most active isomer. In the presence of a chiral copper catalyst efficient addition of the diazoacetate (15) to the diene (16) occurs to give the chrysanthemate ester (14) which contains an 80:20 *trans*:*cis* mixture and with an enantiomeric excess of 90% (80% of the R isomer).[13] Obviously the chiral catalysts cost more than the simple copper bronze catalysts but if they can be designed to give large excesses of the right isomer with high turnover numbers, then the increased cost will be more than offset by the increased biological activity of the product.

13.7 Summary

As can be seen from the few examples given above, very high levels of selectivity can be achieved in catalytic processes. Our knowledge of organometallic chemistry is still restricted compared to what is known of organic chemistry itself, and so it is not surprising that in the synthesis of speciality chemicals the more classical synthetic routes dominate. However, as the demand for new speciality chemicals (and in particular biologically active materials), increases we can expect selective catalysis to play an increasingly important role in this area.

Potentially the most useful area is the synthesis of pure optical isomers from non-chiral starting materials, an area which until recent years was dominated by enzymic catalysis. To date only asymmetric hydrogenation has achieved commercial success, but it is to be hoped that new processes will emerge from work in this important area. Another area where organometallic catalysts should play an increasingly important role is in the synthesis of heterocyclic compounds, particularly nitrogen heterocycles, where it has already been demonstrated that compounds can be synthesised in good yield under mild conditions.[2] In the long term, if cellulosic products become an important feedstock for the chemical industry, then there will be ample scope for the use of organometallic catalysts to compete with enzymic catalysts to produce useful chemical products.

REFERENCES

1. (a) K. P. C. Vollhardt and C. Peter, *Acc. Chem. Res.*, **10**, 1, 1977;
(b) J. Tsuji, *Advances in Organometallic Chem.*, **17**, 182, 1979.
2. H. Bonnemann, *Angew. Chem. Internat. Edn.*, **17**, 505, 1978.
3. P. Chabardes, E. Kuntz and J. Varagnat, *Tetrahedron*, 1775, 1977; U.S. Pat. 3,925,485 and 3,920,751.
4. T. Onishi, Y. Fujita and T. Nishida, *Chem. Lett.*, 765, 1979.
5. K. B. Sharpless and R. C. Michaelson, *J. Amer. Chem. Soc.*, **95**, 6136, 1973.
6. *Homogeneous Hydrogenation in Organic Chemistry*, F. J. McQuillin, Reidel, 1976; *Homogeneous Hydrogenation*, B. R. James, Wiley-Interscience, 1973; *Catalytic Hydrogenation in Organic Chemistry*, M. Friefelder, Wiley, 1979; *Homogeneous Hydrogenation Catalysts in Organic Synthesis*, A. J. Birch and D. H. Williamson, *Organic Reactions*, **24**, 1, 1977.
7. D. R. Fahey, *J. Org. Chem.*, **38**, 80, 1973.
8. S. Nishimura, T. Ichino, A. Akimojo and K. Tsuneda, *Bull. Chem. Soc. Japan*, **46**, 279, 1973.
9. D. Valentine Jr., and J. W. Scott, *Synthesis*, 329, 1978.
10. J. D. Morrison, W. F. Masler and M. K. Neuberg, *Adv. Catal.* **25**, 81, 1976; R. Pearce, Chem. Soc. Spec. Period. Report, *Catalysis*, **2**, 176, 1978.
11. W. S. Knowles, M. J. Sabachy and B. D. Vineyard, *Ann. New York Acad. Sci.*, **214**, 119, 1973; W. S. Knowles, M. S. Sabachy, B. D. Vineyard and D. J. Weinkauff, *J. Amer. Chem. Soc.*, **97**, 2567, 1975; **99**, 5966, 1977.
12. Stauffer Chemical Company, British Patent 1,306,191.
13. T. Aratani, Y. Yoneyoshi and T. Nagase, *Tetrahedron Lett.*, 1707, 1975.

Additional Reviews

1. Catalysis and organic synthesis, *Adv. Organometallic Chem.*, **17**, 1979.
2. *Transition Metal Organometallics in Organic Synthesis*, H. Alper (Ed.). Academic Press, Vol. I, 1976; Vol. II, 1978.
3. Applications of organometallic compounds to organic synthesis, A. P. Kozikowski and H. Wetter, *Synthesis*, 561, 1976.
4. Organopalladium intermediates in organic synthesis, B. M. Trost, *Tetrahedron*, **33**, 2615, 1977.
5. Organic synthesis by means of transition metal complexes, J. Tsuji, *Topics Current Chem.*, **28**, 41, 1972.
6. *Catalysis in Organic Synthesis*, P. N. Rylander and H. Greenfield, eds., Academic Press, New York, 1976.

CHAPTER FOURTEEN

ENZYMIC CATALYSIS

D. J. THOMPSON

14.1 Introduction

Enzymes are catalysts that have been used for centuries in a variety of food fermentations to produce bread, alcoholic beverages, cheese and vinegar, although, of course, the nature and function of the enzyme was not understood. Only at the beginning of this century were enzymes shown to be the agents responsible for all fermentation processes. It is now recognised that enzymes are proteins made up of long chains of amino acids. The primary structure, the sequence in which the amino acids are assembled, has been determined in many cases. The three-dimensional spatial structure, which influences catalytic ability, is also known for some enzymes.

Despite this knowledge the essential details of the mechanism are often obscure since further information, for example the binding mode of the substrate, is difficult to obtain. The active site of the enzyme may or may not involve a metal ion. Enzymic reactions involving small molecules (for example O_2, CO_2 and N_2) usually involve a metal centre. Interestingly, metals which are generally regarded as being catalytically very active, for example rhodium, platinum and palladium, have not as yet been implicated in enzymic catalysis, whereas zinc, copper, iron and molybdenum are commonly found in this role.

14.2 Enzyme types and uses

Enzymes have been classified into six major groups based on their catalytic functions (Table 14.1).[1] They offer a number of advantages as catalytic species.

(1) They are highly versatile, catalysing a wide spectrum of reactions. In fact there is an enzyme-catalysed equivalent to almost every type of organic reaction, but some industrially important transformations, for example cracking, olefin polymerisation and CO/H_2 reactions, have no enzymic counterparts. Enzymes exist only where the catalysis of a given reaction is possible in nature. However an important feature of enzymes is that they can selectively catalyse reactions where no suitable industrial analogue is known, for example the 11-hydroxylation of steroids.[2]

Table 14.1 Types of enzymes

Catalytic function	*Name*	*Example*
Oxidation-reduction	Oxidoreductases	C—H → C—OH CH(OH) ⇌ C=O CH—CH ⇌ C=C
Group transfer	Transferases	Transfer of a group from one molecule to another, e.g. acyl, sugar, phosphoryl
Hydrolysis	Hydrolases	Hydrolysis of esters, amides, anhydrides, glycosides
Group removal	Ligases	Formation of C=C, C=O, C=N bonds
Isomerisation	Isomerases	Various types of isomerisation including racemisation
Joining of molecules	Ligases	Formation of C—O, C—S, C—N bonds

(2) Enzymes are effective under mild conditions, often at room temperature and neutral pH.
(3) The catalytic activity of enzymes is very high. Rates of enzyme-catalysed reactions can be faster than those of corresponding non-enzymic processes by a factor of 10^9 to 10^{12}, but these high rates may be obscured by the fact that conversion, and not rate, is the major limitation in practice.
(4) Enzymes are usually very selective with respect to the type of reaction they catalyse and the structure and stereochemistry of the substrate and product.

However, in spite of these advantages, to date only 20 of the 1500 known enzymes are in large-scale use. There are a number of reasons for this. Isolating and purifying enzymes is expensive, and many enzymes are unstable when removed from the living cell. Most enzymes have been used in soluble form in dilute aqueous media, thus making it difficult and expensive to recover them and usually limiting their use to batch operations followed by disposal of the spent enzyme-containing solvent. Furthermore, many enzymes require a non-protein co-enzyme in order to be catalytically active, particularly in the case of oxidoreductases.[3] Co-enzymes are in reality co-substrates since they undergo chemical transformation during reaction and therefore must be constantly reconverted to their active form in order for catalysis to continue. This is not a problem when using growing micro-organisms because all the co-factors required are present in the medium and are continuously produced during the fermentation process. However, with purified or immobilised enzymes, maintaining a sufficient concentration of co-enzyme is a major problem. Finally, while enzyme selectivities are higher than for other catalysts the penalty for these high activities and selectivities is the large volume required to accommodate these sites.

Enzymes can be used in a number of forms: the enzyme may be contained in whole cells, metabolising or resting, or in artificially immobilised cells; alternatively, isolated free enzymes or immobilised enzymes may be used.

Metabolising cells are those present during conventional fermentation processes, and before 1940 a few companies were using such processes to convert carbohydrates into a range of industrial chemicals, pharmaceuticals and food ingredients. Among these products were ethanol, acetone, *n*-butanol, citric and gluconic acids. Over the years most of these fermentations producing simple organic products have been replaced by chemical processes, but in today's economic climate, the supply of surplus cellulosic material at a favourable cost may make fermentation processes a viable alternative to oil-based processes. Already in Brazil schemes are under way for providing ethanol from sucrose as a fuel for motor vehicles.

As the fermentation of simple chemical products decreased, it was replaced by the production of complex molecules which cannot easily be prepared chemically, examples of which include antibiotics, enzymes, vitamin B_{12} and amino acids[4]. The further use of fermentation processes using metabolising cells was limited because

(1) the processes are not very efficient—the catalysts are used for just one batch reaction;
(2) enzymes and cells are relatively unstable and lose activity during fermentation;
(3) conventional recovery methods are either expensive or cause denaturation and loss of catalytic activity.

Whole cells are potentially cheap, but they contain a large collection of different enzymes. If the cells could be modified so that only one enzyme was active, for example by selective mutation or breeding, then they could become of more interest to chemists.

If enzymes are to be used effectively then stability must be improved and inexpensive non-destructive recovery methods must be developed. Immobilisation of enzymes on an inert support offers a way of achieving both (Table 14.2). Each of the methods shown in the table has advantages and

Table 14.2 Immobilised enzymes

Advantages

(1) Enzymes can be re-used
(2) Continuous processes become practical
(3) Stability of enzyme is improved
(4) Purity of product greater
(5) Enzyme activity sometimes enhanced

Methods of immobilisation

(1) Adsorption at a solid surface
(2) Trapping in a gel
(3) Cross-linking of enzyme using a bifunctional agent
(4) Covalent binding to a support

disadvantages both in terms of cost and ease of application, as is discussed in detail in various reviews.[5]

Although the immobilisation of enzyme was first described in 1908[6] it was not until after World War II that serious study was undertaken, and work in the area showed a spectacular growth in the mid-1960's. The immobilisation of enzymes has led to a whole new era of enzymic catalysed reactions and several have found use industrially (Table 14.3).

Table 14.3 Current and potential applications of immobilised enzymes in industry

Enzyme	*Application*
Amino acid acylase	resolution of DL-amino acids
Glucoamylase, α-acylase	conversion of starch to glucose
Glucose isomerase	conversion of glucose to fructose
Glucose oxidase	gluconic acid production
Penicillin amidase	production of semi-synthetic penicillins
Steroid-modifying enzymes	steroid modification

A variety of reactors can be used to perform chemical conversions catalysed by immobilised enzymes: stirred-tank, fixed-bed, fluidised-bed or open tubular.[7] Of these the fixed-bed type of reactor is most common. Various kinds of solid support have been used for enzyme immobilisation: among the common ones are inorganic materials such as glass, ceramics, carbon and alumina and organic matrices such as cellulose, agarose, nylon, collagen, sepharose, acrylamide and polystyrene polymers.[8] The kinetic behaviour of immobilised enzyme is discussed in detail in several reviews.[7b,9].

14.3 Amino acid synthesis

Amino acids are used as food supplements, cosmetic additives, medical agents and as raw materials for the synthesis of other compounds. For food supplementation and most biological applications, only the L-isomers of the amino acids are biologically active. Although there are efficient chemical syntheses of amino acids, the products are racemic DL-mixtures, and while these mixtures can be resolved chemically the process is expensive. The first process that utilised an immobilised enzyme was for the resolution of racemic amino acids and was introduced by the Tanabe Seizaku Company of Japan (Table 14.4).[10] The *N*-acetyl derivatives of the racemic acid mixture are prepared chemically, and after preferential hydrolysis of the acetyl group of the L-isomer the free L-amino acid and unreacted *N*-acetyl D-amino acid are easily separated by crystallisation. The unreacted D-isomer can then be chemically racemised to a DL-mixture and recycled back to the enzyme reactor. The enzyme is bound only by ion exchange to a Sepharose bed support so leaching of the enzyme occurs. The catalyst half-life, however, is 65 days and by periodic regenerations the column can be

Table 14.4 Features of a Tanabe–Seizaku process for the production of L-amino acids

Column capacity	100 litres
Carrier	DEAE Sephadex
Amount of acylase	333 IU/ml
Activity of bound enzyme	157 IU/ml
Operating temp.	50°
Operating pH	7.0
Catalyst half-life	65 days
L-Amino acid yield (L-methionine)	715 kg/day
Column regeneration	once in 35 days

used for 2 years without loss of binding capacity or physical deterioration. This process gives a cost saving of 40% compared with a batch process using free enzyme.

14.4 Glucose isomerisation

Glucose can be isomerised to the related sugar, fructose, which is much sweeter and is used in a wide range of food products. Chemical isomerisation is not specific, giving objectionable coloured materials and unwanted sugars such as psicose. In contrast, glucose can be converted into a syrup containing a high proportion of fructose using an immobilised glucose isomerase enzyme as shown below. Some details of such a process are shown in Table 14.5.

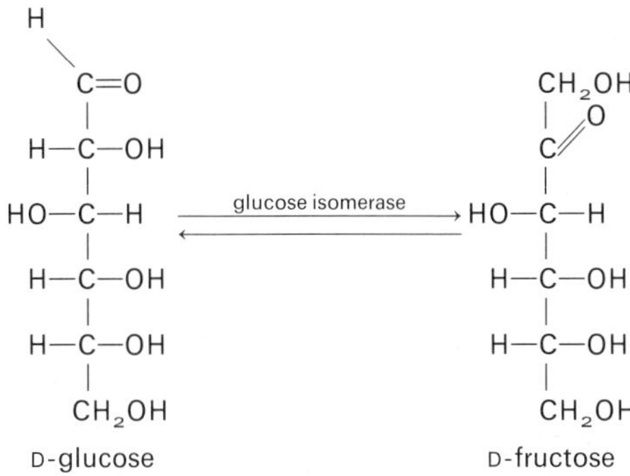

Table 14.5 Glucose isomerase technology

Worldwide scale of operation	2 million tpa (dry syrup)
Reactor size (packed bed)	4 m^3
Price of immobilised enzyme* (1977)	£12/kg
Catalyst half-life*	1000 hr
Total reactant converted*	2500 kg/kg immobilised enzyme
Optimum reactant concentration	40–45% D-glucose in water

* Novo Industri A/S enzyme cross-linked with glutaraldehyde.

14.5 Future uses

The prospect of a world protein shortage has spurred research into the production of cells of micro-organisms to meet some of these needs. The simplest conversion is into micro-organisms for animal feed (single-cell protein). British Petroleum has devised a process utilising *n*-alkanes, a Shell process uses methane and an ICI process uses methanol as feedstock. Of these only the latter is currently in production, but it is thought that an *n*-alkane-based process may be operating in Russia.

As petroleum becomes more expensive, there is a brighter prospect that biotechnology may produce our bulk chemicals, but if so it is likely that we will have to find new uses for readily-available enzymic products rather than trying to find new enzymic routes to currently attractive feedstocks.

It can be envisaged that enzymic catalysis will, in the first instance, come into its own in shortcutting current long or tedious processes. As oil becomes relatively expensive, enzymic processes could make inroads into more established processes, especially where they offer increased feedstock efficiency, for example the oxidation of benzene to phenol and the formation of propylene oxide from propylene.[11] Ultimately we might also hope to see significant advances in the economic conversion of cellulose, particularly for processing food wastes into more useful compounds.

REFERENCES

1. *Enzyme Nomenclature*, Elsevier, 1973.
2. G. S. Fonken and R. A. Johnson, *Chemical Oxidations with Microorganisms*, Dekker, 1972.
3. J. B. Jones, *Tech. Chem.*, **10**, 1, 1976; D. Metzler, *Biochemistry*, Academic Press, 1976.
4. D. Perlmann, *Chemtech*, 210, 1974; S. J. Gutcho, *Chemicals by Fermentation 1973*, Noyes Data Corp., New Jersey, 1973.
5. O. R. Zaborsky, *Immobilized Enzymes*, CRC Press, 1973; J. C. Johnson, *Industrial Enzymes—Recent Advances*, Noyes Data Corp., New Jersey, 1977; K. Mosbach ed., *Methods in Enzymology*, Vol. XLIV, Academic Press, 1976; A. J. Gutcho, *Immobilized Enzymes; Preparation and Engineering Techniques*, Noyes Data Corp., New Jersey, 1974; H. H. Weetall ed., *Immobilized Enzymes, Antigens, Antibodies and Peptides; Preparation and Characterization*, Dekker, 1975; G. P. Royer, *Catalysis Rev.*, **22**, 29, 1980.
6. L. Michaelis and M. Ehrenreich, *Biochem. Z.*, **10**, 283, 1908.
7. (a) W. R. Vieth and K. Venkatasubramanian, *Chemtech*, 434, 1974.
   (b) M. Engasser, in *Enzymic and Non Enzymic Catalysis*, ed. P. Dunnill, A. Wiseman and N. Blackebrough, Ellis Horwood Ltd., Chichester, 1980.
8. R. A. Messing, *Immobilized Enzymes for Industrial Reactors*, Academic Press, 1975, p. 63; L. Goldstein and G. Manecke, *Applied Biochemistry and Bioengineering*, eds. L. N. Wingard, E. Katchalski, and L. Goldstein, Academic Press, 1976, Vol. 1, p. 23; W. R. Vieth and K. Venkatasubramanian, *Chemtech*, 47, 1974.
9. C. Buckle, *Topics in Enzyme and Fermentation Biotechnology*, ed. A. Wiseman, Ellis Horwood, Chichester, 1977, Vol. 1, p. 147.
10. I. Chibata, T. Tosa, T. Sato, and T. Mori, *Methods of Enzymology*, Vol. XLIV, ed. K. Mosbach, Academic Press, 1976, p. 746.
11. S. L. Niedleman, *Hydrocarbon Proc.*, **135**, Nov. 1980; *Chem. Week*, **41**, March 4th, 1981; *Europ. Chem. News, New Technol. Suppl.*, **4**, Dec. 22nd, 1980.

Additional Reviews

1. Immobilized enzymes, C. J. Suckling, *Chem. Rev.*, 215, 1977.
2. *Immobilized Enzymes*, O. R. Zaborsky, CRC Press, Cleveland, 1973.
3. Enzymes in organic synthesis, C. J. Suckling and K. E. Suckling, *Chem. Rev.*, 387, 1974.
4. *Handbook of Enzyme Biotechnology*, ed. A. Wiseman, Wiley, New York, 1975.
5. Immobilized enzyme principles; *Applied Biochemistry and Bioengineering*, Vol. 1; ed. L. B. Wingard, E. Katchalski-Katzir and L. Goldstein, Academic Press, New York, 1976.
6. Metalloenzyme catalysis, J. J. Villafranca and F. M. Russell, *Adv. Catalysis*, **28**, 324, 1979.
7. Oxygenations with micro-organisms: R. A. Johnson *Oxidation in Organic Chemistry*, W. S. Trahanovsky, Academic Press, Part C, 1978, p. 131.

Index

Sources of technocommercial information used in this book are:
Chem. Eng. News, *Europ. Chem. News*, *Chem. Age*, *Chem. Week*, *Oil/Gas J. Hydrocarbon Processing*, *Chem. Marketing Reporter*, Federal Tariff Commission data, CIA data, OECD Ann. Rep., Europ. Chem. Market Research Assoc. data, Weissermal and Arpe's *Industrial Organic Chemistry*, Faith, Keyes, and Clark's *Industrial Chemicals*, and industry sources.